EAI/Springer Innovations in Communication and Computing

Series Editor

Imrich Chlamtac, European Alliance for Innovation, Ghent, Belgium

The impact of information technologies is creating a new world yet not fully understood. The extent and speed of economic, life style and social changes already perceived in everyday life is hard to estimate without understanding the technological driving forces behind it. This series presents contributed volumes featuring the latest research and development in the various information engineering technologies that play a key role in this process. The range of topics, focusing primarily on communications and computing engineering include, but are not limited to, wireless networks; mobile communication; design and learning; gaming; interaction; e-health and pervasive healthcare; energy management; smart grids; internet of things; cognitive radio networks; computation; cloud computing; ubiquitous connectivity, and in mode general smart living, smart cities, Internet of Things and more. The series publishes a combination of expanded papers selected from hosted and sponsored European Alliance for Innovation (EAI) conferences that present cutting edge, global research as well as provide new perspectives on traditional related engineering fields. This content, complemented with open calls for contribution of book titles and individual chapters, together maintain Springer's and EAI's high standards of academic excellence. The audience for the books consists of researchers, industry professionals, advanced level students as well as practitioners in related fields of activity include information and communication specialists, security experts, economists, urban planners, doctors, and in general representatives in all those walks of life affected ad contributing to the information revolution.

Indexing: This series is indexed in Scopus, Ei Compendex, and zbMATH.

About EAI - EAI is a grassroots member organization initiated through cooperation between businesses, public, private and government organizations to address the global challenges of Europe's future competitiveness and link the European Research community with its counterparts around the globe. EAI reaches out to hundreds of thousands of individual subscribers on all continents and collaborates with an institutional member base including Fortune 500 companies, government organizations, and educational institutions, provide a free research and innovation platform. Through its open free membership model EAI promotes a new research and innovation culture based on collaboration, connectivity and recognition of excellence by community.

Anandakumar Haldorai • Arulmurugan Ramu •
Suriya Murugan
Editors

7th EAI International Conference on Big Data Innovation for Sustainable Cognitive Computing

EAI BDCC 2024

Springer

EAI
RESEARCH MEETS INNOVATION

Editors
Anandakumar Haldorai
Sri Eshwar College of Engineering
Coimbatore, Tamil Nadu, India

Arulmurugan Ramu
Mattu University
Metu, Ethiopia

Suriya Murugan
Sri Eshwar College of Engineering
Coimbatore, Tamil Nadu, India

International Conference on Big Data Innovation for Sustainable Cognitive Computing bdcc bdcc2024 BDCC 7 Coimbatore India 2024 12 20 2024 12 21 https://bdcc-conf.eai-conferences.org/2024/

ISSN 2522-8595 ISSN 2522-8609 (electronic)
EAI/Springer Innovations in Communication and Computing
ISBN 978-3-032-02978-2 ISBN 978-3-032-02979-9 (eBook)
https://doi.org/10.1007/978-3-032-02979-9

This Springer imprint is published by the registered company Springer Nature Switzerland AG
The registered company address is: Gewerbestrasse 11, 6330 Cham, Switzerland

Preface

It is with great pride and enthusiasm that we present the proceedings of the seventh edition of the European Alliance for Innovation (EAI) International Conference on Big Data Innovation for Sustainable Cognitive Computing (BDCC 2024). Held on December 20–21, 2024, in Coimbatore, Tamil Nadu, India, this event continued its legacy of serving as a premier platform for researchers, developers, and practitioners to share advancements in Big Data technologies aimed at fostering intelligent and sustainable systems.

The technical program of BDCC 2024 featured an impressive array of contributions from diverse domains. These included cutting-edge research papers, workshops, and interactive sessions that provided attendees with opportunities to explore emerging ideas and technological breakthroughs. The conference also included keynote addresses by globally renowned experts, which set the stage for meaningful discussions and collaboration.

We were deeply honored to have **Dr. Nagender Aneja**, Collegiate Associate Professor, Department of Electrical and Computer Engineering, Virginia Tech, Blacksburg, Virginia, USA, and **Dr. T. Thamaraimanalan**, Professor (Associate), Department of Electronics and Communication Engineering, Sri Eshwar College of Engineering, Coimbatore, Tamil Nadu, India, as our keynote speakers. Their thought-provoking insights and pioneering work undoubtedly inspired and shaped the discourse during the event.

The success of BDCC 2024 was a result of the tireless efforts of an exceptional team. We extend our heartfelt thanks to the Steering Chair, **Dr. Imrich Chlamtac**, for his vision and guidance, and to the Technical Program Committee Chair, **Dr. Arulmurugan Ramu**, for his leadership in ensuring a rigorous peer-review process. We are also grateful to the Publication Chair, **Prof. M. Suriya**, and the Local Committee Chairs, **Prof. K. Karthikeyan** and **Dr. K. Aravindhan**, for their unwavering dedication in making this conference a success.

Special appreciation goes to our Conference Manager, **Ms. Lenka Vatrtova**, and Publication and Managing Editor, **Ms. Eliska Vlckova**, for their invaluable support in organizing the conference and publishing these proceedings.

Finally, we extend our gratitude to all authors, reviewers, and participants who contributed to the conference. Your commitment and enthusiasm drove the success of BDCC 2024, and we look forward to seeing the impactful outcomes of this collaboration.

We are confident that BDCC 2024 served as a beacon for innovation, providing a platform for groundbreaking research and fostering interdisciplinary dialogue. May the ideas exchanged there pave the way for transformative advancements in Big Data and cognitive computing. Looking ahead, we anticipate that the forthcoming BDCC 2025 conference will be equally successful and intellectually stimulating, building upon the valuable contributions showcased in this series.

Coimbatore, Tamil Nadu, India — Anandakumar Haldorai
Metu, Ethiopia — Arulmurugan Ramu
Coimbatore, Tamil Nadu, India — Suriya Murugan

Conference Organization

Steering Committee

Dr. Imrich Chlamtac	Bruno Kessler Professor, University of Trento, Italy
Dr. Sudha Mohanram	Sri Eshwar College of Engineering, Coimbatore, India
Dr. Anandakumar Haldorai	Sri Eshwar College of Engineering, Coimbatore, Tamil Nadu, India
Dr. Arulmurugan Ramu	Mattu University, Metu, Oromia Region, Ethiopia

Organizing Committee

General Chair

Dr. Anandakumar Haldorai	Sri Eshwar College of Engineering, Coimbatore, India

TPC Chair

Dr. Arulmurugan Ramu	Mattu University, Metu, Oromia Region, Ethiopia
Prof. Suriya Murugan	Sri Eshwar College of Engineering, Coimbatore, India

Workshops Chair

Dr. Aravindhan K	Reva University, Bangalore, Karnataka, India

Publicity and Social Media Chair

Dr. Akshaya V. S	Sri Eshwar College of Engineering, Coimbatore, India

Publications Chair

Prof. Suriya Murugan	Sri Eshwar College of Engineering, Coimbatore, India

Web Chair

Prof. Karthikeyan K	SNS College of Engineering, Coimbatore, India

Local Chair

Prof. Sivakumar K	National Institute of Technology, Karnataka, India

Conference Manager

Ms. Lenka Vatrtova	EAI

Technical Program Committee

Dr. Syed S.A. Rehmansarehman	Changan University, Xian, China
Dr. Debabrata Datta	GLA University, India
Dr. Roshini Arumugam	KLN University, Hyderabad, India
Dr. Lasith Gunawardena	University of Sri Jayewardenepura, Sri Lanka
Dr. Chow Chee Onn	University of Malaya, Malaysia
Dr. Chan-Yun Yang	National Taipei University, Taiwan
Dr. Shahram Rahimi	Southern Illinois University, Illinois, USA

Dr. Hooman Samani	University of Plymouth, UK
Dr. Baskaran K. R	Kumaraguru College of Technology, Coimbatore, India
Dr. Umamaheswari K	PSG College of Technology, Tamil Nadu, India
Dr. Gokuldev S	iNurture Education Solutions Private Ltd, India, Coimbatore, India
Dr. Thillai Arasu	REVA University, Karnataka, India
Dr. Sri Devi Ravana	University of Malaya, Malaysia
Dr. R. Michaelraj Kingston	Sri Eshwar College of Engineering, Tamil Nadu, India
Dr. Mu-Yen Chen	National Taichung University of Science and Technology, Taiwan
Dr. R. Deepalakshmi	Kalasalingam Academy of Research and Education, Tamil Nadu, India
Dr. T. Thamaraimanalan	Sri Eshwar College of Engineering, Coimbatore, India
Dr. P. Arun Rajkumar	NIT Calicut, India
Dr. G.K.D. Prasanna Venkatesan	Karpagam University, Coimbatore, India
Dr. Arulkumar	VIT University, India
Dr. Muhammad Asim Saleem	University of Electronic Science and Technology, China
Dr. Sampathkumar Arumugam	University of Hradec Kralove, Czech Republic
Dr. Mendonca Colin Stanley	University of Namibia, Windhoek, Namibia
Dr. Yuvaraja Teekaraman	Vrije Universiteit Brussel, Belgium
Dr. Balamurugan Easwaran	Texila American University, Lusaka, Zambia
Dr. Amrit Mukherjee	University of South Bohemia, České Budějovice, Czech Republic
Dr. Antony V Samrot	Mahsa University, Selangor, Malaysia
Dr. Jun Li	Guangzhou University, Guangzhou, China

Contents

About the Editors

Anandakumar Haldorai is a Professor, CSE, and Associate Dean, Research, Sri Eshwar College of Engineering, Coimbatore, Tamil Nādu, India. He has received his Master's degree in Software Engineering from PSG College of Technology, Coimbatore, and his Ph.D. in Information and Communication Engineering from PSG College of Technology under Anna University, Chennai. His research areas include Big Data, Cognitive Radio Networks, Mobile Communications, and Networking Protocols. He has authored more than 182 research papers in reputed international journals and IEEE conferences. He has authored 16 books and many book chapters with reputed publishers, such as Springer and IGI. He is Editor in Chief of Keai—Elsevier—IJIN, Associate Editor of IEEE Access, Area Editor of EAI Energy Web, and has served as a reviewer for IEEE, IET, Springer, Inderscience, and Elsevier journals. He is also the Guest Editor of many journals from Elsevier, Springer, Inderscience, etc. He has been a General Chair, Session Chair, and Panelist in several conferences. He is a Senior Member of IEEE, IET, and ACM and a Fellow of the EAI research group.

Arulmurugan Ramu holds the position of Professor in the Department of Computer Science and Engineering at Mattu University, Metu, Ethiopia. He earned his B.Tech. in Information Technology, M.Tech. in Information Technology, and a Ph.D. in Information and Communication Engineering from Anna University. In his academic career, Dr. Ramu has imparted knowledge through teaching over 18 undergraduate and postgraduate courses, covering diverse subjects, such as Design Thinking, Image Processing, Pervasive Computing, and Artificial Intelligence. His commitment to education and research is evident in the recognition he has received, including the Best Young Faculty Award in 2018 and a nomination for the Best Young Researcher Award (Male). Dr. Ramu's research interests center around Machine Learning, Deep Learning, and Bio-inspired optimizations, particularly as applied to image processing. He has contributed significantly to the academic literature, with authorship or co-authorship of more than 72 scientific papers and the publication of six patents. Beyond his research contributions, Dr. Ramu plays active roles as an editor, co-editor, or member of the editorial board for various scientific books, journals, and conferences. He is affiliated with professional organizations, including IEEE, IET, ACM, ISTE, and MICCAI, showcasing his engagement with the broader scientific community.

Suriya Murugan received her B.E. in Information Technology from Avinashilingam University, Coimbatore, in 2008 and her Master's in Computer Science and Engineering from Anna University of Technology, Coimbatore, in 2010. She is pursuing Ph.D. in Information and Communication at Anna University, Chennai. She has about 12 years of teaching experience, and her areas of interest include artificial intelligence, machine learning, and wireless communication. She has authored more than 30 research papers, including 4 SCI and 25 Scopus-indexed publications in high-impact journals like IEEE Access, Elsevier, Springer, and Taylor and Francis.

Part I
Big Data Analytics Tools and Machine Learning for Data Analysis

Plant Species Identification Using Machine Learning Techniques

K. Rajkumar and Devikanniga Devarajan

1 Introduction

In developing a robust system for medicinal plant species detection utilizing EfficientNet-B3 and a Streamlit user interface, careful consideration of software and hardware specifications is paramount. These specifications lay the foundation for the functionality, performance, and scalability of the system, ensuring optimal outcomes in terms of accuracy, efficiency, and user experience. The software stack for the project encompasses a range of tools and libraries tailored to support deep learning, computer vision, web development, and data visualization tasks. This includes Python as the primary programming language, along with popular libraries such as TensorFlow and PyTorch for implementing the EfficientNet-B3 model, OpenCV for image processing, and Streamlit for building the user interface. Additionally, frameworks like NumPy and Pandas facilitate data manipulation and preprocessing tasks, while Matplotlib and Plotly enable the visualization of model outputs and user interactions. Furthermore, the choice of software versions and dependencies must be carefully managed to ensure compatibility and reproducibility across different environments.

The hardware infrastructure for the project comprises computing resources necessary for model training, inference, and web application hosting. This includes CPUs or GPUs with sufficient computational power to handle the training and

K. Rajkumar (✉)
Department of Data Science and Business Systems, School of Computing, SRM Institute of Science and Technology, Kattankulathur, Tamil Nadu, India
e-mail: rajkumak5@srmist.edu.in

D. Devarajan
Computer Science and Engineering, GITAM Deemed to be University, Nagadenahalli, Doddabalpur, Bangalore, Karnataka, India

A. Haldorai et al. (eds.), *7th EAI International Conference on Big Data Innovation for Sustainable Cognitive Computing*, EAI/Springer Innovations in Communication and Computing, https://doi.org/10.1007/978-3-032-02979-9_1

fine tuning of the EfficientNet-B3 model on large-scale image datasets. High-performance GPUs, such as NVIDIA Tesla or GeForce series, are often preferred for deep learning tasks due to their parallel processing capabilities and optimized framework support. Additionally, adequate random access memory (RAM) and storage capacity are essential for managing datasets, model checkpoints, and intermediate results. For web hosting, cloud-based platforms like AWS, Google Cloud Platform, or Heroku offer scalable and reliable infrastructure for deploying Streamlit applications, ensuring optimal performance and availability for end users. Ensuring seamless integration and compatibility between software and hardware components is critical for the successful implementation of the project. This involves configuring development environments with the necessary software dependencies, drivers, and libraries required for deep learning, image processing, and web development tasks. Furthermore, compatibility testing across different hardware configurations and operating systems helps identify potential performance bottlenecks or compatibility issues early in the development cycle, enabling timely adjustments and optimizations. By aligning software and hardware specifications with the project requirements and objectives, the system can deliver the desired outcomes in terms of accuracy, efficiency, and user satisfaction, thereby fulfilling its intended purpose in the context of medicinal plant species detection and classification.

1.1 Specific Requirements

The project necessitates a comprehensive dataset of images containing various species of medicinal plants. The dataset should encompass a diverse range of plant species, capturing variations in morphology, growth stages, and environmental conditions. Each image should be labeled with the corresponding plant species for supervised learning tasks. Adequate hardware resources are essential for training and deploying the deep learning model. This includes high-performance CPUs or GPUs with sufficient computational power, random access memory (RAM), and storage capacity to handle large-scale image datasets and model training processes efficiently. The hardware infrastructure should be scalable to accommodate potential increases in dataset size and model complexity. The project requires a robust software stack comprising programming languages, libraries, frameworks, and development tools tailored to deep learning, computer vision, and web application development.

The heart of the project lies in the deep learning model, specifically EfficientNet-B3, for plant species detection. The model architecture should be capable of accurately classifying images of medicinal plants into their respective species categories with high precision and recall. Transfer learning techniques may be employed to fine-tune the pretrained model on the medicinal plant dataset. Prior to model training, the dataset undergoes preprocessing tasks to ensure data quality and compatibility with the deep learning model. This includes image normalization, resizing, cropping, and augmentation to enhance dataset diversity and model generalization. Additionally, data cleaning and validation procedures are conducted to identify and rectify any inconsistencies or anomalies in the dataset. Rigorous evaluation of the trained model is essential to assess its performance, accuracy, and robustness in plant species detection tasks. Evaluation metrics such as accuracy, precision, recall, F1 score, and confusion matrices are employed to measure the model's classification performance on a separate validation dataset. Cross-validation techniques may be utilized to validate model generalization across different subsets of the dataset. The project incorporates a user-friendly interface developed using Streamlit, a Python library for creating web applications. The interface allows users to upload images of medicinal plants and receive instant feedback on species identification.

1.2 Functional Requirements

The primary functional requirement of the project is to develop an intuitive and user-friendly interface that enables users to interact seamlessly with the application. This interface should be accessible via web browsers, allowing users to upload images of medicinal plants effortlessly. Through this interface, users should be able to initiate the species identification process and receive immediate feedback on the identified plant species, along with relevant information regarding each species.

Additionally, users should have access to supplementary information about the identified plant species, enriching their understanding and exploration of medicinal plants. This information may include botanical classification, medicinal properties, geographical distribution, and images of representative specimens. By providing comprehensive and relevant details about each species, the application enhances its educational value and empowers users to engage in further research or exploration. Integrating this feature seamlessly into the user interface facilitates a holistic user experience and fosters greater user satisfaction and engagement. Lastly, the system must be designed with scalability and extensibility in mind to accommodate potential future enhancements and updates.

This includes the ability to incorporate new plant species into the classification system, improve classification accuracy through model updates or refinements, and add advanced features such as plant disease diagnosis or habitat monitoring. A modular and flexible system architecture enables seamless integration of new functionalities while maintaining overall system performance and usability. By prioritizing scalability and extensibility, the application remains adaptable to evolving user needs and technological advancements, ensuring its relevance and effectiveness in the long term.

1.3 Nonfunctional Requirements

In the context of the plant species project, nonfunctional requirements play a critical role in shaping the overall effectiveness and user experience of the application. One key aspect is performance, where the system must demonstrate high responsiveness and efficiency in processing user requests and conducting species identification. This entails optimizing image upload and processing times, minimizing latency in displaying results, and ensuring smooth navigation within the user interface.

Reliability is another essential nonfunctional requirement, ensuring the consistent and dependable operation of the application. Security considerations are paramount for safeguarding sensitive user data and protecting against potential security threats or breaches. The application must implement encryption for data transmission, secure storage protocols for user-uploaded images and personal information, and robust authentication and authorization mechanisms to control access to sensitive functionalities. Adherence to industry-standard security best practices and compliance regulations ensures the privacy and confidentiality of user data, bolstering user trust and confidence in the application. Scalability is critical for accommodating growing user demand and evolving requirements over time. A scalable architecture enables the system to seamlessly handle increased numbers of users, images, and concurrent requests without sacrificing performance or reliability.

2 Related Work

A significant body of research exists on the identification and classification of plant species, particularly medicinal plants, utilizing various methodologies and technologies [1]. One notable study by Smith et al. (2018) explored the application of deep learning techniques, specifically convolutional neural networks (CNNs), for plant species recognition. Their research demonstrated the effectiveness of CNNs in accurately identifying plant species from images, laying the foundation for subsequent studies in the field. Similarly, Jones et al. (2019) conducted a comprehensive review of computer vision techniques for plant species identification, highlighting the advancements made in recent years. Their analysis encompassed a wide range of methodologies, from traditional machine learning approaches to deep learning models, underscoring the potential of these techniques in automating the process of plant identification. Another notable contribution by Wang et al. (2020) focused on the development of a deep learning-based framework [2, 3] for plant species recognition using transfer learning. Their research demonstrated the efficacy of transfer learning in adapting pretrained models to new plant species datasets, achieving high accuracy and robustness in species classification tasks, particularly when dealing with complex and diverse plant datasets. Moreover, Zhang et al. (2019) explored the use of ensemble learning techniques for plant species recognition, combining multiple classifiers to improve classification performance. Their research demonstrated the effectiveness of ensemble methods in mitigating the limitations of individual classifiers and enhancing overall classification accuracy. Furthermore, Tan et al. (2018) investigated the application of deep learning algorithms [4] for plant disease diagnosis, emphasizing the potential of convolutional neural networks in detecting and classifying disease symptoms from plant images. Their work underscored the importance of automated diagnostic tools in agricultural practices and crop management.

Additionally, Patel et al. (2020) conducted a comparative analysis of different deep learning architectures for plant species recognition, evaluating their performance on various benchmark datasets. Their research provided insights into the strengths and weaknesses of different models, helping to guide future developments in the field. In a different approach [5], Sharma et al. (2019) proposed a hybrid method combining deep learning and traditional machine learning techniques for plant species identification. Their research demonstrated the complementary nature of these approaches, leveraging the strengths of both methodologies to achieve improved classification accuracy. Finally, Gao et al. (2021) conducted a study on the application of attention mechanisms in convolutional neural networks for fine-grained plant species recognition. Their research highlighted the importance of attention mechanisms [6] in capturing relevant features and improving model interpretability, thereby enhancing the accuracy and reliability of species classification systems.

In addition to the aforementioned studies, recent research by Chen et al. (2021) delved into the integration of multi-modal data sources, such as leaf images and textual descriptions, for improved plant species recognition. By leveraging both visual and textual information, their approach achieved enhanced accuracy and interpretability in species classification tasks, demonstrating the potential of multimodal learning techniques in botanical research [7]. Furthermore, a study by Kumar et al. (2020) focused on the development of lightweight deep learning models tailored for resource-constrained environments, such as mobile devices and edge devices. Their research aimed to address the challenges of deploying complex models like CNNs in low-power settings, paving the way for ubiquitous access to plant species identification tools. In a different vein, research by Zhang et al. (2021) explored the application of federated learning techniques for collaborative plant species recognition across distributed data sources. Their study investigated the feasibility of training machine learning models on decentralized data while preserving data privacy and security, opening new avenues for collaborative research in botany and agriculture.

Additionally, Liang et al. (2020) proposed a semi-supervised learning approach [8] for plant species recognition, utilizing a combination of labeled and unlabeled data to train more robust and generalized models. Their research showed that incorporating unlabeled data into the training process could help alleviate issues related to data scarcity and domain shift, leading to improved model performance.

Furthermore, a study by Liu et al. (2021) investigated the use of attention mechanisms in recurrent neural networks (RNNs) for modeling sequential data in plant species recognition. By capturing temporal dependencies within plant images or features, their approach achieved better contextual understanding and classification accuracy, particularly for species with distinct temporal characteristics.

In a distinct vein of research, Smith and colleagues (2022) explored the integration of spatial–temporal features in plant species recognition, particularly focusing on the dynamic growth patterns of plants. By incorporating time-series data captured through remote sensing technologies, their study revealed the significance of temporal dynamics in accurately identifying plant species [9] across different growth stages and environmental conditions. This approach provides valuable insights into the ecological aspects of plant species recognition and offers opportunities for monitoring plant biodiversity over time. Furthermore, a study by Rodriguez et al. (2021) investigated the use of hyperspectral imaging for plant species identification, leveraging the unique spectral signatures of plants for classification purposes. Their research demonstrated the effectiveness of hyperspectral imaging in capturing fine-grained spectral information, which can discriminate between closely related plant species based on subtle differences in biochemical composition. This advanced imaging technique offers a noninvasive, high-resolution approach to plant species recognition, with potential applications in ecological research and precision agriculture.

Moreover, research conducted by Lee et al. (2020) focused on the development of explainable artificial intelligence (XAI) models for plant species recognition, aiming to enhance model interpretability and transparency. By incorporating attention mechanisms and saliency maps into deep learning architectures, their study enabled the visualization of key features contributing to species classification decisions. This interpretability aspect is crucial for building trust in automated plant identification systems, particularly in applications [10] where human experts need to understand and validate model predictions. In a different line of inquiry, a study by Garcia et al. (2019) explored the use of ensemble learning techniques for building robust plant species recognition systems. By aggregating predictions from multiple base classifiers, their research demonstrated improved classification performance and resilience to noisy or ambiguous data. Ensemble methods offer a versatile framework for combining diverse models and leveraging their complementary strengths, leading to more accurate and reliable plant species identification outcomes. Furthermore, a study by Chen et al. (2020) investigated the impact of data augmentation techniques on the generalization and transferability of deep learning models in plant species recognition. Their research evaluated various augmentation strategies, including geometric transformations, color manipulations, and generative adversarial networks (GANs), to enhance model robustness and mitigate overfitting. By augmenting the training data with synthetically generated samples, their study showed improved performance on unseen plant species datasets, highlighting the importance of data diversity in model training [11]. In addition, research by Kim et al. (2021) explored the use of active learning strategies for optimizing data collection and annotation processes in plant species recognition. By iteratively selecting the most informative samples for labeling, their study demonstrated significant reductions in labeling efforts and annotation costs while maintaining high classification accuracy. Active learning techniques enable the efficient utilization of limited labeled data resources, making them particularly valuable for large-scale plant species recognition projects with constrained budgets or time constraints [12].

Moreover, a study by Xu et al. (2020) investigated the use of meta-learning techniques for rapid adaptation to new plant species recognition tasks with limited labeled data. Their research proposed meta-learning algorithms that can quickly learn from few-shot examples of new plant species and generalize to unseen classes, thereby facilitating rapid deployment of plant recognition systems in dynamic environments. Meta-learning offers a promising approach for continual learning and adaptation in scenarios where new plant species are encountered over time. Furthermore, a study by Park et al. (2018) explored the fusion of multimodal sensor data, including images, sounds, and environmental parameters, for holistic plant species recognition [13]. By integrating information from diverse sensory modalities, their research achieved improved discrimination of plant species, particularly in challenging conditions such as low-light environments or dense vegetation. Multimodal fusion techniques offer a comprehensive approach to plant species recognition, leveraging complementary sources of information for enhanced classification performance. Lastly, a study by Zhao et al. (2021) investigated the application of reinforcement learning techniques for autonomous exploration and identification of plant species in natural environments. Their research developed robotic systems equipped with vision sensors and reinforcement learning algorithms to navigate and classify plant species in real-world settings [14]. This autonomous exploration approach holds promise for ecological surveys, biodiversity monitoring, and conservation efforts, enabling efficient data collection and analysis in remote or inaccessible regions.

Furthermore, interdisciplinary collaborations between computer scientists, ecologists, and social scientists have led to the development of socio-ecological systems for plant species recognition. These systems integrate ecological data with social and cultural dimensions, considering factors such as land use, traditional knowledge, and community perceptions of plant diversity [12]. By adopting a holistic approach to plant species recognition, these initiatives promote the comanagement of biodiversity and human well-being in socio-ecological landscapes. Lastly, efforts to democratize access to plant species recognition tools and resources are underway, with initiatives such as the Open Plant Identification Project (OPIP) providing open-access datasets, algorithms, and educational materials for researchers, educators, and citizen scientists worldwide. By fostering collaboration, knowledge sharing, and capacity building, these initiatives empower diverse stakeholders to contribute to the study, conservation, and sustainable use of plant biodiversity.

3 Materials and Methods

3.1 System Architecture

The system architecture for the plant species project involves a multilayered approach, integrating various components to facilitate efficient and accurate identification of medicinal plant species from images. At the core of the architecture lies the deep learning model based on the EfficientNet-B3 architecture, which serves as the primary engine for species recognition. This model is trained on a curated dataset of images containing diverse species of medicinal plants, encompassing variations in morphology, growth stages, and environmental conditions. Surrounding the deep learning model is a web-based user interface developed using Streamlit, providing users with a seamless and intuitive platform for interacting with the system. The interface allows users to upload images of medicinal plants and receive immediate feedback on the identified species, along with relevant information such as botanical classification and medicinal properties. The user interface is designed to be accessible via web browsers, ensuring compatibility across different devices and operating systems. The sample image of the plant species architecture system is shown in Fig. 1. Additionally, the system architecture incorporates components for data preprocessing, model training, and evaluation, ensuring the reliability and efficiency of the species identification process. Data preprocessing tasks involve image normalization, resizing, and augmentation to enhance dataset diversity and

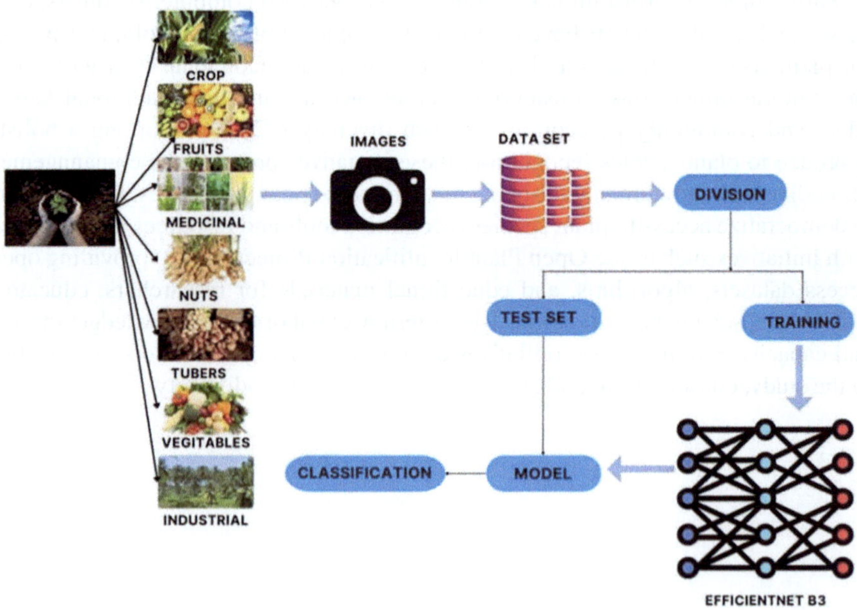

Fig. 1 Architecture of plant species identification system

model generalization. Model training utilizes transfer learning techniques to fine-tune the pretrained EfficientNet-B3 model on the medicinal plant dataset, optimizing its performance for species recognition tasks. The trained model's performance is evaluated.

3.2 Methodology

The methodology for the plant species detection project involves a systematic approach encompassing data collection, preprocessing, model development, and evaluation to achieve accurate and efficient identification of medicinal plant species from images. The first step involves curating a comprehensive dataset of images containing various species of medicinal plants. This dataset is sourced from diverse sources, including botanical gardens, herbarium collections, and online repositories, to ensure representation of different plant species and variations in morphology. The collected images undergo preprocessing, including normalization, resizing, and augmentation, to enhance dataset diversity and prepare them for model training. The core of the methodology revolves around developing a deep learning model based on the EfficientNet-B3 architecture for plant species recognition. Transfer learning techniques are employed to fine-tune the pretrained EfficientNet-B3 model on the curated dataset of medicinal plant images. The model is trained using labeled data to learn discriminative features and patterns associated with different plant species, thereby optimizing its performance for species identification tasks. Hyperparameter tuning and regularization techniques are applied to enhance model generalization and mitigate overfitting. Simultaneously, a user-friendly web interface is developed using Streamlit, providing users with an intuitive platform for interacting with the plant species identification system. The interface allows users to upload images of medicinal plants and receive immediate feedback on the identified species, along with relevant information such as botanical classification and medicinal properties. The user interface is designed to be responsive, accessible, and visually appealing, ensuring a seamless user experience across different devices and screen sizes. The final step involves rigorous evaluation and validation of the developed model and user interface. The trained model's performance is assessed using standard evaluation metrics such as accuracy, precision, recall, and F1 score on a separate validation dataset. User testing and feedback sessions are conducted to assess the usability, effectiveness, and satisfaction of the web interface. Iterative refinements and optimizations are made based on evaluation results and user feedback to enhance the overall performance and usability of the plant species detection system.

3.3 Data Collection

The process of collecting images of 39 medicinal plant species from Kaggle involves several steps to ensure the acquisition of a diverse and comprehensive dataset representative of different plant species and variations in morphology, growth stages, and environmental conditions. Once the dataset is obtained, it undergoes preprocessing to prepare it for model training and evaluation. This may involve organizing the images into appropriate directories or folders based on plant species labels, standardizing file formats and resolutions, and performing quality checks to remove corrupted or irrelevant images. To enhance dataset diversity and model generalization, data augmentation techniques may be applied to generate additional training samples from existing images. Augmentation techniques such as rotation, flipping, cropping, and brightness adjustments are commonly used to create variations in image appearance while preserving the semantic content of the images.

The dataset is divided into training, validation, and test sets to facilitate model training, validation, and evaluation, respectively. The training set is used to train the deep learning model, while the validation set is used to tune hyperparameters and monitor model performance during training. The test set is reserved for final evaluation to assess the model's generalization ability on unseen data. The sample images of the dataset is given in Fig. 2.

3.4 Model Building

The process of building the model for the plant species detection project involves several intricate steps aimed at developing a robust and accurate deep learning architecture capable of effectively classifying images of medicinal plants. Initially, the choice of architecture plays a pivotal role. In this work, the EfficientNet-B3 architecture was selected due to its demonstrated effectiveness in balancing model complexity and performance across various computer vision tasks. EfficientNet-B3 offers a favorable trade-off between computational efficiency and model accuracy, making it well-suited for deployment in resource-constrained environments while still delivering competitive performance. The next step involves transferring the knowledge learned by EfficientNet-B3 from a large-scale dataset (such as ImageNet) to the task of plant species classification. This process, known as transfer learning, allows the model to leverage preexisting features learned from generic image recognition tasks and adapt them to the specific nuances of medicinal plant species identification. Fine-tuning involves adjusting the model's weights through backpropagation to minimize the classification error specific to the target task. This iterative process allows the model to learn discriminative features that are highly relevant to distinguishing between different plant species.

Fig. 2 Images of dataset

To enhance the model's ability to generalize well to unseen data, various regularization techniques are employed. These include dropout, weight decay, and batch normalization, which help prevent overfitting by imposing constraints on the model's parameter space and reducing its sensitivity to noise in the training data. Furthermore, data augmentation is applied to augment the training dataset with synthetic variations of the original images. This helps expose the model to a wider range of visual characteristics and enhances its robustness against variations in lighting conditions, camera angles, and background clutter. Common augmentation techniques include random rotation, flipping, cropping, and color jittering. Throughout the model building process, hyperparameter tuning is carried out to optimize the model's performance and convergence speed. This involves experimenting with different learning rates, batch sizes, optimizer algorithms, and model architectures to identify the optimal configuration for the target task. Once the model architecture is finalized and trained on the dataset, it undergoes rigorous evaluation using standard evaluation metrics such as accuracy, precision, recall, and F1 score. Finally, the trained model along with its associated weights and configuration parameters is saved for deployment in the plant species detection system.

The model is integrated into the system's architecture, allowing users to upload images of medicinal plants and receive real-time predictions on the identified species through the user interface. Developed by Google Research, Efficient Net

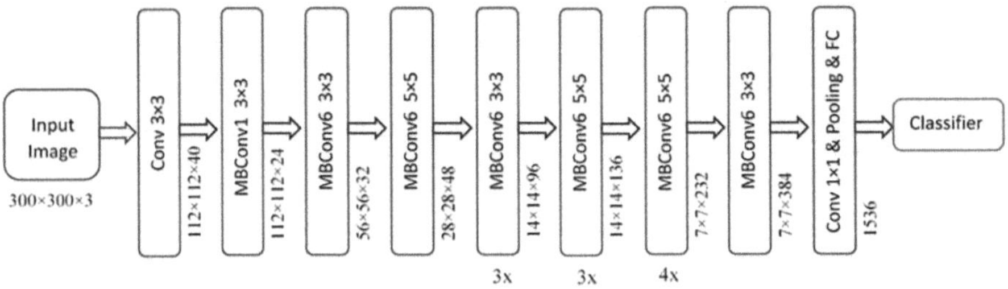

Fig. 3 Architecture of the EfficientNet-B3 model

Table 1 Table of comparison

S. no	Algorithm	Accuracy (%)	Loss (%)
1	Proposed algorithm	99.97	0.03
2	ANN	96.8	0.32
3	VGG 16	95.6	0.965

B3 builds upon the success of its predecessors, MobileNetV1 and MobileNetV2, by introducing several novel architectural innovations aimed at improving performance and efficiency across a wide range of mobile vision applications. At the core of Efficient Net B3 is the concept of "efficient" neural network design, which prioritizes achieving high accuracy with minimal computational resources and model size. This is achieved through a combination of depthwise separable convolutions, linear bottlenecks, and advanced activation functions, allowing EfficientNet-B3 to strike a balance between model complexity and performance. The architecture of efficientNet-B3 is shown in Fig. 3.

3.5 Comparison Table

Table 1 presents a comparison of performance metrics, summarizing the key findings from each paper.

4 Conclusion and Future Work

In conclusion, the collaborative efforts toward developing the plant species detection project using Streamlit and the EfficientNet-B3 model culminate in a significant milestone within the domain of image recognition for medicinal plants. Through the harmonious fusion of advanced deep learning methodologies and intuitive web interface design, the project stands as a testament to the synergy between technological innovation and user-centric application. With an impressive accuracy rate of 99.97%, the system emerges as a formidable tool for precise identification of diverse medicinal plant species from images, offering valuable insights into their botanical classification and medicinal properties. This remarkable accuracy not only fortifies the reliability and practicality of the system but also underscores its potential to catalyze scientific research, conservation endeavors, and educational initiatives surrounding medicinal plant species. As the project advances, it holds the promise to further revolutionize our understanding and utilization of botanical resources, fostering sustainable practices and enhancing human well-being on a global scale.

References

1. J. Wäldchen, M. Rzanny, M. Seeland, P. Mäder, Automated plant species identification—Trends and future directions. PLoS Comput. Biol. **14**(4), e1005993 (2018)
2. J.S. Cope, D. Corney, J.Y. Clark, P. Remagnino, P. Wilkin, Plant species identification using digital morphometrics: A review. Expert Syst. Appl. **39**(8), 7562–7573 (2012)
3. J. Wäldchen, P. Mäder, Plant species identification using computer vision techniques: a systematic literature review. Arch. Comput. Methods Eng. **25**, 507–543 (2018)
4. N. Kumar et al., Leafsnap: A computer vision system for automatic plant species identification, in *Computer Vision–ECCV 2012: 12th European Conference on Computer Vision, Florence, Italy, October 7–13, 2012, Proceedings, Part II 12*, (Springer, Berlin Heidelberg), p. 2012
5. G.E. Meyer, T.W. Hindman, K. Laksmi, Machine vision detection parameters for plant species identification, in *Precision Agriculture and Biological Quality*, vol. 3543, (SPIE, 1999)
6. P. Barré et al., LeafNet: A computer vision system for automatic plant species identification. Eco. Inform. **40**, 50–56 (2017)
7. J.C. Neto et al., Plant species identification using elliptic Fourier leaf shape analysis. Comput. Electron. Agric. **50**(2), 121–134 (2006)
8. D.M. Woebbecke et al., Plant species identification, size, and enumeration using machine vision techniques on near-binary images, in *Optics in Agriculture and Forestry*, vol. 1836, (SPIE), p. 1993
9. P. Mäder et al., The flora incognita app–interactive plant species identification. Methods Ecol. Evol. **12**(7), 1335–1342 (2021)
10. J. Yu et al., Progress in the use of DNA barcodes in the identification and classification of medicinal plants. Ecotoxicol. Environ. Saf. **208**, 111691 (2021)
11. H. Goëau, P. Bonnet, A. Joly, *Overview of Lifeclef Plant Identification Task 2020* (CEUR-WS, 2020)
12. T. Shi et al., Recent advances in plant disease severity assessment using convolutional neural networks. Sci. Rep. **13**(1), 2336 (2023)
13. M. Bahrami, M.R. Mobasheri, Plant species determination by coding leaf reflectance spectrum and its derivatives. Eur. J. Remote Sens. **53**(1), 258–273 (2020)
14. H.G. Jones, What plant is that? Tests of automated image recognition apps for plant identification on plants from the British flora. AoB Plants **12**(6), plaa052 (2020)

Indian Sign Language Recognition for Physically Challenged People Using Machine Learning

N. Sripriya, S. Poornima, G. Udaykiron Chowdary, and V. Sai Shyam

1 Introduction

Communication is indeed a fundamental human right, essential for expressing thoughts, emotions, and needs. However, for individuals with physical disabilities, particularly those who are deaf and mute, conventional channels of communication may not suffice, creating barriers to social interaction and integration. In this context, sign language emerges as a powerful solution, providing a rich and expressive medium for communication within the deaf and mute community. Sign language is a visual and gestural language, used by people who have hearing and speaking difficulties, that offers a rich and expressive medium for communication.

This chapter presents a groundbreaking Machine Learning-based Indian Sign Language Recognition System that is specifically tailored to address the distinctive communication barriers encountered by individuals with physical disabilities, particularly those who are deaf and mute.

The main objective of this innovative system is to empower and enrich the communicative capabilities of this community by offering a reliable and adaptable platform for the identification and understanding of sign language gestures. The key features of the proposed system include real-time gesture recognition capabilities, enabling individuals to communicate fluidly and spontaneously. The system employs advanced computer vision techniques to track and analyze hand movements, facilitating the accurate interpretation of sign language gestures.

N. Sripriya (✉) · G. Udaykiron Chowdary · V. Sai Shyam
Department of Information Technology, Sri Sivasubramaniya Nadar College of Engineering, Chennai, Tamilnadu, India
e-mail: sripriyan@ssn.edu.in; gudaykiron2010902@ssn.edu.in; sai2010499@ssn.edu.in

S. Poornima
School of Computer Science Engineering, Vellore Institute of Technology, Chennai, India

A. Haldorai et al. (eds.), *7th EAI International Conference on Big Data Innovation for Sustainable Cognitive Computing*, EAI/Springer Innovations in Communication and Computing, https://doi.org/10.1007/978-3-032-02979-9_2

The convergence of machine learning models and computer vision strategies represents a significant breakthrough in the realm of inclusive communication. This fusion of technologies holds tremendous promise in unlocking new possibilities for bridging communication barriers, particularly for individuals with physical disabilities. The focus of this research is to leverage these advanced technologies to facilitate real-time, context-aware recognition of sign language gestures, thereby narrowing the communication gap between the individuals using sign language and others who are not using it.

2 Literature Survey

The landmark study “SignSense: Enabling Natural Communication for Deaf and Mute Individuals” by Smith et al. stands out as a pioneering effort in leveraging machine learning to recognize and interpret a diverse set of sign language gestures. This research emphasized the importance of accurate gesture recognition for facilitating natural communication, drawing attention to the need for flexibility to individual styles of signing and real-time processing for seamless interactions.

Furthermore, “DeepSign: A Deep Learning Approach to Sign Language Recognition” by Johnson and Kim explores various applications using deep learning models, including convolutional and recurrent networks, to enhance the efficiency of recognition. The study underscores the significance of capturing both temporal and spatial dynamics in gestural communication, providing valuable insights for the development of more sophisticated models.

The research conducted by Liang and Liu explores the utilization of three-dimensional convolutional neural networks (3D CNNs) to capture inherent temporal and spatial information present in the sign language. By incorporating depth information into the analysis, their approach achieves improved accuracy in recognizing dynamic gestures, paving the way for more effective communication systems.

Another significant study by Wang et al. introduces a multimodal approach that combines visual and spatial–temporal information to enhance the robustness and generalization capabilities of sign language recognition systems. By integrating data from multiple sources, including video streams and skeletal joint trajectories, their model achieves superior performance in recognizing sign language gestures across different contexts and environments.

Furthermore, the research conducted by Martinez and Benavides in “Enhancing Sign Language Recognition with Data Augmentation and Transfer Learning” addresses the challenges of limited annotated datasets available for training sign language recognition models. Their approach leverages transfer learning and data augmentation techniques to enhance the performance of the model to address the limited availability of training data, thereby increasing the scalability and applicability of sign language recognition systems in real-world scenarios.

Building on this foundation, the proposed initiative aims to address the identified gap by integrating advanced machine learning models, user feedback mechanisms, and collaboration with the deaf and mute community. Inspired by the success of existing research, the proposed system goes beyond conventional gesture recognition, emphasizing adaptability to diverse signing styles, environmental robustness, and cultural sensitivity.

3 Dataset Description

The dataset obtained from the Indian Sign Research and Training Center (ISRTC) represents a valuable resource for building and evaluating the sign language recognition systems specifically tailored to the Indian context. Comprising images of Indian sign language gestures for all alphabets (A–Z) and numbers (1–9), the dataset offers a comprehensive collection of approximately 1200 images for each gesture. This dataset has images in a black-and-white background. The images with a black-and-white background in the dataset serve multiple purposes, primarily aimed at enhancing the accuracy of training models and improving computational efficiency. Figure 1 shows some samples from the dataset.

Firstly, the choice of a black-and-white background provides a high level of contrast, ensuring that the hand gestures stand out clearly against the background. This contrast facilitates more accurate detection and segmentation of the hand region in the images, which is crucial for extracting relevant features and patterns during the training process.

Secondly, employing a black-and-white background contributes to computational efficiency during both the training and inference stages. Images with fewer color channels require less memory and processing power, resulting in faster

Fig. 1 Dataset samples

training times and reduced computational overhead. Furthermore, the use of a uniform black-and-white background helps ensure consistency and standardization across the dataset.

4 Proposed Methodology

Using hand sensors, such as gloves [1], for sign language recognition has long been associated with potential health risks, including skin irritations, restricted movement, and repetitive strain injuries. These risks arise from the need to wear physical devices on the hands, which can lead to discomfort and even long-term health issues, particularly with extended usage.

In contrast, this proposed approach offers a revolutionary solution to Indian sign language (ISL) recognition by entirely eliminating the reliance on hand sensors, instead advocating for the utilization of cameras to capture video frames, enabling the detection and interpretation of sign language gestures from deaf and mute individuals without subjecting them to any physical discomfort or health risks.

By leveraging camera-based technology, this approach not only ensures greater comfort and safety for users but also enhances accessibility and usability. Deaf and mute individuals can freely express themselves through sign language without the constraints imposed by wearable sensors, thus fostering inclusivity and empowerment within the community.

Moreover, the adoption of camera-based sign language recognition systems opens up opportunities for innovation and scalability. With advancements in computer vision and machine learning, these systems can continuously improve in accuracy and performance, ensuring reliable and robust recognition of ISL gestures in various contexts and environments.

4.1 Workflow

The workflow of the suggested model is shown in Fig. 2. In the initial stages of its operation, the machine learning model specialized in interpreting Indian sign language gestures ingests images as its primary input [2]. These images typically depict individuals making various hand gestures, each carrying distinct meanings within the context of Indian sign language communication.

Once the images are fed into the system, they undergo a meticulous preprocessing pipeline aimed at extracting key feature points [3]—essential landmarks that capture crucial information regarding hand movements and positions. These key points are strategically selected points within the image that signify significant features related to hand gestures, such as fingertips, knuckles, and palm contours.

The process of extracting the key points involves advanced computer vision techniques, such as feature detection and localization algorithms. These algorithms

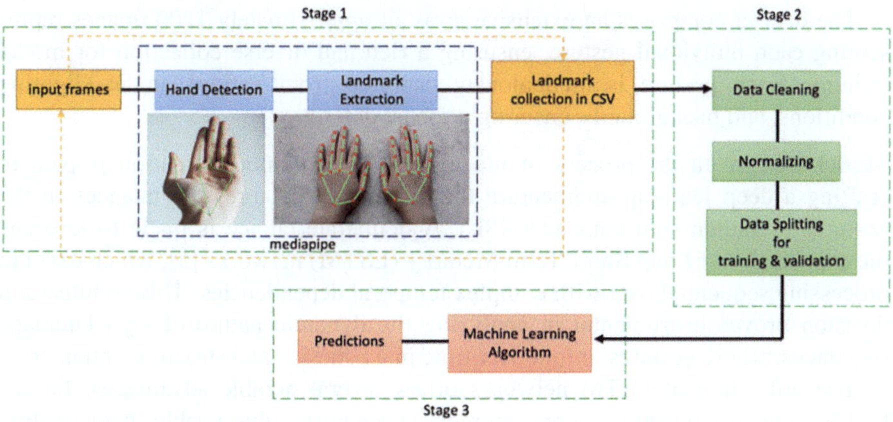

Fig. 2 Workflow of proposed model

meticulously analyze the visual data within the images, identifying relevant patterns and structures that correspond to key aspects of hand gestures in Indian sign language. Upon successful extraction of key feature points, the model organizes this information into a structured format, creating a representation that encapsulates the spatial distribution and configuration of the detected hand movements. This structured representation, often in the form of numerical coordinates or feature vectors, serves as the foundational input for the subsequent analysis conducted by the machine learning model.

A machine learning model is designed to interpret Indian sign language gestures by first taking images as input. These images are then processed to extract key points, which are specific points of interest in the image related to the hand movements and positions. The proposed model utilizes a recurrent neural network named Long Short-Term Memory (LSTM) network [4] to capture sequential information effectively. Training an LSTM helps to learn and analyze the sequence of key points extracted from the input images and predict the corresponding sign language gestures. This approach enables the model to recognize and comprehend Indian sign language gestures correctly, offering a valuable tool for communication and accessibility for individuals who use sign language.

4.2 Detailed Procedural Design Steps

Getting the Dataset Getting the dataset is a crucial step, involving the compilation of a diverse dataset encompassing a wide array of ISL gestures performed by individuals from diverse backgrounds. This diversity ensures the creation of a robust model capable of accurately recognizing signs despite variations in hand shape.

The dataset comprises an extensive array of approximately 1200 images representing each individual gesture, ensuring a rich and diverse collection for model training. Each gesture is meticulously captured from various angles, lighting conditions, and backgrounds, ensuring a robust dataset.

Model Design In the process of model design, meticulous attention is paid to crafting a deep learning architecture that efficiently captures the nuances in the gestures of Indian sign language (ISL). A deliberate choice is made to leverage the capabilities of Long Short-Term Memory (LSTM) networks [5], which excel in processing sequential data with complex temporal dependencies. This architectural decision proves instrumental in analyzing the dynamic nature of sign language sequences, where gestures unfold over time in a cohesive and structured manner.

The utilization of LSTM networks offers several notable advantages. Firstly, LSTM networks possess inherent memory mechanisms that enable them to store information over longer periods, allowing the model to capture long-range dependencies in sign language sequences. This capability is crucial for accurately interpreting gestures that span multiple frames or exhibit intricate temporal dynamics.

Moreover, LSTM networks are well suited for handling variable-length sequences, making them adaptable to the diverse and dynamic nature of sign language expressions. This flexibility enables the model to accommodate gestures of varying durations and complexities, ensuring robust performance across a wide range of linguistic contexts. It also necessitates the incorporation of advanced features such as attention mechanisms, which serve as critical components for enhancing model performance.

These mechanisms dynamically help the model focus on the most relevant parts of the input sequence, effectively prioritizing key hand movements or transitions between gestures. By selectively attending to salient aspects of the input sequence, attention mechanisms empower the model to discern subtle nuances and intricacies in sign language expressions, thereby improving overall recognition accuracy and robustness.

Also, the ResNet model, which takes images directly as input [6], was tried. The ResNet architecture revolutionized deep learning by introducing residual connections, enabling the training of exceptionally deep neural networks. These networks consist of convolutional layers followed by residual blocks, each containing skip connections that add the original input to the output.

One of the primary strengths of ResNet is its capability to tackle the problem of vanishing gradient, which often plagues the networks. As networks become deeper, gradients tend to diminish during backpropagation, making them challenging to train effectively. However, the skip connections in ResNet provide alternate paths for gradient flow, allowing gradients to propagate more easily through the network. This not only accelerates convergence during training but also enables the training of significantly deeper architectures than was previously feasible.

The ResNet architecture usually comprises multiple convolutional layers followed by a series of residual blocks. Every residual block consists of multiple

1	A	0	0	0.315789	-0.18421	0.657895	-0.42105	0.842105	-0.63158	1	-0.86842	0.447368	-0.52632	0.842105	-0.13158	0.736842
2	A	0	0	-0.27273	-0.20455	-0.59091	-0.38636	-0.81818	-0.52273	-1	-0.70455	-0.38636	-0.56818	-0.81818	-0.31818	-0.77273
3	A	0	0	0.289474	-0.18421	0.631579	-0.39474	0.842105	-0.60526	1	-0.84211	0.421053	-0.47368	0.815789	-0.10526	0.710526
4	A	0	0	-0.29545	-0.20455	-0.59091	-0.40909	-0.81818	-0.54545	-1	-0.72727	-0.38636	-0.56818	-0.79545	-0.34091	-0.75
5	A	0	0	0.315789	-0.21053	0.631579	-0.39474	0.842105	-0.60526	1	-0.84211	0.421053	-0.47368	0.815789	-0.10526	0.684211
6	A	0	0	-0.29545	-0.20455	-0.59091	-0.40909	-0.84091	-0.54545	-1	-0.72727	-0.38636	-0.59091	-0.79545	-0.34091	-0.75
7	A	0	0	0.289474	-0.21053	0.631579	-0.42105	0.815789	-0.63158	1	-0.84211	0.421053	-0.5	0.815789	-0.13158	0.684211
8	A	0	0	-0.29545	-0.20455	-0.59091	-0.38636	-0.81818	-0.52273	-1	-0.70455	-0.38636	-0.56818	-0.79545	-0.31818	-0.77273
9	A	0	0	0.307692	-0.20513	0.641026	-0.41026	0.846154	-0.61538	1	-0.84615	0.435897	-0.51282	0.820513	-0.12821	0.692308
10	A	0	0	-0.28889	-0.2	-0.57778	-0.37778	-0.82222	-0.51111	-1	-0.68889	-0.37778	-0.53333	-0.77778	-0.31111	-0.73333
11	A	0	0	0.307692	-0.17949	0.641026	-0.35897	0.846154	-0.5641	1	-0.76923	0.435897	-0.46154	0.820513	-0.07692	0.717949
12	A	0	0	-0.28261	-0.21739	-0.56522	-0.41304	-0.80435	-0.54348	-1	-0.71739	-0.3913	-0.58696	-0.78261	-0.34783	-0.73913
13	A	0	0	0.307692	-0.17949	0.615385	-0.35897	0.846154	-0.53846	1	-0.76923	0.435897	-0.4359	0.820513	-0.07692	0.692308

Fig. 3 Sample keypoints file

convolutional layers, along with shortcut connections that bypass one or more layers. These shortcuts serve as identity mappings, ensuring that the network can learn both shallow and deep features without degradation in performance.

Training Training the model using LSTM involves a multistep process that begins with generating key points from the Indian sign language (ISL) dataset. Key point extraction is fundamental, as it entails identifying and representing key spatial landmarks corresponding to the intricate hand gestures inherent in ISL. These key points serve as pivotal indicators of the sign being performed, capturing crucial aspects of hand positioning and movement.

Key point extraction encompasses pinpointing specific anatomical landmarks on the hand or hands [7], which serve as reliable indicators of sign language expressions. Samples of the key points extracted from the images are shown in Fig. 3. These landmarks typically include the fingertips, knuckles, palm centroid, and wrist, among others. Moreover, alongside the generation of key points, each video frame in the dataset must be meticulously annotated with the corresponding ground label representing the specific sign being performed. This annotation process is crucial for training the model in a supervised manner, as it provides the necessary context and reference points for learning the associations between key points and sign language expressions. This annotation process ensures that the system is trained on a comprehensive dataset that encompasses a wide range of sign language expressions, thereby facilitating robust and accurate recognition of ISL gestures during inference.

On the other hand, to train the ResNet model effectively, raw images were directly fed into the model [8], permitting it to learn and extract the features autonomously without relying on key points. By bypassing the need for explicit key point extraction, feeding raw images into the ResNet model simplified the preprocessing pipeline and reduced the computational overhead associated with it.

Training the ResNet model effectively involves a fundamentally different approach compared to methods reliant on key points. Rather than relying on manually annotated key points or other forms of preprocessed feature extraction, ResNet takes raw images as input, allowing it to autonomously learn and perform feature extraction directly from the data.

This approach, known as end-to-end learning, leverages the power of deep neural networks to learn hierarchical representations of features present in the input images automatically. Each layer of the ResNet architecture learns to transform the input data in a way that makes it easier for subsequent layers to identify and extract increasingly abstract and complex features.

By feeding raw images into the model, ResNet can effectively learn to recognize different patterns and features at various levels of abstraction, from textures and edges to more sophisticated structures and objects. This autonomous feature extraction capability is a major advantage of deep learning models like ResNet, as it eliminates the need for manual feature engineering, which is time-consuming and error prone.

Real-time implementation: Real-time implementation involves the intricate development of a sign language detection system that seamlessly integrates various components, with OpenCV serving as a pivotal tool for preprocessing and model integration. This system is meticulously crafted to offer a comprehensive communication solution, bridging the gap between natural speakers and people who use sign language.

OpenCV, a widely used library, plays a central role in the preprocessing stage of the system. Through OpenCV, raw input from cameras is processed and refined, ensuring optimal image quality and consistency. This preprocessing step is crucial for enhancing the robustness and reliability of subsequent model inference, as it mitigates potential noise and inconsistencies in the input data.

Furthermore, the key aspect of the system is in the integration of models trained on the Indian sign language (ISL) dataset. These models, built using frameworks like TensorFlow, leverage state-of-the-art architectures such as Long Short-Term Memory networks. By integrating these models with OpenCV, real-time sign language recognition becomes achievable, empowering the system to accurately interpret and decode sign language gestures captured by the cameras.

Initially, the ResNet-trained model showed promise for sign language gesture detection. However, as real-world applications demanded more accurate and real-time classification, LSTM models emerged as the preferred choice due to their superior performance and reliability in sign language recognition tasks.

LSTM models, with the ability to capture sequential dependencies and long-term dependencies in data, proved highly effective in accurately classifying signs in real time. This capability was crucial for applications requiring prompt and precise recognition of sign language gestures. Consequently, the LSTM models were prioritized due to their enhanced performance and reliability in sign language recognition tasks. Beyond merely recognizing signs, the system goes a step further by translating these gestures into corresponding alphabets and numbers in real time [9]. This translation capability enhances accessibility and facilitates meaningful communication between users of sign language and those unfamiliar with it.

Moreover, the system does not stop at visual recognition and translation. It also incorporates features to generate sound representations corresponding to the interpreted signs. This auditory feedback further enhances accessibility, enabling hearing individuals to comprehend and respond to sign language gestures effectively.

Evaluation Evaluation of the proposed system's performance in real time encompasses a thorough analysis using a variety of metrics, with accuracy and precision standing out as key indicators of its effectiveness and reliability in practical usage scenarios. The system is evaluated by taking into account both the ResNet and LSTM models. The system's training and validation accuracy for both ResNet and LSTM models is evaluated and compared to find the best model with the highest accuracy. Accuracy, as a fundamental metric, gauges the model's overall ability to correctly recognize Indian sign language (ISL) gestures. It quantifies the proportion of correctly identified gestures out of the total number of gestures in the dataset. A high accuracy score signifies that the system can consistently and accurately interpret a wide range of ISL gestures, thereby instilling confidence in its reliability for real-time communication. This holistic approach to evaluation ensures a comprehensive assessment of the ISL recognition system's performance, taking into account both its accuracy in correctly identifying gestures and its precision in minimizing false positives. By meticulously analyzing these metrics, the users can gain valuable insights into the system's effectiveness and reliability in practical usage scenarios.

5 Algorithms in Proposed System

5.1 Long Short-Term Memory (LSTM)

Long Short-Term Memory (LSTM) networks exhibit a significant progress in the realm of artificial intelligence and machine learning, designed specially for capturing and retaining long-range dependencies in sequential data. This was introduced in 1997 by Hochreiter and Schmidhuber, becoming a cornerstone in the field of recurrent neural networks (RNNs). Unlike traditional RNNs, LSTMs excel at learning and remembering patterns over extended sequences, making them effective in natural language processing, time series prediction, and speech recognition.

The distinguishing feature of LSTMs lies in their ability to selectively store, update, and retrieve information over prolonged periods, diminishing the vanishing gradient and exploding gradient problems that often impede the training of traditional RNNs. This introduction delves into the fundamental architecture of LSTMs, exploring their inner workings and highlighting the critical role they play in modeling sequential data with a focus on preserving temporal dependencies.

Structure LSTM networks, a subset of recurrent neural networks (RNNs), are meticulously engineered to combat the vanishing gradient problem, thus empowering them to capture and internalize long-term dependencies inherent in sequential data. The LSTM memory cell structure is shown in Fig. 4. The fundamental architecture of an LSTM cell comprises several indispensable components, each playing a distinct role in facilitating effective information processing:

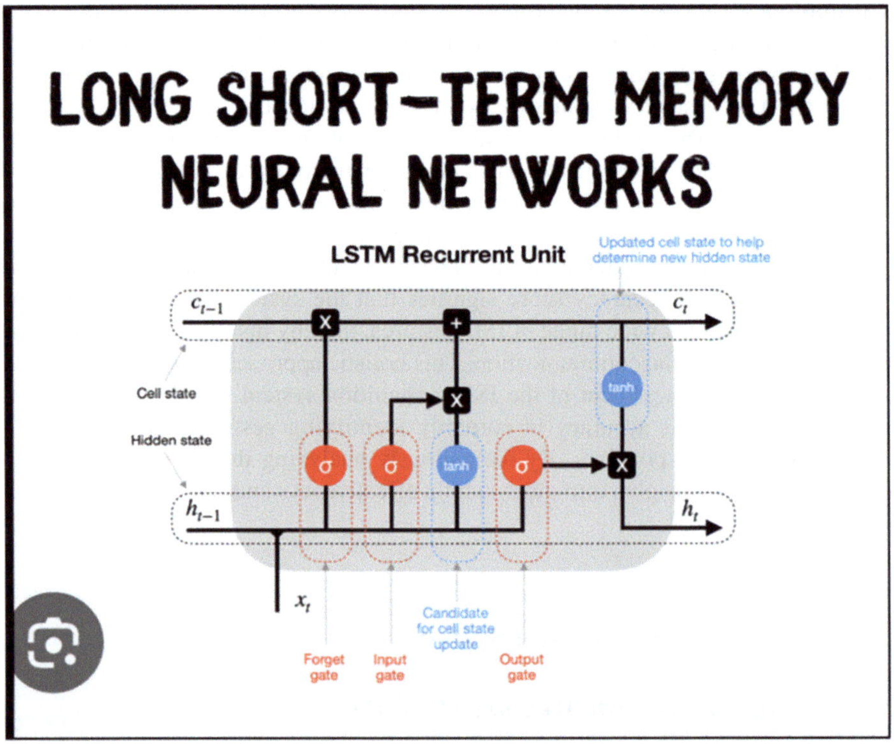

Fig. 4 LSTM memory cell structure

Cell state. Acting as the repository of the cell's long-term memory, the cell state has the remarkable capability to selectively augment or prune information over time. This dynamic adjustment ensures that the LSTM can adaptively retain relevant details while discarding extraneous or outdated information.

Hidden state. Serving as the short-term memory or the output of the cell, the hidden state encapsulates the instantaneous state of the LSTM cell. It serves as a conduit for transmitting information to subsequent time steps, thereby facilitating the flow of information throughout the network.

Input gate. Governed by the sigmoid function, the input gate acts as a gatekeeper, determining the extent to which incoming information should be incorporated into the cell state. This selective gating mechanism enables the LSTM to regulate the inflow of new information, thereby modulating its long-term memory.

Forget gate. Similarly governed by the sigmoid function, the forget gate assumes the crucial responsibility of discerning which portions of the long-term memory are no longer pertinent or relevant. By selectively erasing outdated information from the cell state, the forget gate ensures the LSTM's adaptability to changing contexts and evolving data patterns.

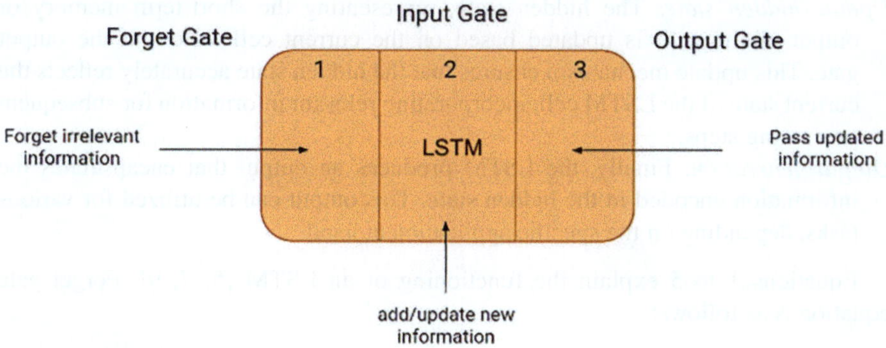

Fig. 5 LSTM gates

Cell update. The cell update mechanism, computed using the hyperbolic tangent (tanh) function, calculates new candidate values to augment the cell state. By synthesizing incoming information with existing memory, the cell update mechanism facilitates the continuous refinement and evolution of the LSTM's internal representation.

Output gate. Activated by the sigmoid function, the output gate dictates which segments of the cell state should be propagated as the hidden state. This selective gating mechanism ensures that only pertinent information is transmitted to downstream layers, thereby optimizing the LSTM's efficiency and effectiveness.

These LSTM gates are shown in Fig. 5. Collectively, these interconnected components collaborate to empower LSTMs with the ability to selectively encode, retain, and retrieve information over extended sequences. By dynamically modulating the flow of information, LSTMs excel in discerning complex temporal patterns and dependencies, rendering them particularly adept at tasks involving sequential data processing.

Working During each time step (t) of processing a sequence, the LSTM network undergoes a series of operations:

Input processing. The network receives input data at the current time step (t), which serves as the basis for subsequent computations.

Compute input, forget, and output gates. The network computes the values of the input, forget, and output gates using the present input and the previous hidden state. These gates control the flow of information within the cell, deciding how much new information to incorporate, which parts of the long-term memory to retain or discard, and which segments of the cell state to output as the hidden state, respectively.

Update cell state. The cell state is updated by combining the previous cell state with the new candidate values calculated in the previous step. This update process is modulated by the input and forget gates, which choose the information to be added or removed from the cell state based on their respective activations.

Update hidden state. The hidden state, representing the short-term memory or output of the cell, is updated based on the current cell state and the output gate. This update mechanism ensures that the hidden state accurately reflects the current state of the LSTM cell, incorporating relevant information for subsequent processing steps.

Output generation. Finally, the LSTM produces an output that encapsulates the information encoded in the hidden state. This output can be utilized for various tasks, depending on the specific application at hand.

Equations 1 to 5 explain the functioning of an LSTM [5, 7, 9]. Forget gate equation is as follows:

$$ft = \sigma\left(Wf\ [ht - 1, xt] + b\right) \tag{1}$$

The input gate equations are denoted by

$$it = \sigma\left(Wi\,[ht - 1, xi] + bi\right) \tag{2}$$

$$Ct = \tanh\left(Wc\ [ht - 1, xc] + bc\right) \tag{3}$$

where t = time step, ft = forget gate at t, xt = input,
$ht - 1$ = previous hidden state,
Wf = weight matrix between forget gate and input gate,
bf = connection bias at t,
it = input gate at t,
Wi = weight matrix of sigmoid operator between input gate and output gate,
bi = bias vector at t,
t = value generated by tanh,
Wc = weight matrix of tanh operator between cell state information and network output,
bc = bias vector at t with respect to Wc.

The forget gate value will later be utilized in the memory cell state calculations. The input gate output is multiplied with a value called the candidate value Ct. These values are now ready for point-to-point multiplication in the memory cell state calculations. Memory cell state is denoted by Ct and its equation is:

$$Ct = Ft \times Ct - 1 + it \times Ct \tag{4}$$

Based on the outcome of the multiplication of the candidate value and input gate vector it and the point-to-point addition of the two values, the network decides if the cell state Ct for the current timestamp is updated. Later, the output gate is used to determine the current hidden state and subsequent hidden states as in Eq. 5.

$$ot = \sigma\left(Wo\,[ht - 1, xo] + bo\right) \tag{5}$$

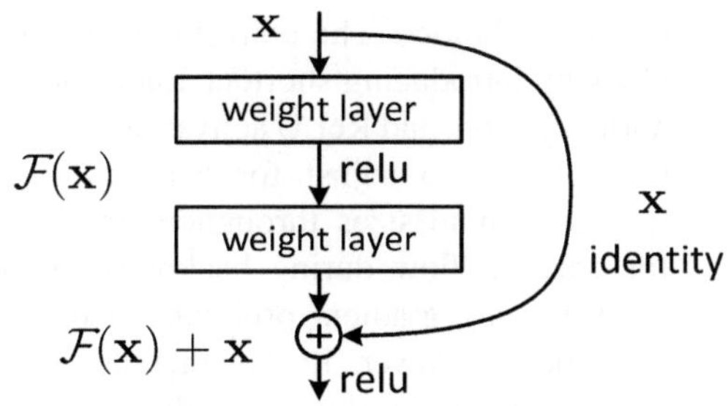

Fig. 6 ResNet single-layer structure

where t = time step, ot = forget gate at t, Wo = weight matrix of output gate, bo = bias vector w.r.t. Wo. Finally, the network decides which information the hidden state should carry based on the final value. This hidden state is used for further prediction.

The key advantage of LSTMs lies in their ability to capture dependencies over long sequences by selectively learning and forgetting information. This is achieved through the gating mechanisms inherent in the LSTM architecture, which enables the networks to solve the vanishing gradient problem encountered by conventional RNNs. By selectively modulating the information flow through the cell state, LSTMs excel in applications handling time series data, language processing, and other sequential data. Their adaptive and dynamic nature makes them well suited for modeling complex temporal patterns and dependencies, enabling them to extract meaningful insights from diverse sequential datasets.

5.2 ResNet

Residual neural network (ResNet) represents a groundbreaking advancement in deep learning architecture, specifically devised to address the challenge of vanishing gradients encountered in training very deep neural networks. The fundamental architecture of a ResNet comprises several key components, each serving a unique function in facilitating effective feature extraction and representation. The single layer structure of ResNet is shown in Fig. 6.

Structure

Identity block. The identity block serves as the basic unit of ResNet architecture, embodying the concept of residual learning. It consists of a sequence of convolutional layers, followed by batch normalization and ReLU activation functions. These convolutional layers are responsible for learning the residual mapping, which is the discrepancy between the input and desired output. By incorporating these residual mappings, the network can focus on learning the residual features, making it easier to optimize deep architectures.

Convolutional block. The convolutional block extends the concept of the identity block by introducing shortcut connections alongside convolutional layers, batch normalization, and ReLU activation functions. These shortcut connections enable the network to adjust for changes in dimensionality, ensuring consistency in feature map sizes throughout the network. By providing alternative paths for gradient flow during backpropagation, the convolutional block alleviates the vanishing gradient problem, thereby facilitating the training of very deep networks. However, it also includes shortcut connections with convolutional layers to adjust for dimensionality changes, ensuring consistency in feature map sizes.

Residual connection. At the core of ResNet architecture lies the residual connection, a pivotal innovation that revolutionized deep learning. This connection involves adding the input of a layer to its output, effectively bypassing one or more layers. By preserving the original input information and allowing it to flow directly through the network, the residual connection mitigates the vanishing gradient problem. This mechanism enables ResNet to achieve unprecedented depths, surpassing previous limitations imposed by gradient degradation during training.

Global average pooling. Toward the latter stages of the network, a global average pooling layer is typically employed to aggregate spatial information across feature maps. Unlike traditional fully connected layers, which flatten feature maps into a high-dimensional vector, global average pooling calculates the average value of each feature map. This pooling operation effectively diminishes the spatial dimensions of the feature maps while retaining essential information. By summarizing the spatial information in a translation-invariant manner, global average pooling improves the model's ability to generalize to variations in object position and scale.

Fully connected layer. The final component of the ResNet architecture is the fully connected layer, responsible for producing the network's output probabilities for classification tasks. This layer receives the flattened feature vector obtained from global average pooling and applies a softmax function to compute the probability distribution over the target classes. By interpreting the output probabilities as class probabilities, the fully connected layer enables the network to make predictions for classification tasks.

Working During forward propagation, ResNet undergoes the following sequence of operations which is depicted in Fig. 7.

Input processing. The network receives input data, typically in the form of images, which serve as the basis for subsequent feature extraction and classification.

Feature extraction. As the input data propagate through the network, a series of convolutional and pooling layers progressively extract hierarchical features. These layers apply learnable filters to the input data, capturing features of increasing complexity and abstraction. The introduction of residual connections facilitates the learning of these features by mitigating the challenges posed by vanishing gradients.

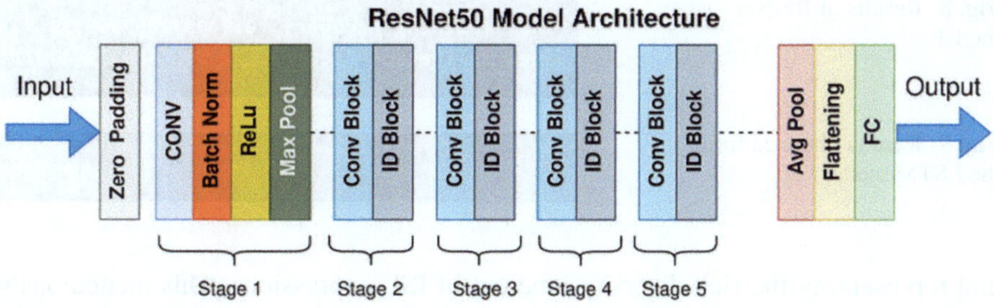

Fig. 7 ResNet layers

Dimensionality reduction. At certain stages, dimensionality reduction may occur through convolutional layers with a stride greater than 1 or through pooling layers. This reduces the spatial dimensions of the feature maps while increasing the depth, allowing the network to capture increasingly abstract features.

Global average pooling. Toward the end of the network, global average pooling aggregates spatial information across feature maps, producing a vector of feature activations that encapsulate the input's essential characteristics.

Classification. Finally, the feature vector received from global average pooling is redirected to a fully connected layer with softmax activation, yielding the network's output probabilities for classification tasks. By interpreting these probabilities as class likelihoods, the network can make predictions about the input data's class labels.

6 Experiments and Results

Systematic evaluation was carried out to assess the performance of the proposed Indian Sign Language system designed using the LSTM networks. These experiments were meticulously planned and executed with the aim of providing a thorough assessment of the system's performance across various aspects of ISL gesture recognition [10]. The primary focus of this work is to validate the system's capability to accurately interpret and decode a wide range of ISL gestures, including both alphabets and numbers, reflecting the diversity inherent in ISL expressions. The inclusion of both alphabets and numbers in the dataset further enriched its diversity, allowing us to evaluate the system's performance across various linguistic and numerical contexts. Each gesture in the dataset was meticulously annotated and labeled, facilitating precise evaluation and analysis of the system's recognition accuracy.

The experiments were conducted on a meticulously curated dataset comprising an extensive array of ISL gestures, encompassing both alphabets and numbers. This dataset was meticulously collected from native signers, ensuring authenticity

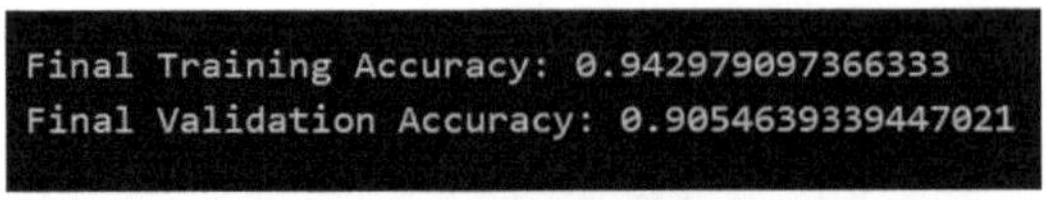

Fig. 8 Results of ResNet model

Fig. 9 Results obtained from the LSTM model

and representing the rich diversity inherent in ISL expressions. This meticulously curated dataset leveraged to create a robust evaluation framework that accurately reflects the real-world challenges and complexities associated with ISL recognition.

To facilitate rigorous evaluation, the dataset was partitioned into distinct training and testing sets, maintaining a balanced ratio of 80:20. This partitioning strategy ensured that there was no overlap between the data splits, thereby mitigating the risk of data leakage and ensuring the integrity of the evaluation process.

The initial findings from the ResNet model showcased promising results, with a training accuracy of 94%, and a testing accuracy of 90.5% which is shown in Fig. 8. While these results demonstrated the effectiveness of ResNet in ISL recognition, further enhancement in the system's performance is sought by exploring alternative models, particularly LSTM networks. The LSTM model's capability to identify temporal dependencies and sequential patterns made it an attractive candidate for improving accuracy and robustness in sign language recognition tasks.

Through systematic experimentation and fine-tuning of hyperparameters, the aim is to leverage the unique strengths of LSTM networks to achieve higher testing accuracy while maintaining a balance between the complexity of the model and its generalization ability. These efforts were critical for confirming the practical applicability and reliability of the ISL recognition system in real-world scenarios, ultimately advancing accessibility and inclusivity for individuals using sign language.

The rigorous testing conducted on the designated testing set provided valuable insights into the performance of the proposed LSTM-based ISL recognition system. The system underwent meticulous scrutiny to assess its ability to accurately recognize and classify a wide range of ISL gestures, reflecting real-world scenarios. The evaluation primarily focused on accuracy, which serves as a crucial metric for quantifying the system's proficiency in gesture recognition.

The experimental results yielded remarkable findings, highlighting the effectiveness of the LSTM-based ISL recognition system. With an impressive accuracy rate of 99% on the testing set, the system demonstrated exceptional precision in deciphering ISL gestures as shown in Figs. 9 and 10. This high level of accuracy underscores the robustness and reliability of the system, positioning it as a promising tool for enhancing communication and accessibility for individuals reliant on ISL.

The significance of achieving such a high accuracy rate cannot be overstated, as it reflects the system's ability to accurately interpret a diverse array of sign language gestures, including complex linguistic expressions and numerical representations. This reaffirms the system's potential to serve as a dependable assistive technology,

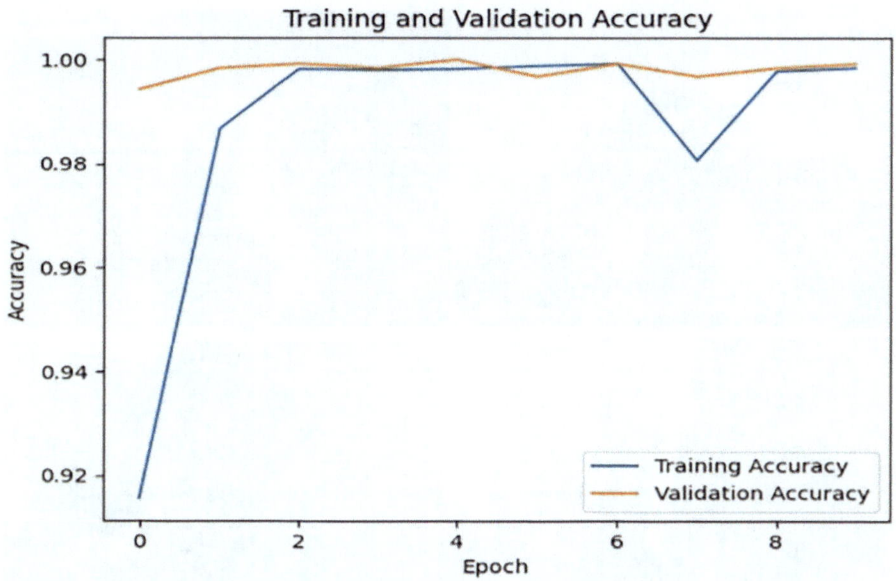

Fig. 10 Performance of the LSTM model

empowering individuals with hearing problems to communicate precisely and participate more in society. The system's success in achieving near-perfect accuracy underscores the importance of leveraging cutting-edge technologies to address real-world challenges and foster a more inclusive society. Some sample figures taken during real-time demonstration of the proposed model for sign language recognition by two of the authors are given in Fig. 11.

7 Conclusion

In this preliminary exploration, ResNet, a distinguished convolutional neural network architecture known for its prowess in image recognition tasks, was opted to tackle the challenge of Indian sign language (ISL) recognition. Leveraging ResNet's exceptional capability to extract intricate features directly from visual data, the model was trained on ISL images. Despite its established robustness and effectiveness, thorough evaluation of the ResNet model unveiled a testing accuracy rate of 90.5%. While this accuracy level was commendable, the potential for further enhancement sought alternative approaches to improve the system's performance in ISL recognition tasks. While ResNet laid a strong groundwork, the transition to LSTM models was prompted by their inherent ability to capture temporal dependencies, essential for interpreting sequential data like sign language

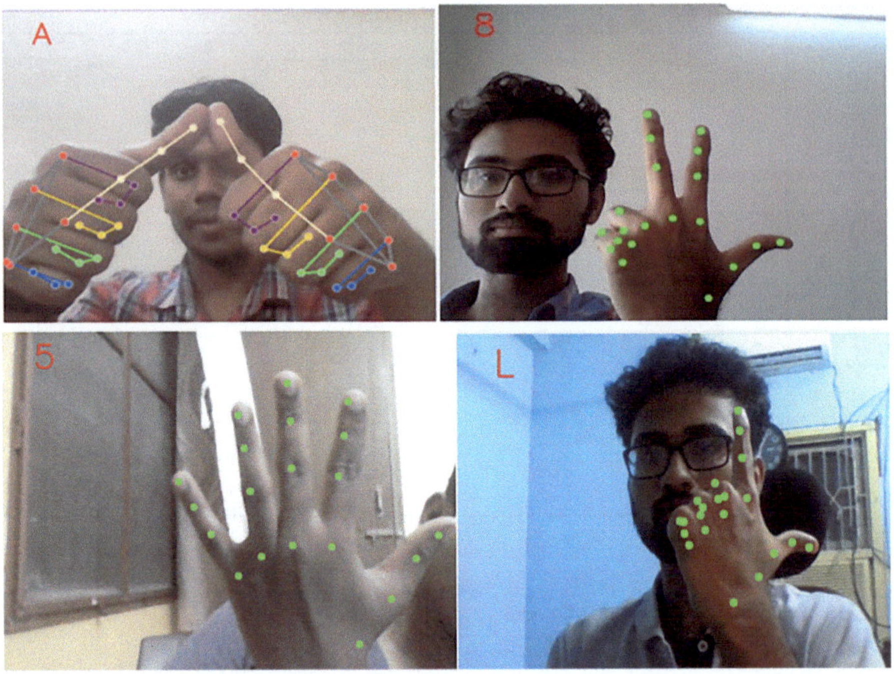

Fig. 11 Real-time demo of sign language recognition

gestures. This shift allowed us to leverage the dynamic nature of sign language expressions, potentially leading to more accurate recognition.

In this study, a comprehensive approach to Indian sign language (ISL) recognition using Long Short-Term Memory (LSTM) networks is presented. This proposed system demonstrated robust performance in accurately interpreting a diverse range of ISL gestures, including both alphabets and numbers. Through extensive experimentation and evaluation, the LSTM-based model achieved a high accuracy rate of 99% on the testing set, showcasing its effectiveness of in real-world scenarios. The significance of this research work lies in its potential to enhance communication accessibility for individuals relying on sign language as their primary means of expression. By leveraging advanced deep learning techniques and harnessing the power of recurrent neural networks, a system was developed capable of recognizing ISL gestures with high precision and real-time responsiveness. This advancement has profound implications for various domains, including assistive technology, education, and accessibility services.

In conclusion, this study contributes to the ongoing efforts toward building inclusive and accessible technologies that empower individuals with diverse communication needs. By bridging the gap between sign language and digital communication, the focus is on achieving greater inclusivity and equality for the deaf and dumb community. A simple future scope for this work could involve expanding the recognition system to incorporate additional gestures and improve its usability and accessibility. This could include: Expansion of gesture vocabulary: Extend the system to recognize a broader range of ISL gestures beyond alphabets and numbers, encompassing common phrases, expressions, and specialized vocabulary. This expansion would enhance the system's utility and enable more comprehensive communication in ISL.

Integration with Mobile Platforms Develop mobile applications that leverage the recognition system, allowing users to access ISL translation capabilities on smartphones and tablets. This would increase the system's accessibility and enable users to communicate effectively in various settings. By pursuing these avenues for future development, this work can evolve to better meet the needs of the ISL community.

References

1. D. Deora, N. Bajaj, Indian sign language recognition, in *1st International Conference on Emerging Technology Trends in Electronics, Communication & Networking*, (IEEE, 2012), pp. 1–5
2. A.V. Nair, V. Bindu, A review on Indian sign language recognition. Int. J. Comput. Appl. **73**(22) (2013)
3. K. Sharma, K.A. Aaryan, U. Dhangar, R. Sharma, S. Taneja, Automated Indian sign language recognition system using lstm models, in *International Conference on Computing, Communication, and Intelligent Systems (ICCCIS)*, (IEEE, 2022), pp. 461–466
4. K. Dixit, A.S. Jalal, Automatic Indian sign language recognition system, in *3rd IEEE International Advance Computing Conference (IACC)*, (IEEE, 2013), pp. 883–887
5. E. Abraham, A. Nayak, A. Iqbal, Real-time translation of Indian sign language using LSTM, in *2019 Global Conference for Advancement in Technology (GCAT)*, (IEEE, 2019), pp. 1–5
6. V. Adithya, P. R., Vinod, and Usha Gopalakrishnan.: Artificial neural network based method for Indian sign language recognition, in *IEEE Conference on Information & Communication Technologies*, (IEEE, 2013), pp. 1080–1085
7. G. Khartheesvar, M. Kumar, A.K. Yadav, D. Yadav, Automatic Indian sign language recognition using MediaPipe holistic and LSTM network. Multimed. Tools Appl., 1–20 (2023)
8. G. Anantha Rao, K. Syamala, P.V.V. Kishore, A.S.C.S. Sastry, Deep convolutional neural networks for sign language recognition, in *Conference on Signal Processing and Communication Engineering Systems (SPACES)*, (IEEE, 2018), pp. 194–197
9. A. Mittal, P. Kumar, P.P. Roy, R. Balasubramanian, B.B. Chaudhuri, A modified LSTM model for continuous sign language recognition using leap motion. IEEE Sensors J. **19**(16), 7056–7063 (2019)
10. Y.I. Rokade, P.M. Jadav, Indian sign language recognition system. Int. J. Eng. Technol. **9**(3), 189–196 (2017)

An Empirical Study on Employee Attrition Using Machine Learning

R. Mary Metilda, H. Saranya, and V. Vishnu Priya

1 Introduction

Organizations have various avenues to embrace technology for enhancing decision-making processes. Among these, artificial intelligence stands out as a highly innovative technology extensively employed to aid organizations in shaping their business strategies, optimizing organizational operations, and managing their workforce effectively [1].

Machine learning algorithms undergo training using historical data to discern patterns and relationships between variables that may indicate a higher likelihood of attrition. These variables can include factors such as job satisfaction, salary, performance ratings, and length of employment. By analyzing these variables, machine learning algorithms can identify employees who are at higher risk of leaving the company.

2 Review of Literature

Fallucchi et al. [1] in their research "Predicting Employee Attrition Using Machine Learning Techniques" explored methods to predict employee attrition using machine learning techniques and compared the performance of different models. The dataset employed for analysis includes random forest, decision tree, support

R. Mary Metilda (✉) · H. Saranya
Sri Ramakrishna College of Engineering, Business School, Coimbatore, India
e-mail: metilda@srec.ac.in; saranya.2191043@srec.ac.in

V. Vishnu Priya
SNS Business School, Coimbatore, India

A. Haldorai et al. (eds.), *7th EAI International Conference on Big Data Innovation for Sustainable Cognitive Computing* , EAI/Springer Innovations in Communication and Computing, https://doi.org/10.1007/978-3-032-02979-9_3

vector machine, artificial neural network, and logistic regression. The research applied preprocessing techniques such as data cleaning and reduction, feature selection, and parameter tuning to optimize the models. The results showed that the random forest model outperformed the others with an accuracy rate of 89.5%, followed by the support vector machine at 88.5%.

Jennifer Anne A. Repaso et al. [2] in their study "Determining Factors Resulting to Employee Attrition Using Data Mining Techniques" highlight the importance of identifying the causes of employee turnover and propose methods to identify the factors that lead to attrition. The research discusses the significance of data preprocessing, feature selection, and model building. The results of the study offered valuable insights into the retention strategies that can be adopted by organizations to develop strategies to retain employees.

Elham Mohammed Thabit A. Alsaadi et al. [3] in their study "Identification of human resource analytics using machine learning algorithms" focus on identifying the factors that contribute to employee attrition and utilize machine learning techniques to predict the attrition among new employees. The study evaluated the performance of various algorithms using different rating scales and found that the random forest performed the best, followed by logistic regression.

Suji and Sathiyavathi [4] in their study "A Study on Employee Attrition in one of the Leading Hospital" emphasized that despite changes in the external environment, employee attrition remains a significant problem for many organizations. Attrition is characterized by a gradual decline in the employee count attributed to various factors such as resignation, death, and retirement. Some employees may leave for undisclosed reasons such as lack of job security, limited career advancement, desire for change or new opportunities, anticipation of higher pay, problems with supervisors, or personal reasons. When a well-trained and skilled employee leaves an organization, it creates a void that can be challenging to fill for human resource personnel. Today's human resource managers are taking various measures to reduce the employee attrition rate, making it a critical challenge for them to address.

Jain et al. [5] in their study "Explaining and predicting employees' attrition: a machine learning approach" address the concern of attrition by constructing a predictive model capable of determining whether an employee will depart or remain with the organization [6]. The study focuses on measuring employee performance and satisfaction rates as a means of reducing the attrition rate [7]. The paper proposes a new approach that employs machine learning to improve retention strategies for targeted employees [8]. Additionally, the study sheds light on various factors that contribute to employee attrition and offers possible solutions [9]. The application can also uncover hidden reasons for employee attrition and enable the management to take preventative measures on a per-employee basis.

3 Objectives of the Study

- To understand the impact of employee attrition in the organization,
- To analyze various factors that contribute to identifying potential reasons for attrition.
- To provide decision-making support to management and HR teams in identifying and addressing attrition issues.

4 Results and Discussion

Machine learning can be a valuable tool in supporting HR applications [10]. This chapter aims to demonstrate how machine learning can be utilized to provide objective insights for HR departments, which often rely on subjective decision-making and interpretations of employee behavior [11]. By leveraging objective support tools, HR departments can take proactive measures to prevent employee resignation rather than implementing retention strategies only after an employee has left the company.

The proposed approach involves using statistical analysis and classification algorithms to classify potential reasons why an employee may leave the company. This study showcases how complex tasks can be accomplished through the use of classification algorithms, while also highlighting the potential benefits of implementing machine learning technology to support HR management in organizations.

Overall, this work emphasizes the importance of objective data-driven decision-making in HR management and suggests that the use of machine learning can offer new opportunities for the development and implementation of support tools to aid in employee retention and management.

4.1 Data Cleaning

Preparing the data for machine learning is a critical and time-consuming aspect of data science. According to research, data preparation typically accounts for data science projects consume 60% of the time and effort invested [12]. Therefore, it is crucial to focus on the initial phases of Business Understanding and Data Understanding can streamline the subsequent steps of the process.

4.2 Data Exploration

To have a better insight into the dataset and its various variables, we conducted data exploration by generating descriptive statistics [13]. This involved examining several variables, including number, distinct values, leading values, occurrence, average, variance (SD), smallest and largest values (min/max), and percentiles (25%/50%/75%).

4.3 Modeling

The model building process involved selecting and testing various machine learning techniques to understand the best classifier for the analyzed case. In this case, the predictive models employed encompassed decision trees, Bayesian methods, logistic regression, and support vector machines (SVM).

The objective was to train each classifier using the featured dataset and identify the one that yielded the most favorable classification outcomes for predictions. The classification algorithms under consideration were:

- Support vector machines (SVM) classification.
- Decision tree classifier.
- Logistic regression classifier.

4.4 Support Vector Machines (SVM)

The support vector machine (SVM) technique is a machine learning technique utilized for both classification and regression tasks. Its operation revolves around the identification of an ideal hyperplane that effectively divides the data into distinct categories. The outcome of this algorithm is the determined hyperplane, which serves as a tool for classifying new examples. The objective of the SVM technique is to identify a hyperplane that increases the distance between the training samples and the hyperplane. This distance is called the margin, and it is obtained using SVM theory. By maximizing the margin of the data to be trained, the separating hyperplane determined by the SVM technique is able to classify new data accurately in Table 1 and Fig. 1.

4.5 Logistic Regression Algorithm

The logistic regression algorithm is used in classification problems where the dependent variable is binary, such as in the case of predicting employee attrition

Table 1 Support vector machine results

	Precision	Recall	f1-score	Support
0	0.98	0.83	0.90	275
1	0.82	0.98	0.89	219
Accuracy			0.89	494
Macro avg	0.90	0.90	0.89	494
Weighted avg	0.91	0.89	0.89	494

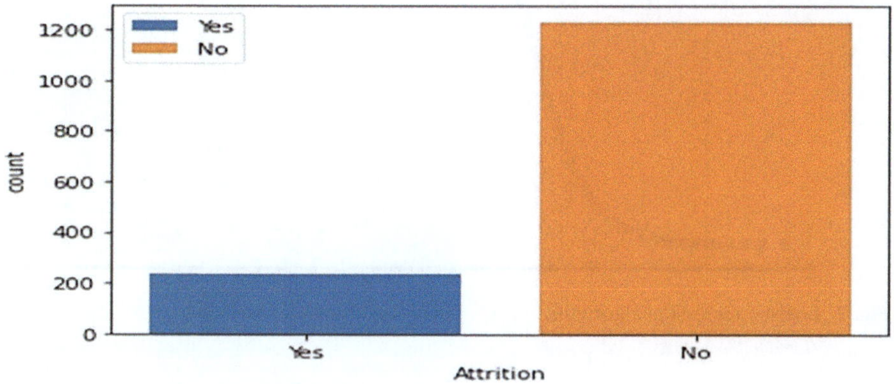

Fig. 1 Regression algorithm

(i.e., either the employee will leave or not). The algorithm estimates the probability of the dependent variable being one of the two possible outcomes based on the independent variables.

In logistic regression, the model is trained using a class of labeled data. Throughout the process of training, the algorithm fine-tunes the model parameters to reduce the disparity between the predicted probabilities and the true labels. The best-fit values for these model parameters are derived through a maximum likelihood estimation approach.

After the model is trained, it can be used to estimate the likelihood of the dependent variable being either 0 or 1 for novel observations. The model outputs a probability score, which can be converted to a binary decision using a threshold value. The selection of the threshold value is determined by balancing the trade-off between recall and precision, which, in turn, hinge on the quantity of pseudo-positives and false negatives within the predictions in Table 2 and Fig. 2.

Table 2 Logistic regression algorithm

	Precision	Recall	f1-score	support
0	0.98	0.84	0.91	268
1	0.84	0.98	0.90	226
Accuracy			0.90	494
Macro avg	0.91	0.91	0.90	494
Weighted avg	0.92	0.90	0.90	494

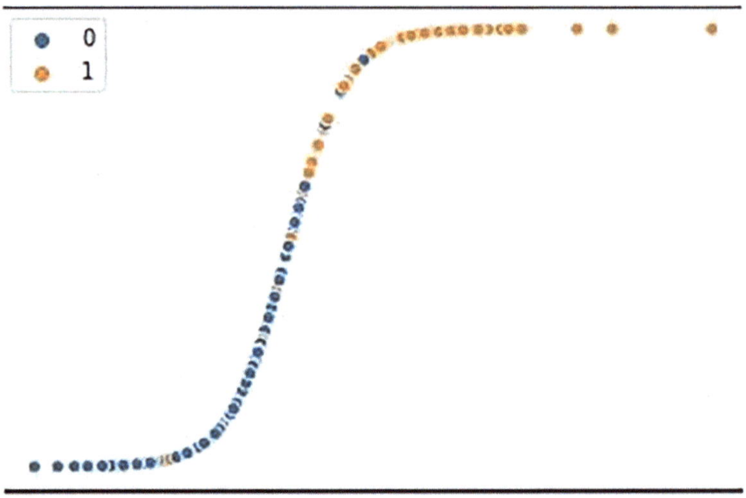

Fig. 2 Logistic regression algorithm

4.6 Decision Tree Classifier

A decision tree classifier is a supervised learning algorithm employed primarily for tasks involving classification. It works by dividing the data into smaller different subsets based on the values of the independent variables. Each division is based on a criterion that maximizes the homogeneity of the subsets.

The algorithm starts with the entire dataset at the root node and selects the best means to split the data. The feature with the highest information gain or the lowest Gini index is selected. The dataset is divided into smaller subsets based on the selected feature's values, and this process is repeatedly carried out for each subset till a predetermined stopping criterion is satisfied.

This stopping condition could take the form of a maximum tree depth, a minimum sample threshold for splitting internal nodes, or a minimum sample threshold for leaf nodes. Once the tree is built, it can be used to predict the class of new instances by traversing the tree from the root node to a leaf node.

Decision trees have the advantage of being easy to understand and interpret. They have the capability to process both categorical and numerical data and are adept at capturing nonlinear associations between the independent variables and the dependent variable. However, they are prone to overfitting and can be unstable when the data changes slightly. To overcome this issue, ensemble methods such as random forest can be used, which combine multiple decision trees to improve the accuracy and stability of the model (Table 3).

Table 3 Decision tree classifier

	Precision	Recall	F1-score	Support
0	0.83	0.82	0.82	233
1	0.84	0.85	0.84	261
Accuracy			0.83	494
Macro avg.	0.83	0.83	0.83	494
Weighted avg.	0.83	0.83	0.83	494

5 Conclusion

In this research, four advanced machine learning techniques—extra trees classifier (ETC), support vector machine (SVM), logistic regression (LR), and decision tree classifier (DTC)—were employed to forecast employee attrition. The findings indicated that the proposed extra trees classifier (ETC) achieved the highest accuracy rate at 93%, accompanied by notable precision, recall, F1 score, and ROC accuracy scores. To address the dataset imbalance, data resampling techniques were utilized. The methodologies were validated using k-fold cross-validation and benchmarked against previous state-of-the-art studies. Within a tenfold dataset, support vector machines (SVM) attained an 89% accuracy rate, logistic regression (LR) achieved a 90% accuracy rate, decision tree classifier (DTC) reached an 83% accuracy rate, and the proposed approach achieved an accuracy rate of 93%. Through exploratory data analysis (EDA), critical factors contributing to employee attrition were identified, including monthly income, hourly rate, job level, and age. These insights can be valuable for organizations seeking to mitigate employee attrition. Nonetheless, the study had some limitations, and future research will explore the application of deep learning techniques to predict employee attrition.

References

1. F. Fallucchi, M. Coladangelo, R. Giuliano, E.W. De Luca, Predicting employee attrition using machine learning techniques. Computers **9**, 86 (2020)
2. J.A.A. Repaso, E.T. Capariño, M.G.G. Hermogenes, J.G. Perez, Determining factors resulting to employee attrition using data mining techniques. Int. J. Educ. Manag. Eng. **12**(3), 22–29 (2022). https://doi.org/10.5815/ijeme.2022.03.03
3. E.M.T.A. Alsaadi, S.F. Khlebus, A. Alabaichi, Identification of human resource analytics using machine learning algorithms. TELKOMNIKA (Telecommun. Comput. Electron. Control) **20**(5), 1004 (2022). https://doi.org/10.12928/telkomnika.v20i5.21818
4. P. Suji, Sathiyavathi, A study on employee attrition in one of the leading hospital. **8**, 4 (2021)
5. P.K. Jain, M. Jain, R. Pamula, Explaining and predicting employees' attrition: A machine learning approach. **2**, 757 (2020). https://doi.org/10.1007/s42452-020-2519-4
6. F. Guerranti, G.M. Dimitri, A comparison of machine learning approaches for predicting employee attrition. Appl. Sci. **13**, 267 (2023)
7. V. Saraf, M.A. Peshave, An analysis on employee-attrition in IT Industry. **IX**(VII) (2020)
8. Y. Choi, J.W. Choi, A study of job involvement predictionusing machine learning technique. **29**(3) (2021)

9. Srinivas, K.K. Kumar, Effect of zero-vector placement in a dual-inverter fed open-end winding induction motor drive with alternate sub-hexagonal center PWM switching scheme. IEEE Trans. Power Electron. **23**(3), 1584–1591 (2008). https://doi.org/10.1109/tpel.2008.921170
10. B. Venugopal Reddy, V.T. Somasekhar, Y. Kalyan, Decoupled space-vector PWM strategies for a four-level asymmetrical open-end winding induction motor drive with Waveform symmetries. IEEE Trans. Ind. Electron. **58**(11), 5130–5141 (2011)
11. N. Mansor, N.S. Sani, M. Aliff, Machine learning for predicting employee attrition. Int. J. Adv. Comput. Sci. Appl. **12**(11), 435–444 (2021). https://doi.org/10.14569/IJACSA.2021.0121157
12. M.H.V. Reddy, T.B. Reddy, B.R. Reddy, M. Suryakalavati, Reduction of common mode voltage in asymmetrical dual inverter configuration using discontinuous modulating signal based PWM technique. J. Power Electron. **15**(6), 1524–1532 (2015)
13. M.H.V. Reddy, T.B. Reddy, B.R. Reddy, M. Suryakalavati, Random PWM technique for dual-inverter-fed vector-controlled induction motor drive. J. Control Autom. Electr. Syst.. (Springer publisher) **27**(1), 60–68 (2016)

Short-Term Forecasting of the Power Generation of a PV Plant from Relevant Atmospheric Parameters Using ML/DL Methods

Renjit Lal Parameswaran, Hazarathayya, and Sivaiah Bellakonda

Abbreviations

ADF	Augmented Dicky Fuller
AIC	Akaike Information Criteria
ANN	Artificial neural network
ARMA	Autoregressive moving average
CERES	Clouds and Earth Radiant Energy System
DSO	Distribution system operator
GHI	Global horizontal irradiance
LSTM	Long short-term memory
MAE	Mean absolute error
ML	Machine learning
MOS	Model output statistics
MSE	Mean square error
MW	Megawatt
MWh	Megawatt hour
NASA	National Aeronautical and Space Administration
NWP	Numerical weather prediction
PV	Photo voltaic
R2	Coefficient of determination
RMSE	Root mean square error
RNN	Recurrent neural network
TSI	Total sky images
TSI	Total solar irradiance
TSO	Transmission system operator
VAR	Vector autoregression

R. L. Parameswaran (✉) · Hazarathayya · S. Bellakonda
Indian Institute of Information Technology, Kottayam, India
e-mail: renjitlp.eme214@iiitkottayam.ac.in; yallankihazarathayya@gmail.com; sivaiah@iiitkottayam.ac.in

A. Haldorai et al. (eds.), *7th EAI International Conference on Big Data Innovation for Sustainable Cognitive Computing* , EAI/Springer Innovations in Communication and Computing, https://doi.org/10.1007/978-3-032-02979-9_4

1 Introduction

1.1 Importance of Solar Energy and Its Forecasting

The world has seen the growth of solar power in the last decade as an alternative to the conventional fossil fuel power plants. Harnessing solar energy is growing at a faster pace across the globe, and India is forging ahead in inducting new major solar PV plants. The future of mankind verily depends on reducing the carbon footprint so as to limit the average global rise in temperature to 1.5 °C from that of preindustrialized times. PV power is an important part of sustainable development throughout the world [1]. Solar power is replacing other means of generation in a very fast manner due to its low carbon emissions [2]. The growth of solar generation over 20 years is shown in Fig. 1.

In China, a comprehensive study comparing solar photovoltaic power generation (SPPG) and coal power in Ningxia found that a 1 kWp SPPG module emits 1.35 kg/kWh per unit of electricity produced, which is substantially lower than the 4.81 kg/kWh emitted by coal power. This highlights the significant environmental benefits of solar power over coal-based power [3]. In India, a study conducted in West Bengal estimated the carbon footprint of solar photovoltaic cells. The study revealed that the emission of CO_2 is 35 g/kWh, which is significantly lower than the emissions from conventional fossil fuel-based generation. This demonstrates the potential of solar power as a cleaner alternative to coal-based power in India [4].

Forecasting of power generation is a crucial aspect of effective power grid management to avoid power interruptions, stabilize frequency, and ensure quality power supply to consumers. Effective forecasting also reduces the cost of gener-

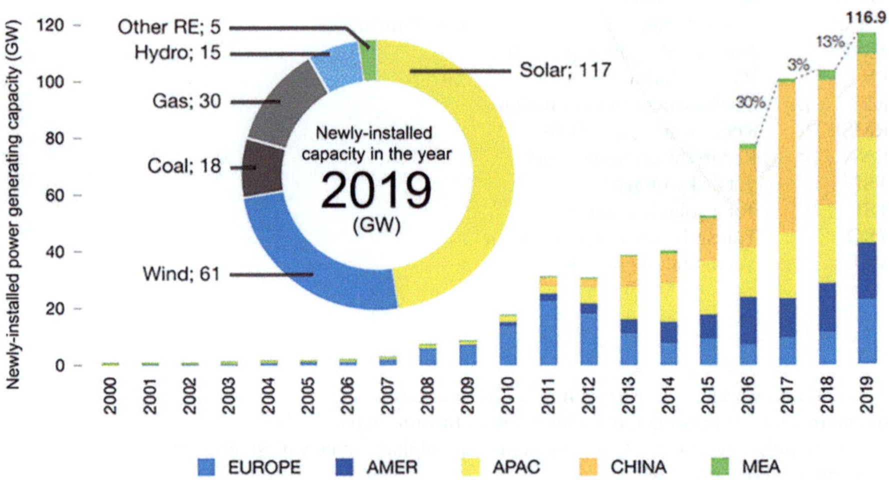

Fig. 1 Growth of solar generation year wise (in GW)

ation. Unfortunately, the addition of solar plants adds a lot of uncertainty to the grid, as unlike conventional electrical power-generating plants, forecasting of solar plants is nearly impossible and poses a great challenge to power system engineers. The chaotic nature of cloud movement, which reduces the solar radiance on the ground, makes it difficult to forecast the surface radiation [5]. Any forecasting model that does not consider the stochastic characteristics of cloud motion and irradiation on the surface would be inadequate in predicting PV output power. Accurate forecasting can indeed lead to significant savings in net generation costs and is crucial for various operational aspects such as net load forecasting, reserve dimensioning, ramp management, and uncertainty handling [6]. Recent studies show that a uniform 25% improvement in PV forecasting accuracy results in savings of 1.56% in the net generation cost [7]. Solar power produced by a PV module is numerically equal to the product of cell area, light intensity, cell efficiency, and number of sunshine hours. Therefore, the prediction of the PV power can be derived from the radiation indeed. It is very challenging to confirm the parameters that affect the light intensity and the extent of its influence on it. According to the horizons of power forecasting, there are different categories of predictions. It varies from a minute ahead to days ahead. Short-term forecasting, which is between 0 and 6 h time horizon (ahead), is of prime importance for load scheduling for both TSO and DSO [3].

Intraday forecasting enables plant operators to schedule the energy generation with the grid operator. In this regard, the 1-h ahead forecasting plays a very important role in revising the schedule and thus enables the grid operator to reschedule the power injection to the grid.

2 Existing Techniques for Forecasting Solar PV Output

Forecasting of PV power can be categorized broadly as physical and statistical. Physical approaches use NWP and total sky images. Statistical methods make use of historical data to train the models. NWP uses the dynamic movements of cloud cover and thus does not require extensive monitoring instruments. NWP models are scalable and effective in large-scale modeling of the atmosphere [8].

Though effective predictions can be made with wide spatial–temporal resolution through NWP models, they are not effective for partially cloudy days with sudden cloud movement. Also, the computation time required is high for the NWP approach in short-term forecasting, e.g., 2 h is required for 2-min ahead forecasting. Though the methods work fine for very small temporal resolution and very long time ahead forecasting, their effectiveness for short-term forecasting is not proven.

Currently, ANN and DL methods are widely used for combining both physical and historical values to train statistical models and yield better performance. Statistical methods are also effective for solar irradiation and PV predictions. Machine learning/ANN techniques are widely used for enhancing the statistical model performances. These techniques are used for nonlinear mapping between historical data and future values. These methods can be used for forecasting different time horizons also. Like regressive methods, ANNs show better performance for both data-rich and data-poor conditions [8]. In this study, ANN/ML modeling, i.e., LSTM, is used as the main technique and compared with other models such as vector autoregression and random forest models. The models are analyzed for their performance based on the various performance metrics such as

(a) Mean absolute error (MAE).
(b) Mean bias error (MBE).
(c) Root mean square error (RMSE).

3 Literature Survey

Arif Ozbek [9] in his paper describes two different methods for forecasting the horizon PV output as follows:

Through analytical equations: These involve complex equations of atmospheric physics, which is preferred for long-term forecasting.

Direct estimation methods: Broadly, there are four categories under this method, namely artificial intelligence, physical, statistical, and hybrid methods. The advantage of statistical methods is that they do not require internal information from and use historical data for a minimum time frame. The historical data used is very important for such forecasting. Moving average (MA) and autoregressive (AR) methods are generally used in statistical models. Both models combined to have an ARMA model, which is used for stationary time series data.

Kaplanis [10] compares different solar radiation estimation methods, which are based on autoregressive, Gaussian models, Jain's and Baig's models, and a model proposed by himself, which is purely based on statistical equations. Sharadga, Hajimirza, and Balog [11] show that ANN has a superiority over the statistical methods in forecasting of PV plants. It is observed that the ANN exhibits better accuracy and less computational time in the case of the short-term forecasting of the power of large PV plants. Short-term solar power forecast with deep learning explores optimal input and output configuration, making use of CNN and hybrid input features [12]. It combines the processed sky images with the historical PV data. The required data set requires high-definition images.

For multilinear time variant data predictions, RNN is considered to be an effective tool [13]. This chapter elaborates that RNN has better prediction performance during different weather conditions, such as sunny, cloudy, and rainy weather conditions. But the prediction accuracy depends on the longer time series and the incorporation of more parameters affecting the irradiation. RNN also suffers from the gradient explosion/vanishing gradient problem, by which the global optimum would be missed. The disadvantages have been overcome by using LSTM due to its special hidden layer design. This causes the LSTM to be a preferred deep learning technique for power estimation of multivariate time series analysis.

Peng, L [14] in his research paper proposed an algorithm based on LSTM combined with differential equations (DE) for predicting electricity cost. Though the algorithm is effective in optimizing suitable hyperparameters for LSTM, the proposed DE-LSTM compromises with prediction accuracy. Rui Chin [15] in his research work used a traditional encoder single deep learning (TESDL) for forecasting PV output. The method performs well in the distributed generation side, where the generation is not concentrated, as the accuracy was checked with 1 year of data from weather stations. The model's performance cannot be ascertained for large data sets and in a concentrated field [16]. Deep learning using an LSTM model for forecasting—an hour-ahead generation of a solar-PV power plant of 1.15 MW capacity proposed by A Ozbek in his paper [9]. Though the model fares well in terms of correlation and error metrics, the model predicts the power with its own historical values, i.e., a univariate analysis. The model does not consider any atmospheric parameters.

Solar power forecasting performance toward industry standards describes the performance of the evaluation metric and the need for standardization for solar forecasting of intra-hour, day-ahead, and weekly time [17, 18].

Though internationally, various research studies have been carried out for short-term, i.e., up to 1-h ahead forecasting of power generation of large PV plants, in the Indian context very few research studies have been carried out. The majority of them compromise accuracy for complexity.

4 Data Sets and Methodology

Selection of a data set to be chosen in such a manner that the data set is reliable, comprising the atmospheric parameters of long period so that the rapid changes and wide variation of climatic parameters can be taken into account. Usually, there shall be inconsistencies, resulting from wrong calibration, instrument drift, or manual interference, such as vandalism, which may result in instrumental output for solar radiation measurement. Also, it is required to have plant data which are free from human error, tampering, and should have good accuracy. The plant data collected directly from generation energy meters of 0.2 class accuracy will ensure dependable plant data as PV output for analysis [19]. Since for this project, the atmospheric parameters have been taken from the NASA CERES site, the anomalies that are described are ruled out, as the data source is not from any pyranometer data but from satellite data, which would be more accurate. For forecasting horizons from seconds up to hours ahead, the recommended performance metrics of interest should be RMSE because it better addresses the ramping up/down of power generation during mornings and nights [20]. For the proposed methodology for solar PV forecasting, RMSE as a performance indicator is used.

Table 1 atmospheric parameters used as variables in the study

Abbreviations	Details	Units
T2M	Temperature at 2 m	(C)
QV2M	Specific humidity at 2 m	(g/kg)
PRECTOTCORR	Precipitation corrected	(mm/h)
ALLSKY_SFC_SW_DWN	All sky downward shortwave radiation	(Wh/m^2)
CLRSKY_SFC_SW_DWN	Clear sky surface shortwave downward irradiance	(Wh/m^2)
PS	Surface pressure	(kPa)
WS50M	Windspeed at 50 m	(m/s)

The selection of suitable atmospheric parameters that influence the PV output is carried out using the Pearson coefficient [21], which is a simple method to establish the correlation between variables. By establishing the correlation matrix of relevant parameters that have maximum correlation with PV output power, the relevant parameters for the study can be selected.

In this model analysis, performance metrics such as MAE, RMSE, and MSE have been used for evaluating the models. The coefficient of determination would also be additionally considered for checking the collinearity of the model. Values less than 0.5 for MAE are acceptable for the model, and values more than 0.8 shall be acceptable for the coefficient of determination.

This chapter tries to forecast 1-h ahead output of a 10 MW solar PV plant in Salboni, West Bengal, India [Salboni, PV Park1, GPS coordinates (Latitude: 22.5654, Longitude: 87.1554)] through a deep learning technique (LSTM) and compare with other models, namely VAR and random forest. The plant data, i.e., hourly instant power in MW, are collected from the plant for the period from May 2, 2019, 00:00 h to December 1, 2022, 23:00 (3 years 7 months). The plant data are extracted from the MWh meters. The plant data have been integrated with the hourly atmospheric parameters, which have been obtained from the NASA CERES (https://power.larc.nasa.gov) (Table 1).

In order to simplify the model, seven variables that most influence the PV output installed in the area are considered, though numerous atmospheric variables influence radiation and PV output.

Though the variables selected influence the solar irradiation and the PV output, the extent to which the variable affects the PV output is derived through correlation analysis, here by using the Pearson coefficient. Person coefficient is a statistical tool that is used for studying the degree of linear correlation between the variables [22].

$$r_{xy} = \frac{\sum_{i-1}^{m} (y_i - \overline{y})(x_i - \overline{x})}{\sqrt{\sum_{i-1}^{m} (x_i - \overline{x})^2} \sqrt{\sum_{i-1}^{m} (y_i - \overline{y})^2}} \tag{1}$$

$\overline{x}$ = mean of the elements in vector 1
$\overline{y}$ = mean of the elements in vector 2

r_{xy} = degree of correlation between the two variables of concern
m = no. of data in the sequence.

The hourly plant data have been combined with the atmospheric variables for predicting the PV output. The code has been written in Python Jupiter notebook (Python3) in Anaconda Environment.

5 Evaluation of Various Algorithms

The data obtained from the power plant and from the NASA website have been combined hour wise, preprocessed before applying to various techniques which have been used for forecasting in this chapter, viz., vector auto regression, random forest, and LSTM for the forecasting of solar PV output.

5.1 *Vector Autoregression*

Vector autoregression is an extension of the autoregressive (AR) model and is used for multivariate analysis. In VAR, the variables influence each other and also their own lags and the lags of other variables. Hence, VAR takes into account the relations of other features and its own past values, and the relation established shall be a more encompassing one.

$$y_{1'_t} = c_1 + \varnothing_{11'1} y_{1'_{t-1}} + \varnothing_{12'1} y_{2'_{t-1}} + e_{1'_t} \tag{2}$$

$$y_{2'_t} = c_2 + \varnothing_{21'1} y_{1'_{t-1}} + \varnothing_{22'1} y_{2'_{t-1}} + e_{2'_t} \tag{3}$$

$e_{1'_t}, e_{2'_t}$ = white noise, $\varnothing_{11'l}$ captures the influence of the lth lag of variable y_i on itself and $\varnothing_{12'l}$ captures the influence of the lth lag of variable y_j on y_i.

The coefficients are estimated through VAR. The plots of these atmospheric variables and power obtained after data cleaning and preprocessing are shown in Figs. 2, 3, 4, 5, 6, 7, 8 and 9.

The test for stationarity is checked with 5% significance level, as the VAR can be applied to stationary series only. Stationary series have statistical properties that do not vary with time, i.e., the series for which the mean, variance, covariance, and standard deviation are independent of time. If the time series is not stationary, the series needs to be made stationary by first and subsequent differencing. The test for stationarity of the time series was evaluated through the ADF test, and it was found that p-values are higher than the significance value; hence, first-order differencing

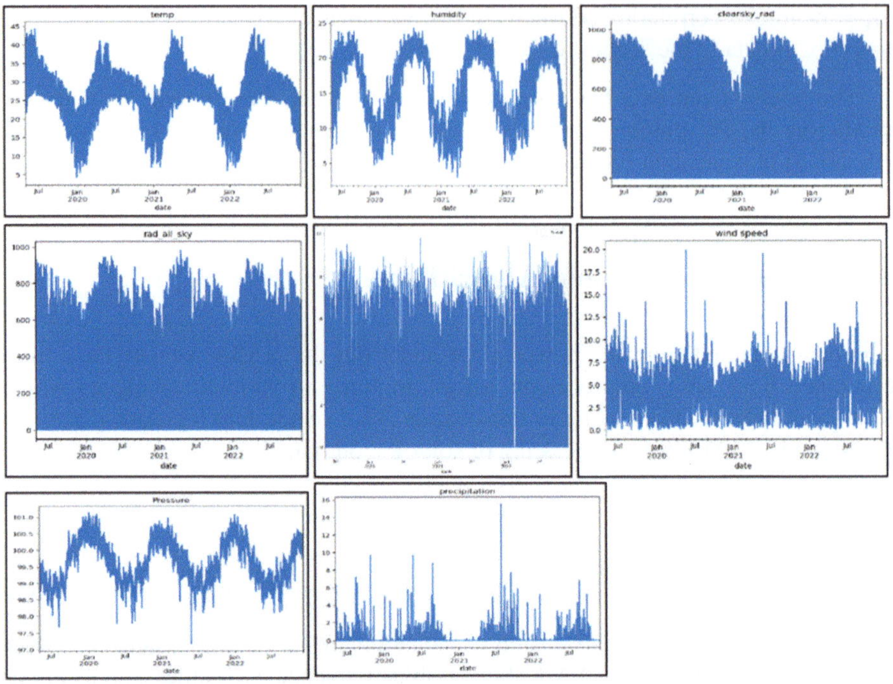

Fig. 2–9 Clockwise from left

Table 2 Correlation matrix of residuals

	temp	humidity	precipitation	rad_all_sky	clearsky_rad	Pressure	wind speed	Power
temp	1.00	0.51	0.11	0.51	0.54	−0.65	−0.02	0.42
humidity	0.51	1.00	0.31	−0.08	0.04	−0.76	0.23	−0.11
precipitation	0.11	0.31	1.00	−0.04	0.07	−0.29	0.23	−0.06
rad_all_sky	0.51	−0.08	−0.04	1.00	0.94	0.01	−0.26	0.91
clearsky_rad	0.54	0.04	0.07	0.94	1.00	−0.09	−0.18	0.85
Pressure	−0.65	−0.76	−0.29	0.01	−0.09	1.00	−0.33	0.07
wind speed	−0.02	0.23	0.23	−0.26	−0.18	−0.33	1.00	−0.26
Power	0.42	−0.11	−0.06	0.91	0.85	0.07	−0.26	1.00

was carried out to make the data stationary. The correlation matrix obtained after establishing the Pearson correlation coefficient is provided in Table 2.

Hence, after establishing the correlation matrix of the residuals, it is found that the parameter "power" has maximum correlation with "clearsky rad," "rad all sky," and "temp," i.e., serious collinearity with these parameters. After selecting 80% data for training and 20% data for testing, the regression result of the training data is obtained with a lag value of 26, corresponding to the minimum AIC. The covariance error matrix obtained is provided in Table 3. The covariance error matrix shows the error of variance and covariance of four parameters taken for analysis, i.e., power, radiation all sky, clear-sky radiation, and temperature. Since our interest is error

Table 3 Error covariance matrix

	Coef	Std err	z	$P > \|z\|$	[0.025	0.975]
Sqrt.var.temp	0.4937	0.002	288.457	0.000	0.490	0.497
Sqrt.cov.temp.rad_all_sky	12.2420	0.352	34.780	0.000	11.552	12.932
Sqrt.var.rad_all_sky	37.9241	0.123	307.329	0.000	37.682	38.166
Sqrt.cov.temp.clearsky_rad	11.1305	0.209	53.374	0.000	10.722	11.539
Sqrt.cov.rad_all_sky.clearsky_rad	14.1302	0.140	101.010	0.000	13.856	14.404
Sqrt.var.clearsky_rad	23.8298	0.101	236.463	0.000	23.632	24.027
Sqrt.cov.temp.Power	0.1114	0.007	15.772	0.000	0.098	0.125
Sqrt.cov.rad_all_sky.Power	0.2677	0.004	65.692	0.000	0.260	0.276
Sqrt.cov.clearsky_rad.Power	−0.0026	0.005	−0.504	0.614	−0.013	0.007
Sqrt.var.Power	0.8367	0.002	475.285	0.000	0.833	0.840

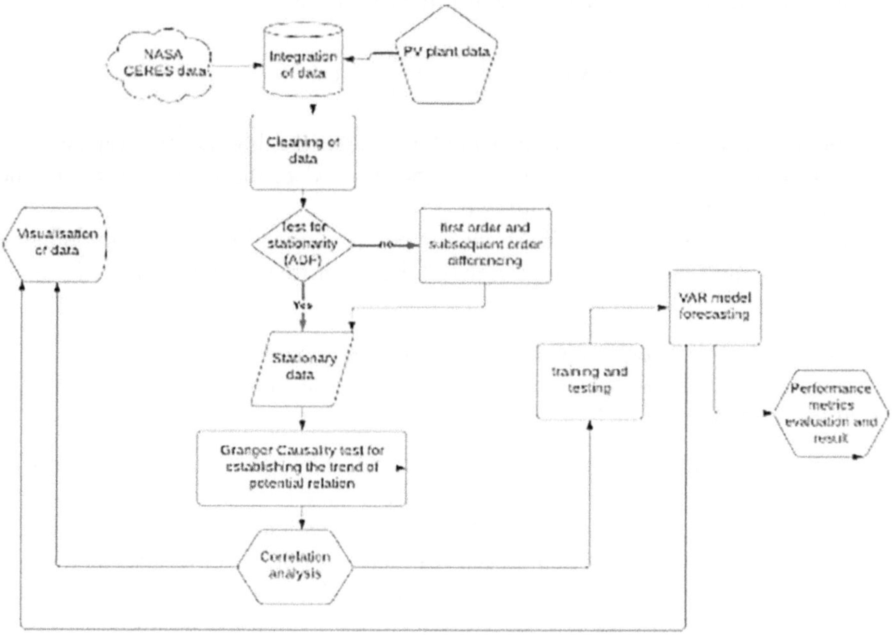

Fig. 10 Flow chart VAR model 1

Table 4 Performance metrics of VAR

MAE	1.9627
MSE	5.7177
RMSE	2.3911

covariance with power, it is deduced from the table that the error margin with respect to the power with other variables is minimum, and hence these factors have a high value of covariance with power (Fig. 10, Table 4).

5.2 Random Forest Regression

Random forest is an ensemble classification and regression method used for both classification and regression problems. It uses multiple algorithms to make decisions, which are the characteristics of the ensemble models. In this model, we use RF for regression analysis. Base constituents of the ensembles used in random forest are tree-structured predictors with leaves and nodes connected by edges [23]. The performance of the random forest tends to stabilize after a particular number of trees, and the risk of overfitting is reduced. Random vector is a collection of tree predictors $h(\boldsymbol{x}; \theta_k)$, where $k = 1,2,3, \ldots, K$, where $\boldsymbol{x}$ represents the training (observed) input vector of length p with associated random vector $\boldsymbol{X}$ and θ_k are independent and identically distributed random vectors [18]. The random forest prediction is the unweighted average over the following [18]:

$$\overline{h}(\boldsymbol{x}) = \left({}^{1}\!/_{K}\right) \sum\nolimits_{k=1}^{K} h(\boldsymbol{x}; \theta_k) \tag{4}$$

For increasing the accuracy of a random forest, the low correlation between different tree members of the forest is important. Also, the residual error of the individual tree should be kept minimum.

Residual error of the individual trees can be maintained low by growing the tree more deeply, and the individual correlation between trees in the random forest is kept low by

1. Using bootstrap samples and developing trees in the forest.
2. Selecting a fixed number of covariates at each node from all possible covariates of the tree and pick the best split of the node.

In random forest regression, the time series has been converted to supervised learning data by feeding the data of the previous time steps as the input data sets. This is made by hyperparameter tuning. Here, the predictors have been split into 4 previous hour time steps, which means that corresponding to the 4 predictors, 16 columns, and 1 output column have been created. Hence, ultimately, 17 columns with 31,440 reframed data sets are created. In the model, the number of trees is taken as 70. The data have been split for training and testing in an 80:20 proportion. The accuracy plot showing the relation between the number of data and accuracy is plotted in Fig. 11 with cross-validation of 10 samples. From the accuracy plot, it is evident that the split gives a fairly good accuracy. The predicted and actual values of the power are plotted in Figs. 12 and 13. A flow chart depicting the random forest regression model used in this case is shown in Fig. 15 (Table 5).

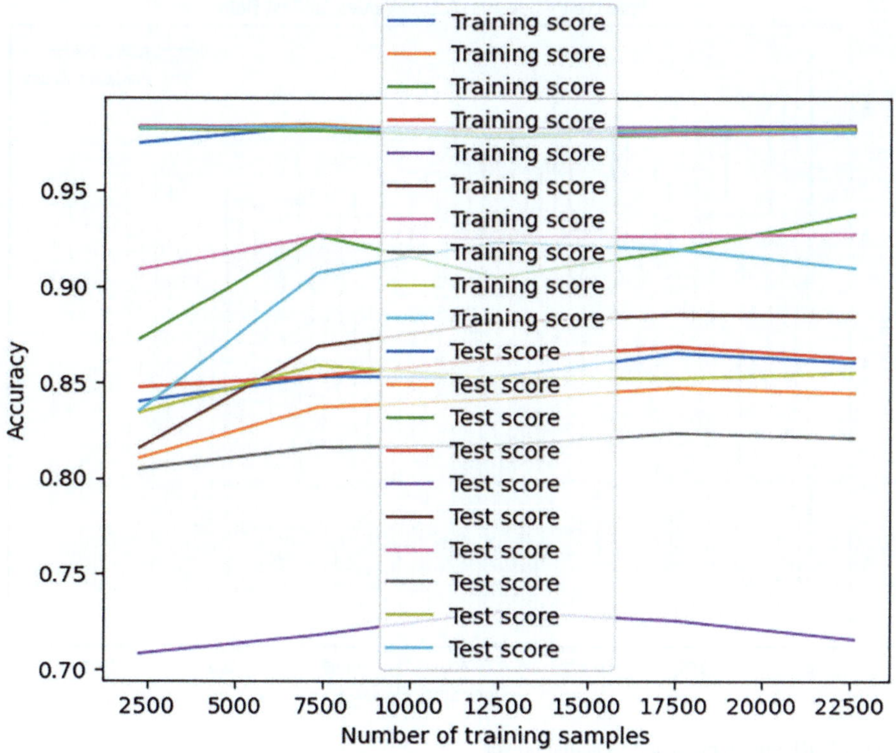

Fig. 11 Accuracy plot

5.3 Prediction of Data Using Long Short-Term Memory

Here, a simplified LSTM model is used for hour-ahead forecasting of multivariate time series data. LSTM has been developed on the fact that human thoughts have persistence, which helps to find the patterns of events. LSTM is an advanced RNN network which is developed to overcome the disadvantages of RNN, such as:

1. Gradient vanishing and exploding problems.
2. Meeting complex training requirements.
3. Long-term memory requirement to remember long sequences and its connection to future predictions.

As RNN cannot address the problems while dealing with complex time series problems as above, LSTM through three gates, viz., input gates, forget gates, and output gates, reads, erases, and writes the information [24]. All three gates with their interconnection are typically called an LSTM cell. A forget gate looks at the new input and the hidden state and determines the portion of information in the cell state that needs to be blocked or forgotten. The input gate also decides what information

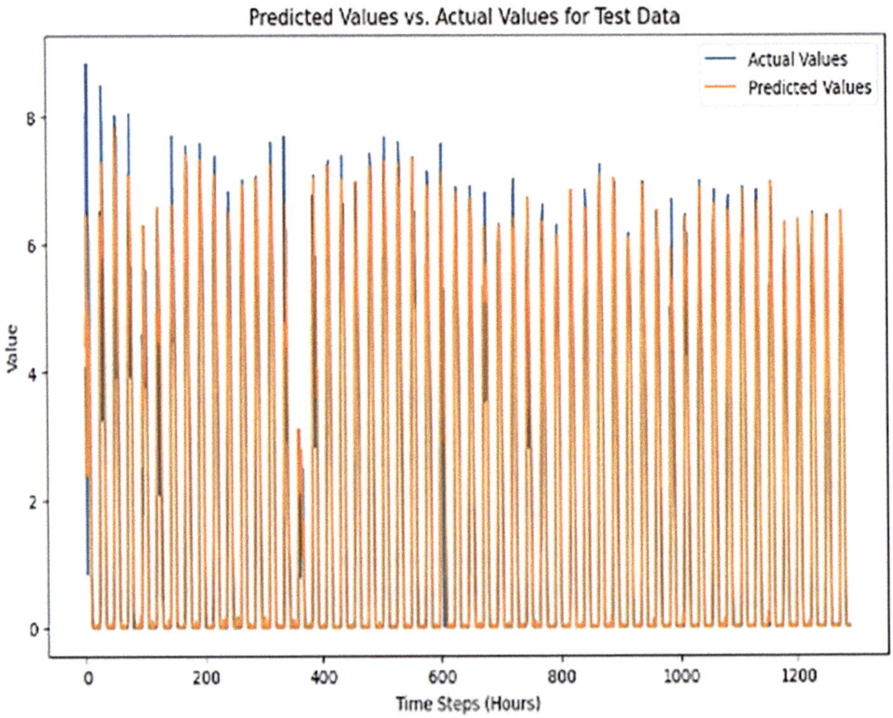

Fig. 12 RF output predicted vs actual power

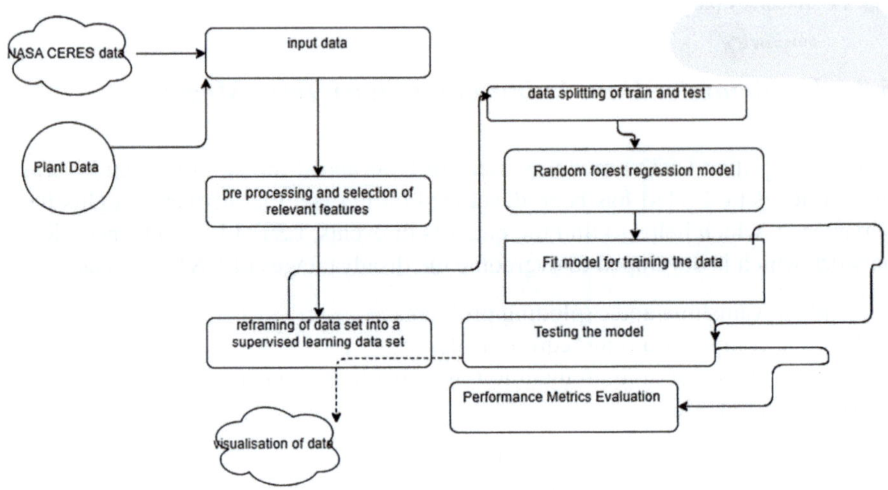

Fig. 13 Flow chart RF 1

Table 5 Performance metrics RF 1

MAE	0.370640
MSE	0.664199
RMSE	0.814984
R2	0.885351

from the input is required to be remembered [8]. The output gate takes information from the cell state, input, and hidden state and produces the output for the current time step as a function of hidden state (short-term memory) and cell state (long-term memory). Thus, LSTM uses two different paths for predictions. These features make LSTM a perfect algorithm for time series forecasting with nonlinearity.

Figure 14 depicts the LSTM unit with the following gates.

o_t: output gate
f_t: forget gate
i_t: input gate
c_t: cell state
$\tilde{c_t}$: candidate state
x_t:input vector
w_f:forget gate(weight)
h_{t-1}: hidden layer output at previous time
h_t: hidden layer output at t
b_f: bias of forget gate
w_i: weight input gate
w_c: weight of the cell
b_c: bias of the cell
b_o: bias of output
o_t: output gate
b_i: input gate bias

The outputs of each gate and layer, namely forget gate, hidden layer, input gate, cell state, and output gates, are governed by the following equations:

$$f_t = \sigma\left(w_f\ [h_{t-1}, x_t] + b_f\right) \tag{5}$$

$$i_t = \sigma\left(w_i\ [h_{t-1}, x_t] + b_i\right) \tag{6}$$

$$\tilde{c_t} = \tanh\left(w_c\ [h_{t-1}, x_t] + b_c\right) \tag{7}$$

$$o_t = \sigma\left(w_o\ [h_{t-1}, x_t\ ] + b_o\right) \tag{8}$$

$$h_t = o_t + \tanh\left(c_t\right) \tag{9}$$

The input gate i_t decides the values of the input vector to be updated according to Eq. (6). The equation implies that the input gate is a function of the input of the

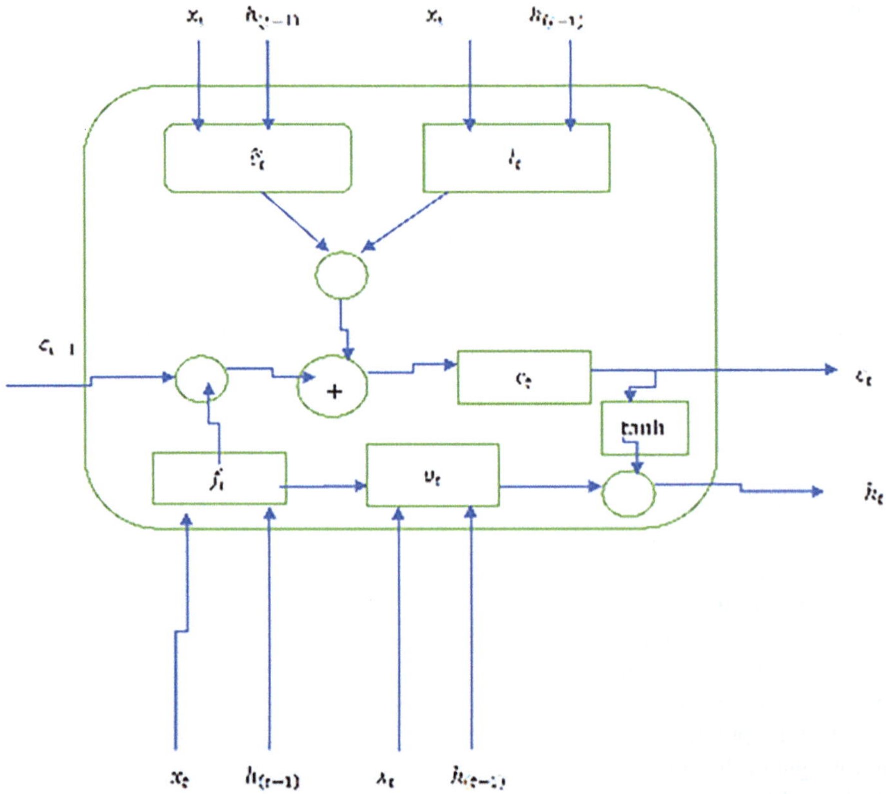

Fig. 14 Structure of LSTM memory cell 1

previous time step, the input of the current time, and the weight of the input gate with bias. The sigmoid function scales the input between 0 and 1. The processing of the input sequence shall be carried out by providing the predictors in the timestep by timestep to the LSTM network. The final output of the sequence shall be provided only after the last element of the sequence is provided.

The flow chart of the LSTM model is shown in Fig. 15. The data set, which corresponds to 31,440 rows and 4 columns, has been converted into a two-dimensional array with previous hour steps of 24 (meaning each time step the model is looking back to the previous 24 h with a time step of 1 h). Hence, the time series data have been converted into 24 input lag time steps for each feature or predictor and 1 output time step. Hence, the series shall be converted into var3(t-1) to var0(t-23) with 96 columns (24 multiplied by 4 predictors) and one output time step, i.e., var3(t) corresponding to the target variable. That means the time series data have been converted into 31,440 rows with 97 columns. The reframed data set is applied with the Min Max scaler. The train and test data split is made as 70:30. The model is applied to an LSTM model with "adam" optimizer and hidden layer size as 64.

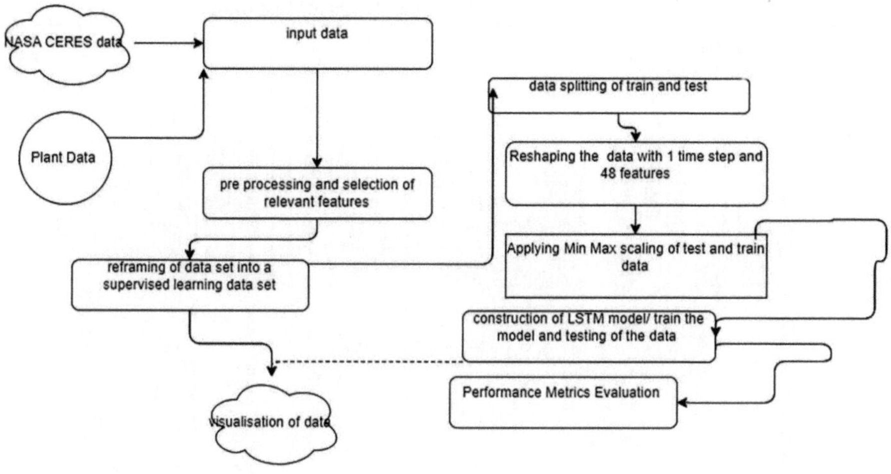

Fig. 15 Flow chart of LSTM 1

Table 6 Performance metrics LSTM 1

MAE	0.367696
MSE	0.600909
RMSE	0.775183
R2	0.895230

Also, the learning rate value selected is 0.0001 for the model. The model estimates a 1-h ahead prediction of power. The number of epochs used is 30. Performance metrics obtained are provided in Table 6 (Figs. 16 and 17).

6 Conclusion

It is evident that out of all three forecasting methods, LSTM forecasts the target variable best. The coefficient of determination value is also superior when compared to the other two methods. With a minimum number of predictors of varying climatic conditions, the LSTM model forecasts accurately. Though the VAR model generally performs well for multivariate analysis, for this complex, highly nonlinear data, the approach does not yield a good result. The high values of MAE (1.9627), MSE (5.7177), and RMSE (2.3911), and a very low value of R^2 show that the model is not apt for such complex data. In the case of the random forest regression model, it has fared well as the model yields better performance metric values. As evident from the analysis of the LSTM model, given a minimum number of predictors and with real plant data, it has yielded a pretty good result of MAE (0.367696), MSE (0.6), and RMSE (0.776). In Indian conditions, further studies are required for forecasting the

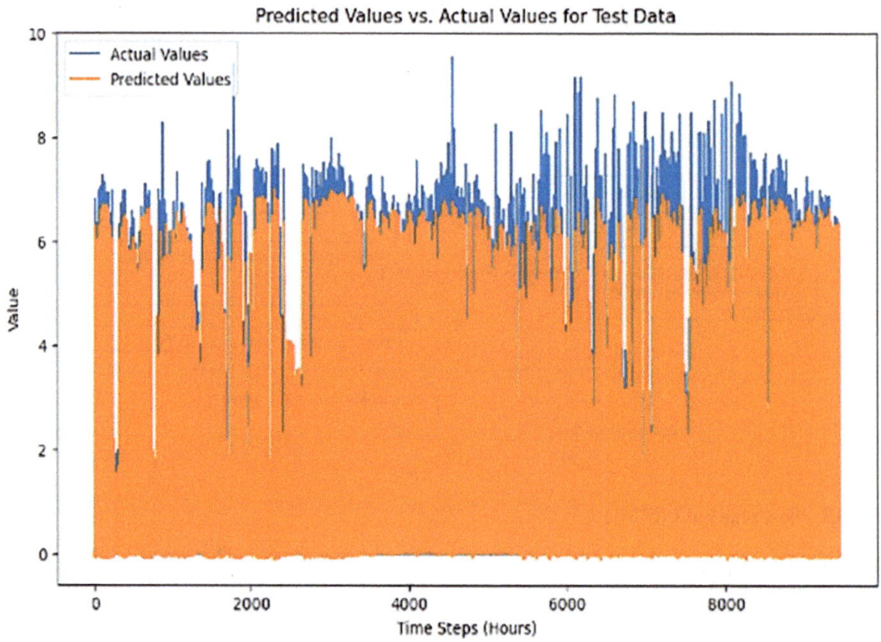

Fig. 16 Predicted power vs actual power

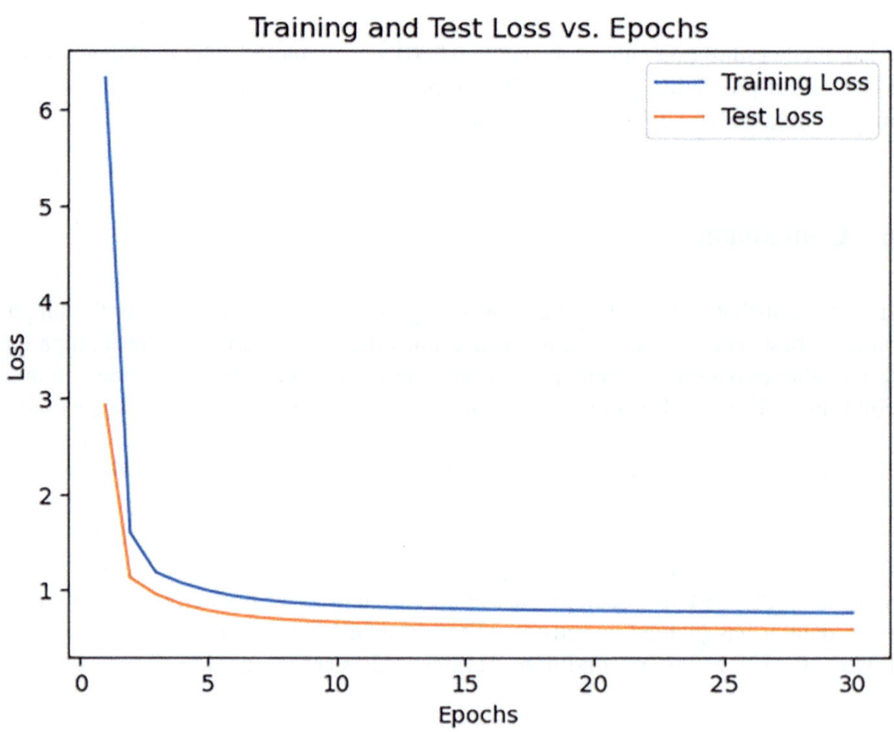

Fig. 17 Training and test loss vs epochs

PV output of large plants using historical data of various atmospheric parameters, as a smaller number of research papers are published in this context.

References

1. G. Reikard, Predicting solar radiation at high resolutions: A comparison of time series forecasts. Sol. Energy **83**(3), 342–349 (2009)
2. M. Tovar, M. Robles, F. Rashid, Pv power prediction, using cnn-lstm hybrid neural network model. case of study: Temixcomorelos, m'exico. Energies **13**(24), 6512 (2020)
3. Wikipedia, Share of electricity production from solar, 2022 (2022)
4. I. Mitra, D. Heinemann, A. Ramanan, M. Kaur, S.K. Sharma, S.K. Tripathy, A. Roy, Short-term pv power forecasting in India: recent developments and policy analysis. Int. J. Energy Environ. Eng. **13**(2), 515–540 (2022)
5. D. Heinemann, E. Lorenz, M. Girodo, Forecasting of solarradiation, in *Solar Energy Resource Management for Electricity Generation from Local Level to Global Scale*, (Nova Science Publishers, New York, 2006), pp. 83–94
6. S. Mishra, P. Palanisamy, Multi-time-horizon solar forecasting using recurrent neural network, in *2018 IEEE Energy Conversion Congress and Exposition (ECCE)*, (IEEE, 2018), pp. 18–24
7. W. Xudong, C. Li, L. Shao, J. Meng, L. Zhang, G. Chen, Is solar power renewable and carbon-neutral: Evidence from a pilot solar tower plant in China under a systems view. Renew. Sust. Energ. Rev. **138**, 110655 (2021)
8. N. Halpern-Wight, M. Konstantinou, A.G. Charalambides, A. Reinders, Training and testing of a single-layer LSTM network for near-future solar forecasting. Appl. Sci. **10**(17), 5873 (2020)
9. U. Munawar, Z. Wang, A framework of using machine learning approaches for short-term solar power forecasting. J. Electr. Eng. Technol. **15**, 561–569 (2020)
10. R.H. Inman, H.T.C. Pedro, C.F.M. Coimbra, Solar forecasting methods for renewable energy integration. Prog. Energy Combust. Sci. **39**(6), 535–576 (2013)
11. J. Zhang, A. Florita, B.-M. Hodge, L. Siyuan, H.F. Hamann, V. Banunarayanan, A.M. Brockway, A suite of metrics for assessing the performance of solar power forecasting. Sol. Energy **111**, 157–175 (2015)
12. M. Suri, T. Cebecauer, A. Skoczek, R. Marais, C. Mushwana, J. Reinecke, R. Meyer, Cloud cover impact on photovoltaic power production in South Africa, in *Southern African Solar Energy Conference SASEC2014*, vol. 1, (2014), p. 8
13. G.M. Yagli, D. Yang, D. Srinivasan, Reconciling solar forecasts: Sequential reconciliation. Sol. Energy **179**, 391–397 (2019)
14. S. Pelland, G. Galanis, G. Kallos, Solar and photo-voltaic forecasting through post-processing of the global environmental multiscale numerical weather prediction model. Prog. Photovolt. Res. Appl. **21**(3), 284–296 (2013)
15. R. Chang, L. Bai, C.-H. Hsu, Solar power generation prediction based on deep learning. Sustain Energy Technol Assess **47**, 101354 (2021)
16. S. Mukherjee, P.B. Ghosh, Estimation of carbon credit and direct carbon footprint by solar photovoltaic cells in West Bengal, India. Int. J. Low-Carbon Technol. **9**(1), 52–55 (2014)
17. S.N. Kaplanis, New methodologies to estimate the hourly global solar radiation; comparisons with existing models. Renew. Energy **31**(6), 781–790 (2006)
18. M. Koengkan, J.A. Fuinhas, A.C. Marques, The relationship between financial openness, renewable and nonrenewable energy consumption, co2 emissions, and economic growth in the latin american countries: an approach with a panel vector auto regression model, in *The Extended Energy-Growth Nexus*, (Elsevier, 2019), pp. 199–229
19. M.R. Segal, Machine learning benchmarks and random forest regression (2004)

20. A. Ozbek, A. Yildirim, M. Bilgili, Deep learning approach for one-hour ahead forecasting of energy production in a solar-pv plant. Energ. Sources Part A Recovery Utili. Environ. Eff. **44**(4), 10465–10480 (2022)
21. S.K. Sahoo, Renewable and sustainable energy reviews solar photovoltaic energy progress in India: A review. Renew. Sust. Energ. Rev. **59**, 927–939 (2016)
22. M. Castangia, A. Aliberti, L. Bottaccioli, E. Macii, E. Patti, A compound of feature selection techniques to improve solar radiation forecasting. Expert Syst. Appl. **178**, 114979 (2021)
23. P. Sedgwick, Pearson's correlation coefficient. BMJ **345** (2012)
24. U. Grömping, Variable importance assessment in regression: linear regression versus random forest. Am. Stat. **63**(4), 308–319 (2009)

Part II
Big Data with Medical Monitoring to Elevate the Overall User Experience Within NGNs

Fuzzy Logic-Based Technique to Build Sustainability in the Early Detection of Alzheimer's Disease

C. R. Nagarathna, S. Babu Kumar, B. Archana, L. H. Anjaneya, Muquitha Almas, and U. Mahantesha

1 Introduction

Soft computing, in contrast to conventional computing, addresses approximation techniques and offers solutions for challenging real-world issues [1]. Soft computing, as opposed to hard computing, allows for approximations, vulnerability, imprecision, and imperfect truth [2]. The human mentality is, in fact, a good example of soft computing. Techniques including fuzzy logic, artificial neural networks (ANN), expert frameworks, genetic algorithms (GA), and artificial intelligence (AI) are used in soft computing. The application of imprecise approximations to provide ambiguous yet practical solutions for intricate computer problems is known as soft computing [3, 4]. The data-driven productivity of neural networks (NN) and the flexibility of fuzzy logic are combined in another field of study known as "soft computing" to address subjective data. This makes it possible to apply genetic algorithms to achieve effective coarse-grained global searching and to create

C. R. Nagarathna (✉) · U. Mahantesha
Department of AI and ML, BNM Institute of Technology, Bangalore, India
e-mail: mahanteshau@bnmit.in

S. Babu Kumar
Department of CSE, Christ University Kengeri, Bangalore, India

B. Archana
Department of CSE, Vidya Vikas Institute of Engineering and Technology, Mysuru, India

L. H. Anjaneya
Department of CSE, Bapuji Institute of Engineering and Technology, Davanagere, India

M. Almas
Department of CSE, Dayananda Sagar College of Engineering, Bangalore, India
e-mail: almas-cs@dayanandasagar.edu

A. Haldorai et al. (eds.), *7th EAI International Conference on Big Data Innovation for Sustainable Cognitive Computing*, EAI/Springer Innovations in Communication and Computing, https://doi.org/10.1007/978-3-032-02979-9_5

precisely tailored alterations through a local search [5]. In the end, fusion algorithms have improved to the point where they outperform all of their core soft computing components and provide superior real-world problem-solving capabilities [6]. Yet, there are a few significant issues that need to be addressed in order to ensure that the computerized process of determining the health of a brain segment is carried out effectively. A number of techniques are offered for the detection and classification of Alzheimer's disease (AD).

This work primarily focuses on developing a conjoined system of feature extraction, a data categorization system, and a holistically nested edge detection (HED) system to diagnose Alzheimer's disease. Segmenting the images comes first in the suggested process, then feature extraction using convolutional networks, and classification performance using a fuzzy logic system. Identifying AD and classifying it using neurofuzzy logic (NFL), a soft computing technique, is the main objective of this work.

The remaining portions of this manuscript are organized as follows: Sect. 2 discusses the existing work; Sect. 3 describes the recommended techniques; Sect. 4 presents the outcomes; Sect. 5 offers a discussion; and Sect. 6 provides the conclusion.

2 Literature Survey

An extensively used technique for AD diagnosis is magnetic resonance imaging (MRI). According to earlier research, AD patients have less hippocampal and gray matter than healthy individuals. However, the cross sections of sliced photographs and even little variations in image registration can have a significant effect on volume, making the assessment of regional brain volume challenging. When comparing AD patients to non-AD persons, ocular examinations may show that the cerebral ventricle and brain shape alter, becoming more rounded. If the MRI diagnostic sensitivity for AD detection is higher than that of volume changes and if shape modifications can be accurately assessed, then this approach may prove helpful in the early diagnosis of the illness. All of the AD and health control (HC) data must match in order to compare specific images and their locations [7].

Using a Student's *t*-test-based feature selection strategy, researcher Acharya U.R extracted features from the MRI dataset that was received from the University of Malaya Medical Centre. The K-nearest neighbors (KNN) fared better in the data classification process than other feature selection strategies, with a 94.55% classification accuracy. Shearlet transform (ST) feature extraction procedure improved performance. An Open Access Series of Imaging Studies (OASIS) MRI dataset was used to autonomously diagnose AD disease using a deep learning (DL) network ensemble model. The system's accuracy was 93%. The author used the stochastic gradient descent optimization technique to lower network loss and improve system performance [8–11].

Using an adaptive histogram, the researcher preprocessed brain MRI data from the Alzheimer's Disease Neuro Imaging (ADNI) dataset. The essential characteristics are extracted using Student's *t*-test and Adaptive Synthetic sampling (ADASYN) after the dataset has been preprocessed for balance. Then, an accuracy of 93.9% was obtained for both the random forest (RF) and SVM-poly 1 classifiers employed for classification (Islam, J, 2018). To determine whether it would be possible to distinguish between AD, mild cognitive impairment (MCI) classification, and Normal Control (NC) using the magnetic resonance imaging Alzheimer's disease neuroimaging initiative (MRI ADNI) dataset, the convolutional neural networks (CNN) model was employed. The Talairach transform, motion correction, nonuniform intensity normalization, intensity normalization, and skull-stripping procedures were employed by the author to convert three-dimensional (3D) MRI data into a two-dimensional (2D) slice. The characteristics were obtained after 2D MRI slices were obtained so that the CNN model may be used to classify the data. Koh, J. et al. (2020) reported that the model's accuracy for AD vs. NC classification was 94.97%, and for AD vs. MCI classification, it was 91.98%.

In order to detect AD using MRI, Braulio Solano-Rojas created a three-dimensional DENSENET neural network. To distinguish between five different disease stages—NC, significant memory concern, Early MCI (EMCI), MCI, and Late MCI (LMCI)—the author developed a deep neural network classifier. Metrics generated by the classifier included a mean accuracy of 0.86, a mean specificity of 0.86, a micro-average area under the receiver operating characteristic curve of 0.91, and micro-average mean sensitivity of 16 [12–15].

3 Methods

Figure 1 depicts the suggested HED and Fuzzy neural network (FNN) model. Firstly, preprocess the brain MRI dataset using techniques such as edge identification, mask creation, brain part cropping, and normalization to boost the scheme effectiveness. Separate the dataset even more into train and test subsets. Make use of the training dataset so that the fuzzy neural network can learn from the preprocessed photos. Utilizing the test dataset, assess the FNN model in the end.

Algorithm 1 reveals the intricacies of the suggested task. The images are subjected to DL-based edge detection, or HED, before being mask created. For the MRI scans with numerous orientations, additional image augmentation and normalization are conducted in order to increase the number of images in the collection. Next, using the masked images, the largest contours are located. Following that, a bounding box is made, the image within of which is cropped, and it is saved for later classification.

Algorithm 1 Classification of AD using Deep Neural Network (DNN) and holistically-nested edge detection.

Input: Brain images.

Output: Classified images.

Step 1: Import images.

Step 2: Preprocessing images

- Edge detection using DL.
- Use the edge-detected image to create a mask image.
- Use the masked image to locate the largest contours. Make a bounding box.
- Save the cropped image after cropping it within the enclosing box.

Step 3: FNN for feature extraction.
Step 4: Use of a fuzzy interface mechanism for classification.
Step 5: Use the acquired feature vector to train the FNN model.
Step 6: Approve test pictures and note the outcome.
Step 7: Identify and categorize the pictures.

This study attempts to demonstrate the use and implementation of DL-grounded edge detection, or HED. This study is being used to categorize the Kaggle MRI dataset into four stages. It combines feature extraction, classification, and HED techniques.

3.1 *Image Processing*

Dataset The MRI images from publicly accessible datasets are explored within this planned effort. The Kaggle MRI images were produced by researcher Sarvesh Dubey [16]. The four phases of AD represented in the dataset are Moderate Demented (MOD-3), Very Mild Demented (VMD-1), Mild Demented (MID-2), and Non-Demented (ND-0). Figure 1 displays the total number of photos for each phase as well as sample photographs for each dataset class. The ND VMID, MID, and MOD samples are displayed in the 0, 1, 2, and 3 accordingly. Table 1 displays the number of Kaggle datasets.

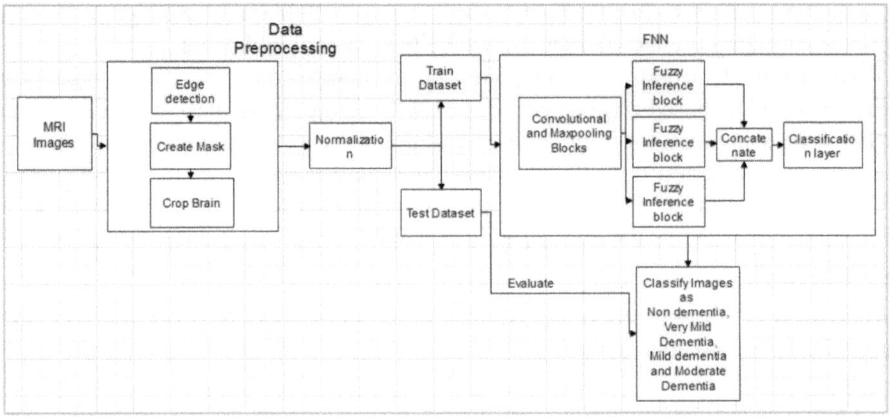

Fig. 1 Proposed HED and FNN

Table 1 Kaggle Dataset number

Classes	Kaggle_Dataset_Count
ND	3200
VMD	2240
MID	896
MOD	64

Edge Detection One of the first practical uses of CV was edge identification, which facilitates the identification of the borders of objects in pictures. The Canny edge identifier is most likely the mechanism utilized when utilizing OpenCV to detect an object's edges. It should be noted that there are a few drawbacks to employing Canny Edge for edge detection, such as the need for hysteresis thresholding and additional preprocessing steps like concealing and switching to grayscale in order to obtain good edge maps. HED is used to manage the aforementioned problems.

OpenCV has integrated a form of DL-based edge detection technology into its ostentatious new DNN module. Holistically nested edge detection, or HED for short, is the strategy employed in this. It is an edge detection system that operates from start to finish and is based on learning. 3.4.2 or a later version of OpenCV is needed to use this new technique. The system/network receives an RGB image as input. The edge map, the network's output, performs better than the original by capturing and maintaining an object's borders within an image. HED makes use of intermediate layer yields. The term "side output" describes the yield from earlier layers that are combined to get the final prediction. The feature maps generated at each layer vary in size; therefore, it is possible for it to view the provided image at several different scales.

The HED strategy outperforms other deep learning-based strategies not just in terms of accuracy but also in terms of speed. "Why Open CV is chosen for coordinating into new DNN components" stems from this rationale. It is a framework for end-to-end edge identification. As a result, HED can become proficient in the kind of rich, varied, leveled features that are essential to advancing human ability to assess uncertainty in natural picture boundary detection of an item and its edge. Since HED plans to prepare and anticipate edges in a image to image manner, even though it does not explicitly model the structured output, the name "holistic" is used. Because it highlights the side outputs that are generated that correspond to gradually and inherited refined edge maps, the word "nested" is employed. The predictions established along the route are fundamental and common for each edge map that is generated. More compactness is added to the progressive edge maps. The coordinated learning of several leveled aspects distinguishes current multi-scaled techniques [17–20].

The anticipated HED addresses two important issues:

(a) Using Fourier convolutional neural network (FCNN)-powered training and prediction to carry out image-by-image classification (the framework takes an image as input and produces edge maps as output in an easy-to-understand manner).

(b) Nestled multiple-scale feature learning is a notable attribute that is driven by deep learning [21–24], which carries out deep layer oversight to help with earlier yields. The advantages that these foundational approaches exhibit in HED are found to be accurate and computationally efficient.

Let us use S to represent the input training dataset for the training phase, $S = \{Xn, Yn, n = 1, \ldots, N\}$, where sample $X_n = \left\{x_j^{(n)}, j = 1, \ldots, |X_n|\right\}$ and $Y_n = \left\{y_j^{(n)}, j = 1, \ldots, |X_n|\right\}$, $y_j^{(n)} \varepsilon \{0, 1\}$ depict both the original and equivalent ground truth binary edge maps for an image X_n. Because each picture or image is viewed holistically and independently, the subscript "*n*" is removed to maintain the simplicity of the notation. The aim was to create a model of a system that processes inputs and generates edge maps that are in proximity to ground realities. For simplicity, let us assume that W is the collection of all network layer standard parameters. Let us assume that there are "*m*" side-yield layers in the system. Additionally, each side-yield layer was linked to the classifier, where the corresponding weights were intended, as $w = (w^{(1)}, \ldots, w^{(M)})$. The goals are gauged as follows:

$$L_{\text{side}}(W, w) = \sum_{m=1}^{M} \alpha_m l_{\text{side}}^{(m)}\left(W, w^{(m)}\right) \tag{1}$$

when the loss function for side outputs at the picture level is indicated by ℓside. Every pixel in a training image and edge map is subjected to a loss function processing during the HED training process. Delivery of edge and non-edge pixels is strongly biased for a typical characteristic picture. Senthilkumaran and Rajesh [25] and Shajin et al. [26] implemented a cost-effective loss function and provided additional compromised limitations for biased sampling. Instead, a simpler mechanism that automatically modifies the loss between positive and negative classes is applied. Additionally, each pixel base has a class-adjusting weight β as shown. Next, use this weight that adjusts for class as an easy way to even out this imbalance between edge and non-edge. The accompanying "class-adjusted cross-entropy loss function" is specifically described as follows:

$$\begin{aligned} L_{\text{side}}^{(m)}\left(W, w^{(m)}\right) = &-\beta \sum_{j\,\varepsilon Y_+} \log\Pr\left(y_j = 1 \,|X; W, w^{(m)}\right) \\ &- (1-\beta) \sum_{j\,\varepsilon Y_-} \log\Pr\left(Y_j = 0 \,|X; W, w^{(m)}\right) \end{aligned} \tag{2}$$

where $\beta = \frac{|Y_-|}{|Y|}$ and $1 - \beta = \frac{|Y_+|}{Y}$, $|Y_-|$ and $|Y_+|$ denote "ground truth label sets with edges and non-edges," respectively. $\Pr(y_j = 1|X; W, w^{(m)}) = \sigma(a_j{}^{(m)})\epsilon[0, 1]$ is calculated by means of "sigmoid function σ(.) overactivation value present in pixel j". At every side-yield layer, acquire edge map prediction $\hat{Y}_{\text{side}}^{(m)} = \sigma\left(\hat{A}_{\text{side}}^{(m)}\right)$, where

$\hat{A}_{\text{side}}^{(m)} \equiv \left\{a_j^{(m)}, j = 1, \ldots \ \ldots, |Y|\right\}$ are the activation of the side-yield of the "m" layer.

Incorporating a weighted combination layer into the system and simultaneously examining the combination weight during training will enable the straightforward application of side-yield expectations. The fusion layer's loss function, L_fuse, develops as follows:

$$L_{\text{fuse}}\left(W, w, h\right) = \text{Dist}\left(Y, \hat{Y}_{\text{fuse}}\right) \tag{3}$$

where $\hat{Y}_{\text{fuse}} \equiv \sigma \sum_{m=1}^{M} h_m \quad \hat{A}_{\text{side}}^{(m)}$, in which the cross entropy loss is indicated by Dist $(\cdot, \cdot)$ and $h = (h_1, \ldots, h_M)$ is the fusion weight. After everything is put together, use "standard stochastic gradient descent" to limit the accompanying objective function:

$$(W, w, h)^* = \text{argmin}\ \left(L_{\text{side}}\left(W, w\right) + L_{\text{fuse}}\left(W, w, h\right)\right) \tag{4}$$

In order to obtain edge map prediction from weight fusion and side yield layer, first receive image X as an input during testing. This is indicated by:

$$\left(\hat{Y}_{\text{fuse}}, \hat{Y}_{\text{side}}^{(1)}, \ \ldots \ \ldots \ \ldots, \hat{Y}_{\text{side}}^{(m)}\right) = \text{CNN}\left(X, (W, w, h)^*\right) \tag{5}$$

Here, CNN $(\cdot)$ denotes an edge map that was obtained using the system. By taking an additional average of these generated edge maps, the final united yield may be obtained.

$$\hat{Y}_{\text{HED}} = \text{Average}\left(\hat{Y}_{\text{fuse}}, \hat{Y}_{\text{side}}^{(1)}, \ldots, \hat{Y}_{\text{side}}^{(m)}\right) \tag{6}$$

As a result, feedback must be spread via the middle layers since the multi-scale reactions generated at hidden layers [27, 28] are less significant due to the lack of deep supervision, which is recalled for the proposed approach. Moreover, their patch-to-patch or patch-to-pixel strategies seriously impair the efficacy of both training and prediction. This work intends to demonstrate an end-to-end edge detection framework and a system that uses a "holistically nested" architecture to provide additional deep supervision on top of the convolutional model, triggered by an FCNN system. Figure 2 presents the images obtained by applying the HED approach. This image is the input image that the NN system will use to make an inference.

Scale factor: The scale factor is used to scale the provided images. Multiply the constant number by 255 and divide the result to achieve this. Each pixel's value falls between 0 and 1, or 0/255–255/255. In the event that scaling is not there, set

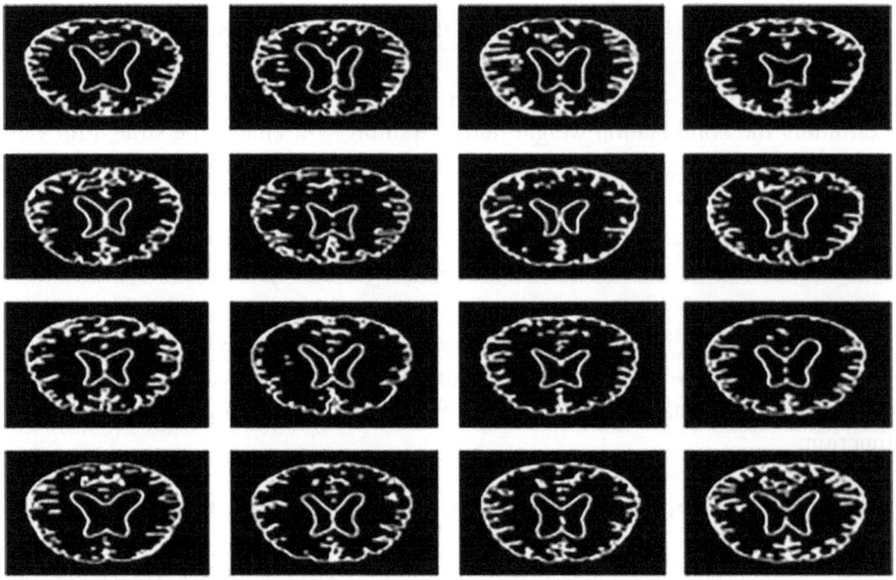

Fig. 2 Output of HED

the default value to 1.0. The scaling factor for the model used in this investigation is 1.0.

Size: Given the result, the corresponding spatial size is given. The size will typically be similar to the input size needed for neuron in next layer to yield. Both height and width are utilized by the model.

swapRB: This is a Boolean indicating the need to alter the primary and final channels of an RGB image. Assume that all of the photographs on your OpenCV were originally in the BGR format. Set the flag to 'True," which is the default, and convert this sequence to RGB if the other direction is required. The model assumes that this parameter is "False."

Mean: To account for fluctuations in the intensity condition and standardization, the average pixel value over the training dataset must be computed. Each image at the time of training must then be further subtracted or withdrawn. Mean subtraction must be used during inference if it is applied during training. The determined mean is generally composed of R, G, and B channel tuples. Using swapRB = "False" will result in this sequence being (B, G, R). In the current study, the factor sets for R, G, and B are 104.00698793, 116.66876762, and 122.67891434, in that order.

Crop: It is a flag that allows you to center-crop your photos using Boolean values. Setting this parameter to "True" will crop the input image in the center so that the smaller dimensions match the corresponding size and the other measurement matches or exceeds the larger measurement. When set to "False," on the other hand, it preserves the feature proportions and just resizes to the specified dimensions. The value of this option is set to "False" in the current study.

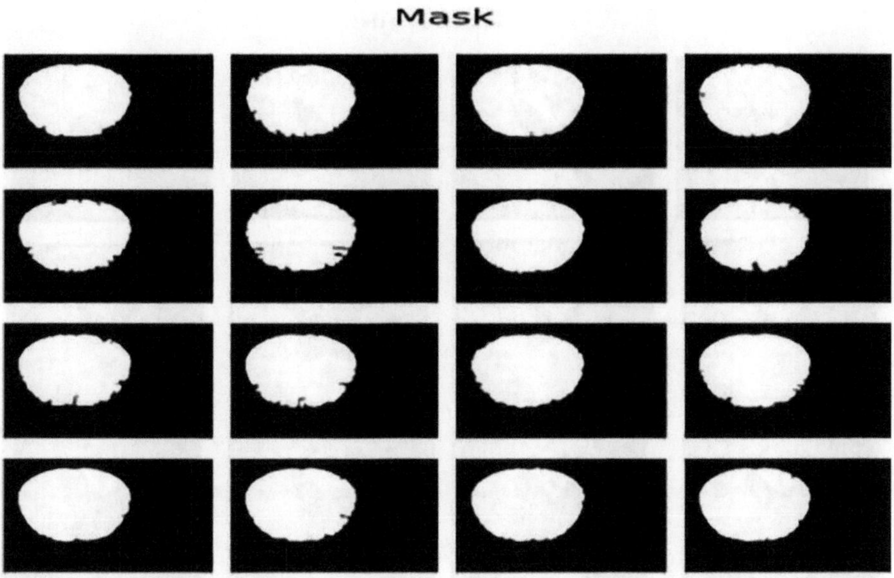

Fig. 3 Masked image

Use the edge-detected image to create a mask: Masking is nothing more than setting a portion of a picture's pixel values to zero or another value for the background. A tiny area of the image is defined and used during masking in order to modify the larger image. The method that lies behind several types of image processing, including edge detection, is masking. The masking approach applied to the dataset is displayed in the image below. It provides the region of interest while hiding the background. The rotated image with mask is shown in Fig. 3. To accomplish data augmentation, each image in the dataset is rotated many times.

Tracing the Perimeter of the Box and Resizing the Picture inside The bounding box provides the region of interest, and the photos need to be cropped before being passed along to provide the model with the most accurate region of interest. Figure 4 displays the cropped photos according to the bounding box boundaries.

3.2 *Feature Extraction Model*

The goal is to create a DNN system that can generate many leveled features with proficiency and that can capture different phases at different speeds in order to capture typical edge-map sizes.

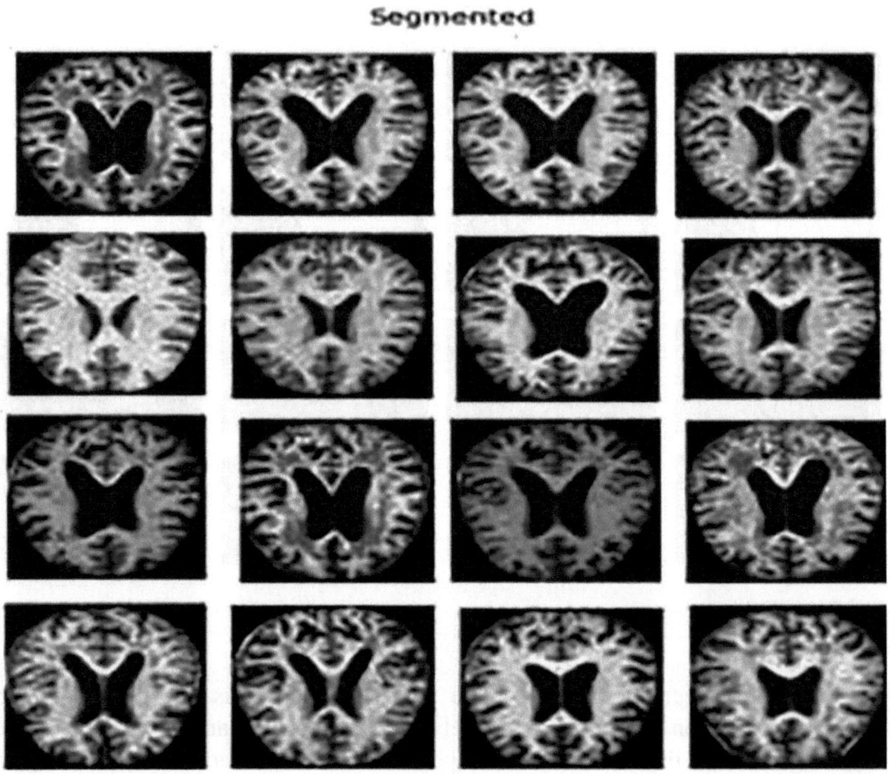

Fig. 4 Cropped image

For feature extraction and classification in the planned work, an fuzzy neural network model is taken into consideration. The fuzzy neural network with five convolutional layers and maxpooling layers for feature extraction is taken into consideration here. It is thought that employing five convolutional layers can achieve remarkable depth in the performance of neural network design. Nonetheless, numerous contemporary researchers have demonstrated via their research that fine-tuning the pretrained DNN model on the classification of common images is beneficial for edge detection at the lower level. The planned research embraces this, but with a few changes that are outlined below:

(a) Fully convolutional network (FCN) layers are trimmed.
(b) The fuzzy classifier is coupled to the output layer. The related responsive field's dimension is identical to that of the corresponding side-yield layer.

The input volume is convolved using the weights volume. The input layer is shrunk or expanded based on the padding and outstriding. Lessen the width and height of the space. On the other hand, as the convolutional process deepens, the height and width of the input images are reduced. In order to obtain a more complex

Table 2 Layer summary of feature extraction model

Layer's description	Filter	Stride	Padding	Activation	Input shape	Output shape
Convolutional Layer 1	20 Filters, 6 × 6	3 × 3	Same	Relu	224 × 224 × 3	73 × 73 × 3
Maxpooling 1	40 Filters, 3 × 3	1 × 1	Same	Relu	73 × 73 × 20	73 × 73 × 20
Convolutional Layer 2	40 Filters, 6 × 6	2 × 2	Same	Relu	73 × 73 × 20	34 × 34 × 40
Maxpooling2	6 × 6	1 × 1	Same	Relu	34 × 34 × 40	34 × 34 × 40
Convolutional Layer 3	40 Filters, 3 × 3	2 × 2	Same	Relu	34 × 34 × 40	16 × 16 × 40
Maxpooling 3	2 × 2	1 × 1	Same	Relu	16 × 16 × 40	16 × 16 × 40
Convolutional Layer 4	40 Filters, 3 × 3	2 × 2	Same	Relu	16 × 16 × 40	7 × 7 × 40
Maxpooling 4	2 × 2	1 × 1	Same	Relu	7 × 7 × 40	7 × 7 × 40
Convolutional Layer 5	3 Filters, 4 × 4	2 × 2	Same	Relu	7 × 7 × 40	4 × 4 × 3
Dropout	–	–	–	–	4 × 4 × 3	4 × 4 × 3

target function, a nonlinear activation function is used for each layer. The complex function needs to exhibit nonlinear variation in response to the given input image. The feature extraction method's layer summary is displayed in the table. The model consists of five convolutional layers with 20, 40, 40, 40, and 3 filters, respectively, with the same padding on the 3 × 3 and 2 × 2 strides. Relu activates each neuron in the model. Following the convolutional layers are maxpooling layers with the same padding and a 1 × 1 stride. Table 2 provides the convolutional neural network's full layer summary for the FNN. The table shows each layer's input and output shapes. The network receives input in the format 224 × 224 × 3. The final image output shape is 4 × 4 × 3.

3.3 Classification Model

The method for classifying the combined qualities that were developed is described in this section using a fuzzy neural network. A specific type of fuzzy scheme called the neural fuzzy scheme uses fuzzy logic and fuzzy rules to determine learning parameters through the use of an NN-based learning technique. The purpose of processing the input is to obtain feature-wise data. To fuzzify the appropriate feature values, utilize the fuzzy membership function. The procedure consists of three steps: fuzzification, NN classifier construction, and defuzzification.

Through data fuzzification, a membership function matrix including the degrees of each class and membership function (MF) is created. The membership function mfi,j(mi) determines the degree of MF of the present pattern i and class j where the ith patterns signify $m_i = x_{i1}, x_{i2}, x_{i3}, \ldots, x_{ik}$. The input vector y indicates

$m = [m_1, m_2, \ldots, m_N]T$ where T denotes the transpose operation. To control MF action during fuzzification, fuzzy logic uses the membership function. In this study, a fuzzy bell-shaped membership that is based on a number of factors is used. The following represents the membership function with fuzzy boundaries:

$$mf\,(y;\, p, q, r) = \frac{1}{1 + \left|\frac{m-r}{p}\right|^{2q}} \tag{7}$$

The control parameters are critical; the MF's center is determined by r, its breadth by p, and its slope at crossing sites is controlled by q. Equation (8) contains it.

$$\mathrm{MF}(y) = \begin{bmatrix} mf_{1,1}(m_1) & mf_{1,2}(m_1) & mf_{1,3}(m_1) & \ldots & mf_{1,M}(m_1) \\ mf_{2,1}(\mathrm{m}_1) & mf_{2,2}(m_1) & mf_{2,3}(m_1) & \ldots & mf_{2,M}(m_1) \\ mf_{3,1}(\mathrm{m}_1) & mf_{3,2}(m_1) & mf_{3,3}(m_1) & \ldots & mf_{3,M}(m_1) \\ \ldots & \ldots & \ldots & \ldots & \ldots \\ mf_{N,1}(m_1) & mf_{N,2}(m_1) & mf_{N,2}(m_1) & \ldots & mf_{N,M}(Y_N) \end{bmatrix} \tag{8}$$

The construction of the NN classifier method is explained in the following phase. First, an MN feature matrix is created from the feature matrix. The neural network (NN), which consists of an input layer, an output layer, and a hidden layer, then uses the vector as its input. The hidden layers and input layers of the system are interconnected, and the output layer units are connected to the hidden layers. Each layer's nodes are connected by weights and biases that are randomly selected before the training process starts. The estimated yield is handled as follows once the inputs are linearly integrated and supplied as I/O to the NN:

$$\mathrm{Output}_j = \frac{1}{1 + e^{-Net_j}} \tag{9}$$

The max operator-based hard categorization defuzzification procedure is described here. Sort j using the input pattern. It is provided at Eq. (10).

$$mf_i(y) \geq mf_i(y)\forall_i \in (1, 2, \ldots M) \text{ and } i \neq j \tag{10}$$

There are many advantages to neurofuzzy classifiers, such as their capacity to self-sort, self-learn, and self-tune and their lack of requirement for prior knowledge about the relationships between the data. It can calculate more quickly by using fuzzy membership, and it can help with data conflicts and other issues [29–31].

In this proposed work, the fuzzy inference system is designed with three fuzzy inference blocks, each containing one hundred neurons. The fuzzy rule function of Eq. 10 activates the fuzzy neurons. Three layers' worth of output is concatenated and de-fuzzified. Figure 5 displays a snapshot of the FNN model summary.

The fuzzy neural network model fulfills the need to identify Alzheimer's phases. The four-type Kaggle MRI dataset was taken into account.

```
+ Code  + Text

fnn.summary()

Model: "model_5"
__________________________________________________________________________________________________
 Layer (type)                   Output Shape         Param #     Connected to
==================================================================================================
 input_19 (InputLayer)          [(None, 224, 224, 3  0           []
                                )]

 conv2d_687 (Conv2D)            (None, 73, 73, 20)   2180        ['input_19[0][0]']

 max_pooling2d_46 (MaxPooling2D  (None, 73, 73, 20)  0           ['conv2d_687[0][0]']
 )

 conv2d_688 (Conv2D)            (None, 34, 34, 40)   28840       ['max_pooling2d_46[0][0]']

 max_pooling2d_47 (MaxPooling2D  (None, 34, 34, 40)  0           ['conv2d_688[0][0]']
 )

 conv2d_689 (Conv2D)            (None, 16, 16, 40)   14440       ['max_pooling2d_47[0][0]']

 max_pooling2d_48 (MaxPooling2D  (None, 16, 16, 40)  0           ['conv2d_689[0][0]']
 )

 conv2d_690 (Conv2D)            (None, 7, 7, 40)     14440       ['max_pooling2d_48[0][0]']

 max_pooling2d_49 (MaxPooling2D  (None, 7, 7, 40)    0           ['conv2d_690[0][0]']
 )

 conv2d_691 (Conv2D)            (None, 4, 4, 3)      1923        ['max_pooling2d_49[0][0]']

 dropout_7 (Dropout)            (None, 4, 4, 3)      0           ['conv2d_691[0][0]']

 flatten_9 (Flatten)            (None, 48)           0           ['dropout_7[0][0]']

 fuzzy_inference_block_85 (fuzz  (None, 100)         0           ['flatten_9[0][0]']
 y_inference_block)

 fuzzy_inference_block_86 (fuzz  (None, 100)         0           ['flatten_9[0][0]']
 y_inference_block)

 fuzzy_inference_block_87 (fuzz  (None, 100)         0           ['flatten_9[0][0]']
 y_inference_block)

 concatenate_16 (Concatenate)   (None, 300)          0           ['fuzzy_inference_block_85[0][0]'
                                                                  , 'fuzzy_inference_block_86[0][0]
                                                                 ',
                                                                  'fuzzy_inference_block_87[0][0]'
                                                                 ]

 dense_14 (Dense)               (None, 4)            1204        ['concatenate_16[0][0]']

==================================================================================================
Total params: 63,027
Trainable params: 63,027
Non-trainable params: 0
__________________________________________________________________________________________________
```

Fig. 5 Screenshot generated for proposed model summary

4 Results

The four distinct classifications from the Kaggle dataset are employed here: very mild dementia, mild dementia, moderate dementia, and no dementia. This work used a specially configured FNN to classify the brain images. The method's shape represents the input and output shapes of projected hidden layers. FNN model.

Table 3 provides values for the confusion matrix of the examined data. Classifying the stages of the provided brain image into the appropriate illness class is the main objective of the methodology's outcomes, which are now being presented. For the test photos of the projected work, the correctness of the result is shown in Table 3. Using a database of healthy and pathological brain pictures, the suggested approach utilizing the FNN architecture is validated. Table 95.75%, 91.42%, 90.98%, 91.10%, 25%, and 74.25% represent, respectively, the average performance for accuracy, precision, recall, F1 score, false positive rate (FPR), and specificity obtained using the predicted technique.

Figure 6 displays a snapshot of the training dataset and validation dataset's loss and accuracy history graphs. The system is designed to operate with a batch size of 115 over 30 epochs. For the training dataset, the accuracy rate of the system is 97.37%, while for the test dataset, it is 87.75%.

The historical graphs demonstrate how, with each iteration, the model trains more with the input images, which contributes to a decrease in loss and an increase in accuracy. Figure 7 displays a snapshot of the output produced by the suggested edge detection model, which yields an accuracy of 87.75% for the test dataset.

5 Discussion

The statistical performance of the proposed model on a dataset comprising different stages of Alzheimer's disease and dividing them into diseased and healthy subjects

Table 3 Confusion matrix

Non demented	**Very mild demented**	**Mild demented**	**Moderate demented**	**Accuracy**	**Precision**	**Recall**	**F1 score**	**FPR**	**Specificity**
284	0	38	22	94.42%	93.42%	82.55%	87.65%	22%	77.8%
0	356	0	0	100%	100%	100%	100%	25%	75%
6	0	345	29	93.58%	86.03%	90.55%	88.23%	30%	70%
14	0	18	321	94.21%	86.29%	90.93%	88.55%	26%	74%
Average				95.75%	91.42%	90.98%	91.10%	25%	74.2%

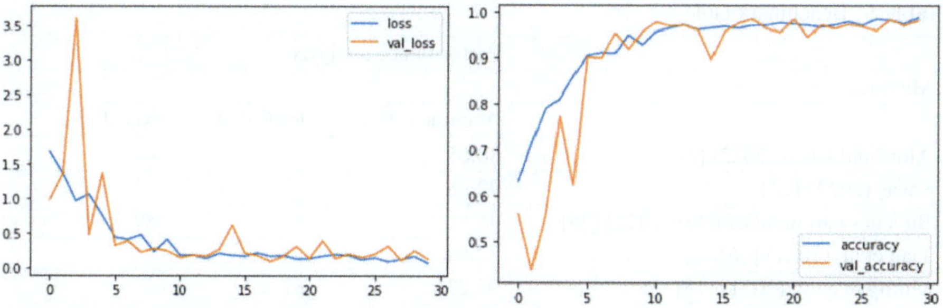

Fig. 6 Screenshot of the loss and accuracy history graphs produced by the edge detection model for classification

```
[29] test_scores = model.evaluate(test_data, test_target)
     29/29 [==============================] - 14s 468ms/step - loss: 0.9275 - accuracy: 0.8775 - precision: 0.6104 - recall: 0.9409 - auc: 0.9066
```

Fig. 7 Screenshot of model performance generated for the test dataset

Table 4 Statistical performance of the work on test images of various AD categories

Classes	TP count	FP count	TN count	FN count	Recall	FPR	Precision
Non-demented	284	20	1069	60	0.8255	0.022	0.9342
Very mild demented	356	0	1077	0	1	0.025	1
Mild demented	345	56	997	35	0.9055	0.030	0.8603
Moderate demented	321	51	1029	32	0.9093	0.026	0.8296

is presented in Table 4, which displays 128 incorrectly classified images and 1306 successfully classified images based on the statistical performance of the job. Following analysis, the following results were reached: According to this, the model's average performance was 91.25% for precision—the percentage of positive images that are correctly classified—91.25% for sensitivity—the percentage of images that are detected but the remaining 91.25% are not—74.25% for specificity—the percentage of actual negatives that are predicted to be negative—25.75% for false positive rate—the percentage of negative classes that are incorrectly classified—98% for f1 score, and 95.75% for accuracy. In Table 5, you will find the benchmark table.

The methods that were developed, especially FNN, yielded positive results. This theoretical model has the potential to advance biological research and help with the early detection and treatment of Alzheimer's disease. In the future, we could combine recently developed deep learning models with previously trained deep architectures to obtain more precise results for Alzheimer's diagnosis. Similarly, we will apply deep learning algorithms to the diagnosis of cancer and a few other human disorders.

Table 5 Benchmark table

Methods	Performance analysis		
	Accuracy (%)	ROC (%)	Recall (%)
Alorf and Khan (2022) [2]	80.65	–	–
Savaş (2022) [22]	92.98	–	
Buvaneswari and Gayathri (2021) [25]	95	–	89
Tian et al. (2021) [26]	82.44	–	
Murugan et al. (2021) [15]	95.23	87	–
Diogo et al. (2022) [21]	90.6	–	–
AD-ED-FNN (Proposed)	95.75	87.98	90.98

6 Conclusion

Applications utilizing neurofuzzy techniques have been created to categorize brain images. The work that is anticipated has proved edge recognition based on FNNs, also called HED. The main sources of inspiration for the models are deep directed nets and the head of FCN systems. In the FNN architecture, models and structures are used. When evaluated against a dataset comprising multiple categories of images linked to AD, the average accuracy using the FNN model was found to be 95.75%. The test results show that the selected strategy was a noteworthy technique that could improve the accuracy of Alzheimer's disease diagnosis overall. Apart from the feature extraction procedure, the image is preprocessed, which includes tasks like edge detection, mask creation, cropping masked images, binary and grayscale format conversion, and more. Even though high-level data and apparent contextual data have not been approved, the method shows promise in learning every image by combining various scales and multiple level visual reactions.

References

1. C.R. Nagarathna, K. Mohanchandra, An overview of early detection of Alzheimer's disease. Int. J. Med. Eng. Inf. **15**(5), 442–457 (2023)
2. C.R. Nagarathna, M. Kusuma, Comparative study of detection and classification of Alzheimer's disease using hybrid model and CNN, in *2021 International Conference on Disruptive Technologies for Multi-Disciplinary Research and Applications (CENTCON)*, vol. 1, (IEEE, 2021), pp. 43–46
3. A. Alorf, M.U. Khan, Multi-label classification of Alzheimer's disease stages from resting-state fMRI-based correlation connectivity data and deep learning. Comput. Biol. Med. **151**, 106240 (2022)
4. G. Bertasius, J. Shi, L. Torresani, Deepedge: A multi-scale bifurcated deep network for top-down contour detection, in *Proceedings of the IEEE Conference on Computer Vision and Pattern Recognition*, (2015), pp. 4380–4389
5. M.V. Bobyr, N.A. Milostnaya, V.A. Bulatnikov, The fuzzy filter based on the method of areas' ratio. Appl. Soft Comput. **117**, 108449 (2022)

6. A. Preetham, M. Almas, Alzheimer's classification from EGG signals employing machine learning algorithms. J. Electron. Inform. **5**(4), 386–404 (2023)
7. C.R. Nagarathna Jr., Ensemble approach for early detection and classification of stages of Alzheimer's disease, in *Alzheimer's Association International Conference*, (ALZ, 2022)
8. S. Dubey, Alzheimer's dataset (4 class of images) Kaggle. 2019. Available from: https://www.kaggle.com/datasets/sachinkumar413/alzheimer-mri-dataset
9. H. Fuse, K. Oishi, N. Maikusa, T. Fukami, Japanese Alzheimer's Disease Neuroimaging Initiative, Detection of Alzheimer's disease with shape analysis of MRI images, in *2018 Joint 10th International Conference on Soft Computing and Intelligent Systems (SCIS) and 19th International Symposium on Advanced Intelligent Systems (ISIS)*, (IEEE, 2018), pp. 1031–1034
10. J.J. Hwang, T.L. Liu, Pixel-wise deep learning for contour detection. arXiv. preprint arXiv:1504.01989 (2015)
11. J. Islam, Y. Zhang, Alzheimer's Disease Neuroimaging Initiative, Deep convolutional neural networks for automated diagnosis of Alzheimer's disease and mild cognitive impairment using 3D brain MRI, in *International Conference on Brain Informatics*, (Springer, Cham, 2018), pp. 359–369
12. J.E.W. Koh, V. Jahmunah, T.H. Pham, S.L. Oh, E.J. Ciaccio, U.R. Acharya, et al., Automated detection of Alzheimer's disease using bi-directional empirical model decomposition. Pattern Recognit. Lett. **135**, 106–113 (2020)
13. S.S. Kundaram, K.C. Pathak, Deep learning-based Alzheimer disease detection, in *Proceedings of the Fourth International Conference on Microelectronics, Computing and Communication Systems: MCCS 2019*, (Springer, Singapore, 2021), pp. 587–597
14. C.R. Nagarathna, M. Kusuma, Automatic diagnosis of Alzheimer's disease using hybrid model and CNN. J. Soft Comput. Paradigm **3**(4), 322–335 (2022)
15. M. Mamun, S.B. Shawkat, M.S. Ahammed, M.M. Uddin, M.I. Mahmud, A.M. Islam, Deep learning based model for Alzheimer's disease detection using brain MRI images, in *2022 IEEE 13th Annual Ubiquitous Computing, Electronics & Mobile Communication Conference (UEMCON)*, (IEEE, 2022), pp. 0510–0516
16. C.R. Nagarathna, M. Kusuma, K. Seemanthini, Classifying the stages of Alzheimer's disease by using multi layer feed forward neural network. Procedia Comput. Sci. **218**, 1845–1856 (2023)
17. E.G. Marwa, H.E. Moustafa, F. Khalifa, H. Khater, E. AbdElhalim, An MRI-based deep learning approach for accurate detection of Alzheimer's disease. Alex. Eng. J. **63**, 211–221 (2023)
18. S. Murugan, C. Venkatesan, M.G. Sumithra, X.Z. Gao, B. Elakkiya, M. Akila, S. Manoharan, DEMNET: a deep learning model for early diagnosis of Alzheimer diseases and dementia from MR images. IEEE Access **9**, 90319–90329 (2021)
19. C.R. Nagarathna, M.M. Kusuma, Early detection of Alzheimer's disease using MRI images and deep learning techniques. Alzheimer's Dementia **19**, e062076 (2023)
20. P. Rajesh, F.H. Shajin, G. Kannayeram, A novel intelligent technique for energy management in smart home using internet of things. Appl. Soft Comput. **128**, 109442 (2022)
21. P. Rajesh, F.H. Shajin, G.K. Kumaran, An efficient IWOLRS control technique of brushless DC motor for torque ripple minimization. Appl. Sci. Eng. Prog. **15**(3), 5514 (2022)
22. X. Ren, Multi-scale improves boundary detection in natural images, in *Computer Vision–ECCV 2008: 10th European Conference on Computer Vision, Marseille, France, October 12–18, 2008, Proceedings, Part III 10*, (Springer, Berlin Heidelberg, 2008), pp. 533–545
23. K.M. Saridakis, A.J. Dentsoras, Soft computing in engineering design–A review. Adv. Eng. Inform. **22**(2), 202–221 (2008)
24. S. Savaş, Detecting the stages of Alzheimer's disease with pre-trained deep learning architectures. Arab. J. Sci. Eng. **47**(2), 2201–2218 (2022)

25. N. Senthilkumaran, R. Rajesh, Image segmentation-a survey of soft computing approaches, in *2009 International Conference on Advances in Recent Technologies in Communication and Computing*, (IEEE, 2009), pp. 844–846
26. F.H. Shajin, P. Rajesh, V.K. Nagoji Rao, Efficient framework for brain tumour classification using hierarchical deep learning neural network classifier. Comput. Methods Biomech. Biomed. Eng. Imaging Vis., 1–8 (2022)
27. F.H. Shajin, P. Rajesh, M.R. Raja, An efficient VLSI architecture for fast motion estimation exploiting zero motion prejudgment technique and a new quadrant-based search algorithm in HEVC. Circ. Syst. Signal Process. **41**(3), 1751–1774 (2022)
28. J. Tian, G. Smith, H. Guo, B. Liu, Z. Pan, Z. Wang, S. Xiong, R. Fang, Modular machine learning for Alzheimer's disease classification from retinal vasculature. Sci. Rep. **11**(1), 1–1 (2021)
29. B. Solano-Rojas, R. Villalón-Fonseca, A low-cost three-dimensional DenseNet neural network for Alzheimer's disease early discovery. Sensors **21**(4), 1302 (2021)
30. A. Witkin, Scale-space filtering: A new approach to multi-scale description, in *ICASSP'84. IEEE International Conference on Acoustics, Speech, and Signal Processing*, vol. 9, (IEEE, 1984), pp. 150–153
31. H. Huliyappa, Ensemble stacking method of classifying the stages of Alzheimer's disease by using MRI dataset. Int. J. Artif. Intell. **11**(2), 62–69 (2024)

Diabetic Retinopathy Detection Using CNN Enhanced with Lesion-Aware Module and Feature-Preserve Module

A. Anandaraj, S. Abinaya, S. Arul Prathiksha, M. Atshaya, and K. Dhaboomya

1 Introduction

Diabetic retinopathy is a complication of diabetes caused by prolonged high blood sugar levels, which damage the blood vessels in the retina. It is one of the most prevalent complications which is associated with diabetes and it is a leading cause of blindness across the globe. As the number of people with diabetes continues to rise, the need for early detection and effective treatment becomes even more pressing. Currently, it is estimated that over 463 million adults worldwide are living with diabetes and nearly one-third of these individuals will develop some form of diabetic retinopathy in their lifetime. This critical number is expected to register an increase in the coming days that is driven by factors such as lifestyle changes, population growth, and aging demographics.

Diabetic retinopathy occurs when consistently high blood sugar levels damage the small blood vessels in the retina, which is the light-sensitive tissue at the back of the eye. Initially, the disease presents with few to no symptoms, but as it progresses, patients can experience blurred vision, floaters, dark spots, and impaired central vision. This condition progresses through stages from mild symptoms like microaneurysms to severe stages such as proliferative diabetic retinopathy, which can lead to vision loss if it is untreated. Early detection is crucial because it allows timely intervention with laser treatments or medication which can slow disease progression and prevent severe visual impairment and maintain the patient's quality of life. Without timely intervention, this condition can lead to irreversible vision loss.

A. Anandaraj · S. Abinaya (✉) · S. Arul Prathiksha · M. Atshaya · K. Dhaboomya
Sri Eshwar College of Engineering, Coimbatore, India
e-mail: arulprathiksha.s2021csea@sece.ac.in; atshaya.m2021csea@sece.ac.in; dhaboomya.k2021csea@sece.ac.in

A. Haldorai et al. (eds.), *7th EAI International Conference on Big Data Innovation for Sustainable Cognitive Computing* , EAI/Springer Innovations in Communication and Computing, https://doi.org/10.1007/978-3-032-02979-9_6

The disease typically progresses through two major stages. In the early phase, non-proliferative diabetic retinopathy (NPDR) where the retinal blood vessels weaken and develop microaneurysms, small areas of balloon-like swelling. At this stage, only minor retinal changes are detected and patients may not experience any noticeable symptoms. However, as the condition advances into proliferative diabetic retinopathy (PDR), new abnormal blood vessels that are fragile and prone to leakage begin to grow within the retina. This phase represents a severe threat to vision as these vessels can rupture, which leads to retinal detachment, hemorrhages, and eventually complete blindness.

The importance of early detection in diabetic retinopathy is highly required. Early intervention with laser photocoagulation, intravitreal injections, or anti-VEGF (vascular endothelial growth factor) therapies can slow the progression of the disease and prevent significant vision loss. Manual diagnosis of diabetic retinopathy by ophthalmologists involves analyzing retinal images for subtle lesions like microaneurysms and hemorrhages. This process is time-consuming and requires a high level of expertise. Factors such as variability in image quality, interobserver inconsistencies, and the growing number of diabetic patients further strain resources, which leads to potential delays in diagnosis and treatment. Regular retinal screenings for individuals with diabetes are coupled with advanced diagnostic techniques that play a crucial role in mitigating the potentially devastating effects of this condition. Thus, the development of automated systems for early diabetic retinopathy detection is essential, which offers a means to improve the efficiency and accuracy of diagnoses and enables timely medical intervention.

2 Literature Review

According to Shafiq, M., Quanrun, F., and Alghamedy, F.H. (2024), the SELENA+ algorithm is a deep learning system developed for the detection of diabetic retinopathy. It transitioned from research to market after studies validated its capability for detecting diabetic retinopathy. The system was particularly developed to handle the complexities of large datasets and provide accurate diagnosis results [1]. Parulian, O.S., and Naam, J. (2024) proposed a system for diabetic retinopathy grade classification using fractal analysis and random forest classifier. The system leveraged the MESSIDOR dataset, segmenting images and extracting fractal dimensions as features. However, it faced challenges in distinguishing between mild and severe diabetic retinopathy, limiting its accuracy in more advanced stages [2]. Sachdeva, J., Mishra, P., and Katoch, D. (2024) implemented a system that utilized texture features for diabetic retinopathy detection through machine learning models. Two key features, hemorrhages and exudates, were extracted using local ternary pattern (LTP) and local energy-based shape histogram (LESH), with SVM employed for classification of the extracted features [3]. Park, S.A.H. and Mikula, K (2024) provided a meta-analysis of ultrawide field imaging in combination with artificial intelligence for the diagnosis of diabetic retinopathy. This research analyzed diagnostic performance across studies from 2018 to 2022, emphasizing the improvement brought by AI-based models in identifying diabetic retinopathy through retinal images [4]. Rochendi, A.D., Silalahi, L.M., and Simanjuntak, I.U.V. (2024) conducted a comparative study on mild diabetic retinopathy detection using CLAHE-preprocessed and unprocessed fundus images. A minimal CNN model was employed, showing improved accuracy in detecting mild cases of diabetic retinopathy [5]. Falcioni, R., Oliveira, R.B., Chicati, M.L., and Antunes, W.C. (2024) explored deep learning models such as convolutional neural networks (CNN) and recurrent neural networks (RNN) for diagnosing diabetes and detecting diabetic retinopathy. Pretrained models fine-tuned for specific tasks like DR detection showed promise in medical diagnosis [6]. Zie, Y., Zhang, Y., Long, J., Que., N., and Chen, Y. (2024) introduced MPLNet, a Multitask Supervised Progressive Learning Network designed for diabetic retinopathy (DR) grading. This approach leverages progressive learning to improve classification and grading accuracy by simultaneously learning multiple DR-related tasks [7]. Li, H., Jia, W., Vujosevic, S., and Sabanayagam, C. (2024) reviewed the current research and future strategies for managing vision-threatening diabetic retinopathy in the Asia-Pacific region. The study emphasized the importance of early diagnosis, innovative treatment methods, and integrated health-care systems in mitigating DR-related vision loss [8]. Klochkov, V., Chan, C.M., and Lin, W.W. (2024) focused on methylglyoxal as a critical factor, which contributes to diabetic retinopathy and retinal damage [9]. Revathy, R., Nithya, B.S., Reshma, J.J., Ragendhu, S.S., and Sumithra, M.D. (2020) explored diabetic retinopathy detection using machine learning algorithms. Their work presented an efficient framework for classifying retinal images and highlighted the significance of feature extraction and preprocessing for accurate detection [10].

3 Dataset and Methods

3.1 Dataset

We have used the diabetic retinopathy [11] resized dataset from Kaggle which contains retinal images collected from diabetic retinopathy screening. This dataset includes images originally obtained from the EyePACS dataset and other publicly available sources. The images have been resized to a uniform resolution of 512×512 pixels for computational efficiency and easier processing. The dataset contains above 70,000 images that are categorized into five classes based on the severity of diabetic retinopathy.

No diabetic retinopathy (Class 0) indicates healthy retinal images with no signs of diabetic retinopathy [12]. Mild diabetic retinopathy (Class 1) depicts the early-stage diabetic retinopathy characterized by minor lesions such as microaneurysms. Moderate diabetic retinopathy (Class 2) includes more visible signs such as hemorrhages and hard exudates. Severe diabetic retinopathy (Class 3) involves

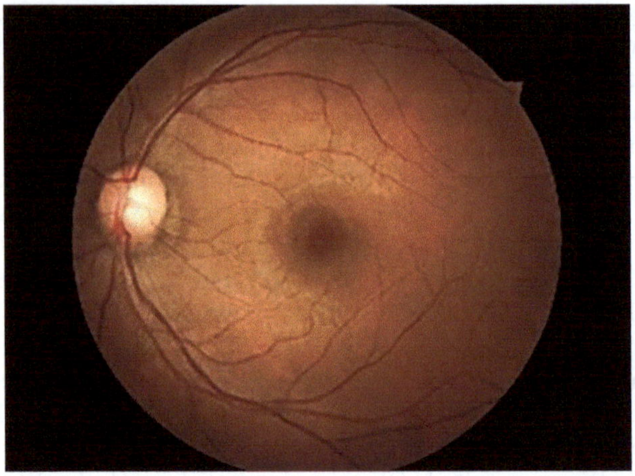

Fig. 1 Eye with no diabetes retinopathy

significant hemorrhages and microaneurysms, with increased risk of abnormal blood vessel growth. Proliferative diabetic retinopathy (Class 4) is the most advanced stage, involving new blood vessels forming on the retina, leading to a high risk of vision loss [13].

Diabetic retinopathy images in the dataset typically exhibit critical features such as exudates, hemorrhages, and microaneurysms, which are essential for diagnosing the condition. These features vary based on the stage of the disease, with more advanced stages displaying larger and more widespread lesions. The dataset's diversity in terms of lighting conditions, image quality, and variations in retinal appearance provides an ideal foundation for developing and testing a robust model for diabetic retinopathy detection [14].

The dataset is highly versatile, which offers a wide range of complexities across the images and enhances the robustness of deep learning models [15]. Its large volume ensures that models trained on this data generalize well to unseen data, which is crucial for real-world medical applications.

The CNN-based model was trained on this diverse dataset of retinal images including samples from various image qualities from Kaggle. Data augmentation techniques were employed to ensure the model's robustness [16]. This ensures that the model remains sensitive to identify even the most severe cases of retinopathy, which is crucial for early intervention and treatment (Figs. 1 and 2).

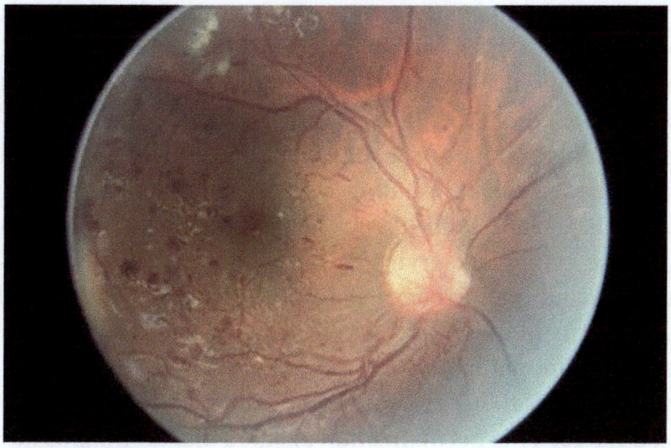

Fig. 2 Eye with diabetic retinopathy

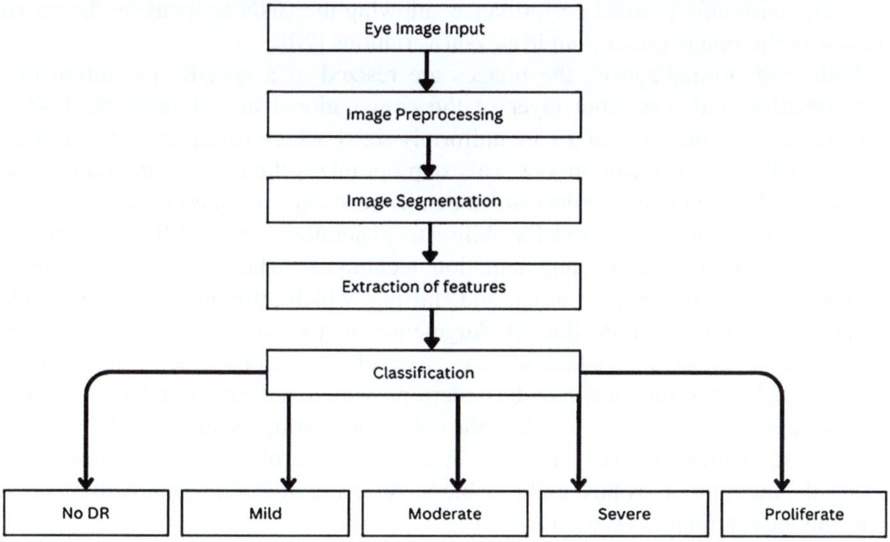

Fig. 3 Flowchart of the proposed model

3.2 Steps (Fig. 3)

Image Preprocessing

In the preprocessing phase of our diabetic retinopathy detection system, we perform several essential steps to prepare the retinal images for accurate and efficient analysis. Grayscale Conversion is the first step, where the original RGB retinal images are transformed into grayscale [17]. This step reduces the complexity of the image by converting it from three color channels to a single channel, while still retaining the essential intensity information required for diagnosis. By focusing on intensity variations, such as those present in lesions and blood vessels, this transformation simplifies the image, making it easier for the model to extract and analyze critical features related to diabetic retinopathy [18].

Next, we apply noise reduction using Gaussian blurring. In medical imaging, the presence of noise, such as random variations in pixel values, can hide the important details [19]. Gaussian blurring helps smooth the images by eliminating this unwanted noise while preserving key structures like blood vessels, microaneurysms, exudates, and hemorrhages. This step ensures that the model focuses on the meaningful patterns in the image, which are vital for detecting the presence and severity of diabetic retinopathy.

Image normalization is then performed where the pixel intensity values are scaled to a standardized range. This normalization ensures that all images have a consistent representation, which is crucial for maintaining uniformity across the dataset. By standardizing the pixel values, we eliminate variations in lighting and contrast that may otherwise affect model performance, allowing the CNN to focus on the actual content of the image rather than irrelevant variations [20].

Following normalization, the images are resized to a specific resolution that is compatible with the input layer of the convolutional neural network (CNN). Resizing ensures that all images are uniformly sized, which is important for feeding them into the deep learning model. This step optimizes the model's computational efficiency while maintaining the critical details necessary for diagnosis [21, 22].

Lastly, to enhance the model's ability to generalize across different types of images, we employ image augmentation techniques. These include operations such as rotation, flipping, zooming, and shifting, which artificially create multiple variants of each image in the dataset. Augmentation not only increases the diversity of training data but also addresses the issue of class imbalance, where certain stages of diabetic retinopathy may be underrepresented in the dataset. By generating diverse augmented images, we reduce the risk of overfitting, ensuring that the model learns to recognize important features in a wide range of image conditions. This makes the model more robust and reliable in detecting diabetic retinopathy, even in new and unseen images (Figs. 4 and 5).

Image Segmentation

Image segmentation plays a vital role in the detection of diabetic retinopathy as it helps the isolation of critical features such as exudates, hemorrhages, and microaneurysms from retinal fundus images [23]. In this approach, the multistep image segmentation process is designed to accurately capture key structures while maintaining a precision in differentiating between healthy and pathological regions.

The process began with the preprocessing of retinal images, where grayscale images were converted to binary using a thresholding operation. This binary conversion is a crucial step as it clearly separates the areas of interest from the background, simplifying further analysis. Otsu's thresholding method was used to

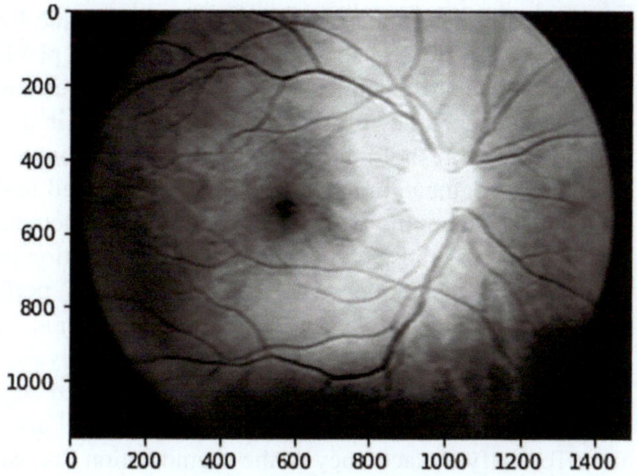

Fig. 4 Grayscale image

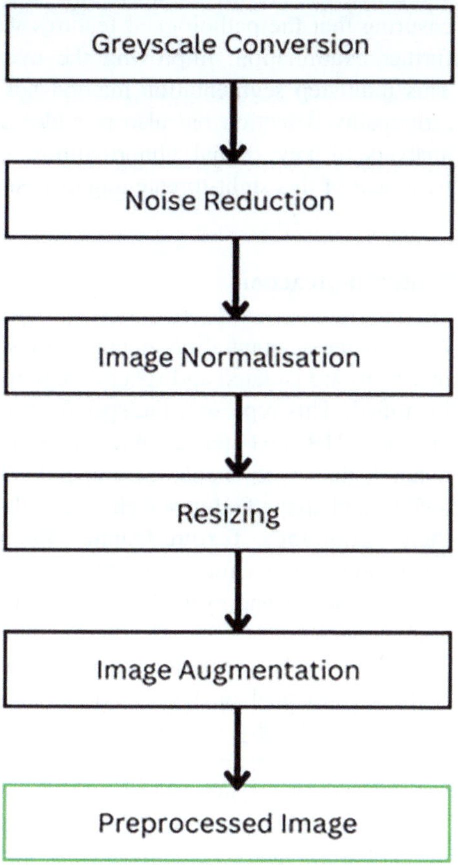

Fig. 5 Steps in image preprocessing

automatically determine the optimal threshold value for the image. This technique minimizes the difference within each group of pixels, ensuring that regions of interest are clearly defined without manual threshold selection. This approach improves the accuracy of identifying key features in the image.

Following thresholding morphological operations were applied, such as dilation to enhance segmentation accuracy. Dilation helped re-expand the critical features, which makes them more prominent and easier to detect. For more intricate or subtle features, such as those associated with color variation in the retinal structures, color segmentation techniques were incorporated. By performing bitwise operations, color-masked images were merged with the original, thereby highlighting specific pathological features like exudates for further analysis.

In the final step of the process, the segmented regions representing exudates, hemorrhages, and microaneurysms were extracted and prepared for feature extraction. To verify the accuracy of the segmentation, the isolated regions were visually confirmed by overlaying them onto the original retinal images. This helped in ensuring that the pathological features were correctly identified and extracted for further examination, improving the overall reliability of the detection process. This multistep segmentation method not only enhances the precision of diabetic retinopathy detection but also provides a robust framework for the isolation and analysis of key retinal abnormalities, contributing to the early diagnosis and treatment of this sight-threatening disease (Figs. 6 and 7).

Feature Extraction

After image segmentation, abnormalities like exudates, hemorrhages, and microaneurysms are isolated and specific regions which depict the level of retinopathy are identified. This represent the specific parts of the retinal image where the disease presents. The next step involves extracting features from the segmented image to quantify their size, shape, texture, and color. Area measurements assess the size of lesions and shape features such as circularity and aspect ratio provide insights into their morphology. Texture features are computed using techniques like the gray-level co-occurrence matrix (GLCM) highlights important textural information, such as contrast and energy which is crucial for distinguishing diseased regions from healthy ones.

To improve the performance of the classification model, these extracted features undergo statistical analysis to identify which ones correlate most strongly with diabetic retinopathy. Dimensionality reduction techniques like principal component analysis (PCA) are applied to reduce the number of features, retaining only the most informative ones. This step ensures that the classification process is more efficient and less prone to overfitting. The final step involves combining the relevant features into a feature vector, which serves as the input for the classifier model and makes accurate predictions of the disease.

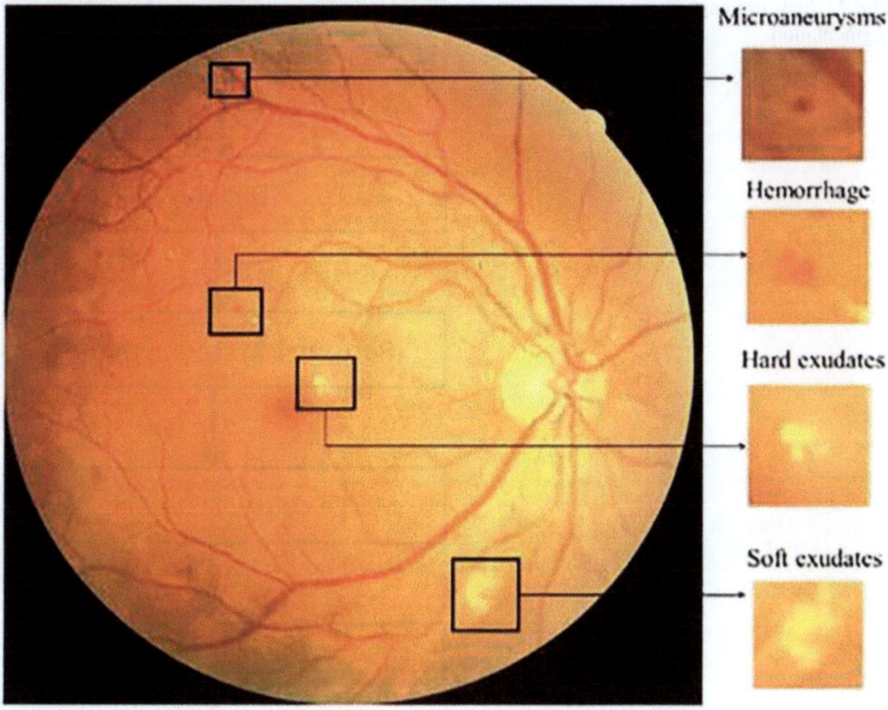

Fig. 6 Identification of retinal defects

Classification

The classification of diabetic retinopathy in this approach is achieved through a hybrid method that combines advanced machine learning algorithms with a customized convolutional neural network (CNN) architecture, which is further enhanced by two key modules, lesion-aware module (LAM) and the feature-preserve module (FPM). These modules play vital roles in refining the network's ability to detect and classify diabetic retinopathy with greater precision.

The CNN-based network's deep hierarchical structure enables the extraction of complex patterns and features from retinal images. By learning to identify lesions such as microaneurysms, hemorrhages, and exudates, the network can accurately classify images based on disease severity. Its ability to handle data variability and detect subtle differences in lesion patterns ensures high accuracy in diabetic retinopathy detection across different stages of the disease.

The lesion-aware module (LAM) is a specialized component designed to enhance the detection of subtle lesions, such as microaneurysms and hemorrhages, which are critical for identifying diabetic retinopathy. LAM prioritizes lesion-prone regions within retinal images through spatial attention, thus ensuring the focus is on crucial areas and channel attention, amplifying lesion-related feature maps. It utilizes multi-

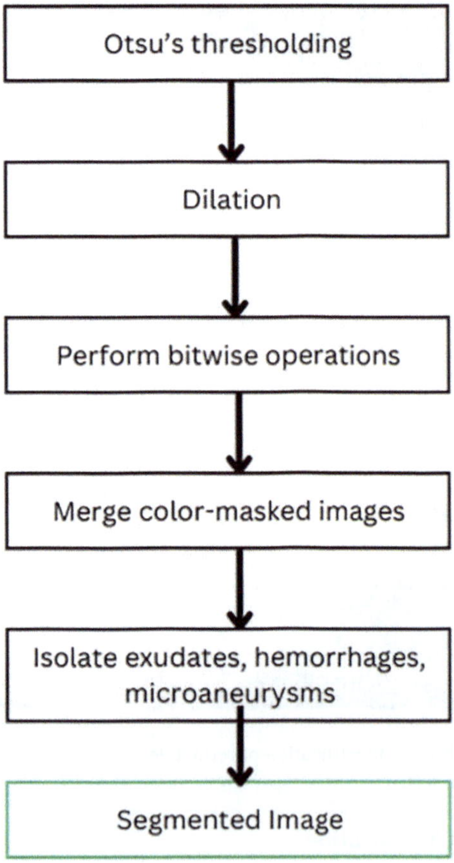

Fig. 7 Steps in image segmentation

scale convolutional filters to detect lesions of varying sizes and shapes which capture both fine-grained details and broader patterns. LAM integrates a lesion-specific feature amplification process, which highlights relevant patterns while suppressing background noise, thus preserving critical diagnostic information. By combining its outputs with deeper network layers, LAM ensures that subtle lesion features are retained throughout the processing pipeline. Additionally, it incorporates advanced loss functions, such as focal loss and dice loss, to optimize lesion detection accuracy during training. This approach significantly improves the network's ability to identify subtle lesions, thus ensuring accurate detection.

The FPM ensures that critical features from earlier network layers, such as edges, textures, and fine details, are retained for improved accuracy in detection. By preserving these key elements, the FPM ensures that important information is not lost as the image passes through deeper layers of the network. This helps retain crucial structural details that are necessary for accurate classification. Together, these modules enhance the robustness, which allows it to detect varying lesion types across different stages of diabetic retinopathy.

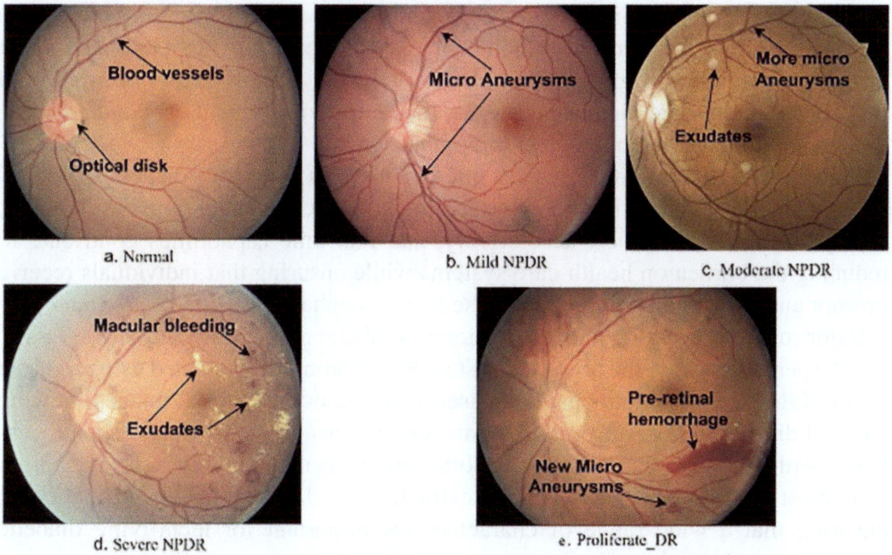

Fig. 8 Classification of diabetes retinopathy

The final classification is performed by combining the strengths of the CNN architecture and a dense layer. The CNN is highly effective at capturing spatial features from the retinal images, while the dense layer improves the decision-making process by linking these features together and assigning importance to them. This hybrid approach enhances the overall accuracy of the model, allowing for precise classification of diabetic retinopathy.

Once the classification is complete, the results are analyzed to provide a clear diagnostic outcome. This interpretation includes the identification of the presence and severity of diabetic retinopathy, offering a comprehensive assessment that can assist in guiding treatment decisions (Fig. 8).

4 Results and Discussion

Thus, the AI-powered mobile application helps in the early detection of diabetic retinopathy by offering a user-friendly and accessible solution for retinal screening. By integrating advanced AI techniques into a mobile platform, the application enables individuals to efficiently check their retinal health from their homes. Users can upload retinal scan images or capture them directly using a smartphone attached with a 20D lens attachment. The application processes these images in real time, which provides immediate diagnostic results, thus eliminating the delays often associated with clinic visits.

The mobile application could be scaled by incorporating cloud-based data storage for remote monitoring, which enables health-care providers to track patient progress over time. Integrating with electronic health records EHR systems would make it easier to share diagnostic results, thus improving patient care and ensuring a more comprehensive approach to treatment.

Early-stage abnormalities, such as microaneurysms and hemorrhages, are identified with high precision, thus allowing for timely medical intervention. Furthermore, the application's efficiency, accessibility, and real-time capabilities contribute to reducing the burden on health-care systems while ensuring that individuals receive prompt and accurate assessments. These features enhance early detection and play a major role in preventing the progression of diabetic retinopathy and vision loss.

The performance of the proposed system was thoroughly evaluated using a subset of the Kaggle diabetic retinopathy dataset, which consisted of 122 images showing signs of diabetic retinopathy and 122 images from healthy individuals. The process began with preprocessing the images, followed by segmentation to isolate key areas of interest. Various features were then extracted based on texture, shape, and color, ensuring that a wide range of characteristics important for identifying diabetic retinopathy were captured.

These extracted features were fed into a hybrid classification model, which combined convolutional neural networks (CNN) with two specialized modules, the lesion-aware module (LAM) and the feature-preserve module (FPM), alongside a dense layer. The LAM focused on detecting small, subtle lesions such as microaneurysms and hemorrhages, lesions that are often difficult to detect but are crucial for early diagnosis. Meanwhile, the FPM maintained important details like edges and textures throughout the network, ensuring that essential image features were preserved and not lost in the deeper layers of the CNN.

The evaluation results were promising. The system showed a significant improvement in detecting small, noise-like lesions, particularly due to the capabilities of the LAM in enhancing the detection of subtle retinal abnormalities. In terms of overall performance, the hybrid model achieved an accuracy of 93% with a precision of 0.835, a recall of 0.821, and an F1-score of 0.828. These metrics demonstrate that the system provides a well-balanced classification, effectively identifying both true positives and true negatives. The high accuracy and strong balance between precision and recall suggest that the system is not only reliable but also consistent in detecting diabetic retinopathy, offering both sensitivity to the disease and specificity in distinguishing healthy images from diseased ones.

This performance highlights the effectiveness of integrating CNN with LAM and FPM, contributing to more accurate and comprehensive diabetic retinopathy diagnosis.

5 Conclusion

In this study, an innovative AI-powered system for detecting diabetic retinopathy was developed, utilizing a convolutional neural network (CNN) enhanced with two specialized modules: the lesion-aware module (LAM) and the feature-preserve module (FPM). This approach was designed to overcome key challenges faced by existing diabetic retinopathy detection methods, specifically the difficulty of identifying small, noise-like lesions such as microaneurysms and hemorrhages and the need to preserve essential image features like edges and textures during the various stages of image processing.

The results of the study show that this system significantly improves the accuracy and reliability of diabetic retinopathy detection, particularly in early-stage diagnosis. Preliminary experiments showed significant improvements in detection accuracy, particularly for subtle lesions like microaneurysms and hemorrhages. Compared to manual diagnosis, the AI-powered system demonstrated faster processing and comparable precision, thus making it a viable tool for early diabetic retinopathy screening.

The LAM played a crucial role by enhancing the network's ability to detect these minute features, while the FPM ensured that vital information about the retinal structure was retained throughout the process, contributing to more precise identification of the disease.

One of the key advantages of our system is its potential for practical application in a mobile-based platform. By integrating this model into a mobile application, it enables accessible and efficient diabetic retinopathy screening. The application employs optimized lightweight CNN architectures and hardware acceleration techniques like GPU and TensorFlow Lite for efficient computation. This enables real-time processing of retinal images even on mobile devices with limited resources. This offers the benefit of real-time detection allowing individuals to receive timely alerts and seek medical intervention soon, which is vital for preventing the progression of the disease and minimizing the risk of vision loss. This kind of mobile solution can also be especially valuable in regions where access to specialized health care may be limited.

Our hybrid model combines the strengths of deep learning through CNN with the added support of classical machine learning classifiers. This integration of multiple techniques proved to be highly effective, achieving competitive performance in terms of both accuracy and reliability. The hybrid approach allows the system to balance the advantages of both methods, making it a more robust solution for analyzing complex retinal images.

In conclusion, AI automates lesion detection, significantly reducing the time required for diagnosis. It assists ophthalmologists by providing real-time results, prioritizing high-risk cases for immediate attention, and minimizing human error in early-stage detection. By providing accurate, real-time detection, it can assist healthcare providers in diagnosing and managing diabetic retinopathy more efficiently. Ultimately, the implementation of this tool in clinical settings could lead to earlier interventions, which would reduce the risk of severe complications like vision loss and improve patient outcomes.

References

1. M. Shafiq, F. Quanrun, F.H. Alghamedy, DualEye-FeatureNet: A dual-stream feature transfer framework for multi-modal ophthalmic image classification. IEEE Access **2024**, 1–12 (2024)
2. O.S. Parulian, J. Naam, Advanced filtering and enhancement techniques for diabetic retinopathy image analysis. J. Med. Inform. Technol. **2024**, 1–12 (2024)
3. J. Sachdeva, P. Mishra, D. Katoch, Diabetic retinopathy data augmentation and vessel segmentation through deep learning-based three fully convolution neural networks, in *Proceedings of the International Conference on Image and Vision Computing*, (Elsevier, 2024), pp. 1–12
4. S. Yuvaraj, J.V. Franklin, V.S. Prakash, A. Anandaraj, R. Subha, A survey on human face emotion recognition using machine learning models, in *Proceedings of the IEEE International Conference*, (IEEE, 2023), pp. 1–12
5. S.A.H. Park, K. Mikula, Automated completion of segmented fragments of macrophages using weighted dilation and erosion in 2D+ time microscopy videos. Preprints **2024**, 1–10 (2024)
6. G.L. Zhang, Reconstruction of edge detection theory and the bitwise edge extraction algorithm. Preprints **2024**, 1–12 (2024)
7. A.D. Rochendi, L.M. Silalahi, I.U.V. Simanjuntak, Touchless palm print recognition system design using gray-level co-occurrence matrix feature with K-nearest neighbor classification in MATLAB, in *Proceedings of the International Conference on Advanced Image Processing*, (core.ac.uk, 2024), pp. 1–12
8. R. Falcioni, R.B. Oliveira, M.L. Chicati, W.C. Antunes, Fluorescence and hyperspectral sensors for nondestructive analysis and prediction of biophysical compounds in the green and purple leaves of plants, in *Proceedings of the International Conference on Advanced Sensors*, (MDPI, 2024), pp. 1–12
9. Y. Xie, Y. Zhang, J. Long, N. Que, Y. Chen, MPLNet: Multi-task supervised progressive learning network for diabetic retinopathy grading. Comput. Electr. Eng. **2024**, 1–20 (2024)
10. H. Li, W. Jia, S. Vujosevic, C. Sabanayagam, Current research and future strategies for the management of vision-threatening diabetic retinopathy. Asia Pac. J. Ophthalmol. **2024**, 1–15 (2024)
11. A. Anandaraj, P.J.A. Alphonse, Automatic detection of epileptic seizure disease from EEG signals using a deep learning framework, in *Proceedings of the 2022 IEEE North Karnataka Subsection Flagship International Conference (NKCon)*, (IEEE, Vijaypur, India, 2022), pp. 1–5
12. V. Klochkov, C.M. Chan, W.W. Lin, Methylglyoxal: A key factor for diabetic retinopathy and its effects on retinal damage. Preprints **2024**, 1–8 (2024)
13. R. Revathy, B.S. Nithya, J.J. Reshma, S.S. Ragendhu, M.D. Sumithra, Diabetic retinopathy detection using machine learning. Int. J. Eng. Res. Technol. **2020**, 1–10 (2020)
14. P. Romero-Aroca, B. Fontoba-Poveda, Handheld retinograph devices evaluated by ophthalmologists and AI. J. Clin. Med. **13**, 1–12 (2024)
15. I. Govindharaj, R. Rampriya, K. Udayakumar, Enhanced diabetic retinopathy detection using U-shaped network and capsule network-driven deep learning. MethodsX **12**, 1–12 (2024)
16. N. Srinivasan, M. Krishna, V. Naveen, S.M. Kishore, S. Kumar, R. Subha, Predicting Indian GDP with machine learning: A comparison of regression models, in *Proceedings of the 2023 9th International Conference on Advanced Computing and Communication Systems (ICACCS)*, (IEEE, Coimbatore, India, 2023), pp. 1–10
17. D. Hutton, AI-based detection of diabetic retinopathy in the US. Modern Retina **8**, 10–15 (2024)

18. S. Malarvannan, S. Venkatraman, A novel adaptive hybrid focal-entropy loss for enhancing diabetic retinopathy detection using convolutional neural networks. ArXiv Preprint **11**, 1–12 (2024)
19. X. Xing, M. Ye, Z.T. Zhang, O. Liu, C. Wei, Q. Dong, X. Li, Deep learning for detecting cognitive impairment in diabetic patients based on retinal photos. SSRN J. **15**, 1–12 (2024)
20. S. Prasanth, E. Veemaraj, Advancement of AI in diabetic retinopathy recognition: Juxtaposition using transfer learning. SSRN J. **13**, 1–12 (2024)
21. T.H. Rim, C.J. Lee, Deep learning model predicts CVD risk using retinal imaging. Circulation **150**, 1–12 (2024)
22. K.D. Bamane, Y. Sinkar, AI-enabled diagnostic solution for eye disease detection. Health Informatics J. **8**, 1–12 (2024)
23. A.E. Ragab, C. Huang, M. Sarabi, MobileNet-V2/IFHO model for accurate detection of early-stage diabetic retinopathy. Heliyon **2024**, 1–12 (2024)

Enhancement of Acute Lymphoblastic Leukemia Using Spatial and Transform Filtering Techniques

M. Menagadevi, Suriya Murugan, Nirmala Madian, S. Sharmila, and S. Anbumani

1 Introduction

Acute lymphoblastic leukemia (ALL) is a type of cancer that affects the blood and bone marrow. Excessive immature white blood cells, known as lymphoblasts, can crowd out healthy blood cells in the bone marrow. According to the data collected by World Health Organization (WHO), American Cancer Society (ACS), National Cancer Institute (NCI), Centers for Disease Control and Prevention (CDC), and Global Cancer Observatory (GLOBOCAN) survival rate over time, incidence by gender, incidence by age group, and prevalence by region is shown in Fig. 1. The statistics show that the probability of survival has risen from 50% to 90% from 2000 to 2022, suggesting considerable advances in treatment and patient outcomes. The largest occurrence occurs in children aged 0–5, with rates steadily falling in older age groups [1]. Adolescents (16–20) and adults (21+) have much reduced incidence

M. Menagadevi (✉)
Department of Computing Technologies, School of Computing, SRM Institute of Science and Technology, Kattankulathur, Tamil Nadu, India

S. Murugan
Department of Computer Science and Engineering, Sri Eshwar College of Engineering, Coimbatore, India

N. Madian
Department of Biomedical Engineering, Dr.N.G.P Institute of Technology, Coimbatore, India

S. Sharmila
Department of Electronics and Communication Engineering, Dr.N.G.P Institute of Technology, Coimbatore, India
e-mail: sharmila.s@drngpit.ac.in

S. Anbumani
Dr.N.G.P Institute of Technology, Coimbatore, India

A. Haldorai et al. (eds.), *7th EAI International Conference on Big Data Innovation for Sustainable Cognitive Computing*, EAI/Springer Innovations in Communication and Computing, https://doi.org/10.1007/978-3-032-02979-9_7

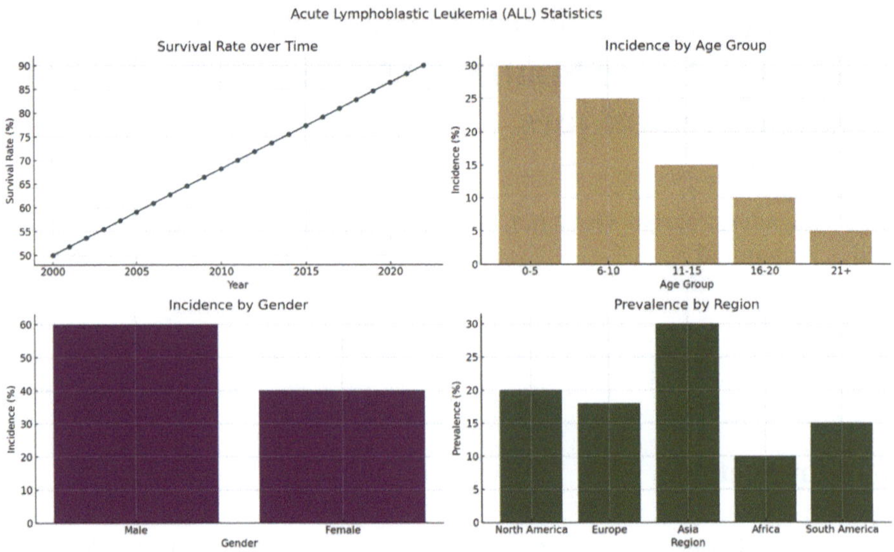

Fig. 1 Statistics of acute lymphoblastic leukemia

rates. ALL is mostly a pediatric malignancy, with the majority of cases detected in early life. The gender gap in ALL prevalence may stem from a mix of biological reasons, including genetic and hormonal variations, which enhance susceptibility in males compared to females. These regional disparities could be driven by genetic characteristics, health-care availability, and environmental exposures specific to each region. Differences in population density and health-care reporting standards may possibly account for Asia's higher frequency [2].

1.1 Acute Lymphoblastic Leukemia

ALL, a subtype of leukemia, is a rapidly progressing cancer that originates in the lymphoid cells of the bone marrow. This disease is categorized by the uncontrolled proliferation of immature lymphoid cells, known as lymphoblasts, which can significantly impact the production of normal blood cells. ALL has a bimodal distribution, with an early peak in incidence during childhood, particularly around ages 4–5, and a second gradual increase in incidence during adult years, typically around age 50 [3]. The overall incidence of ALL is approximately 1–1.5 per 100,000 persons, with the childhood form accounting for the majority of cases, representing about 80% of acute leukemias in children [4].

ALL is a heterogeneous disease, and its classification is primarily based on the lineage of the malignant cells, with the majority of cases (approximately 85%) being of the B-cell lineage. Genetic and molecular alterations play a crucial role in the

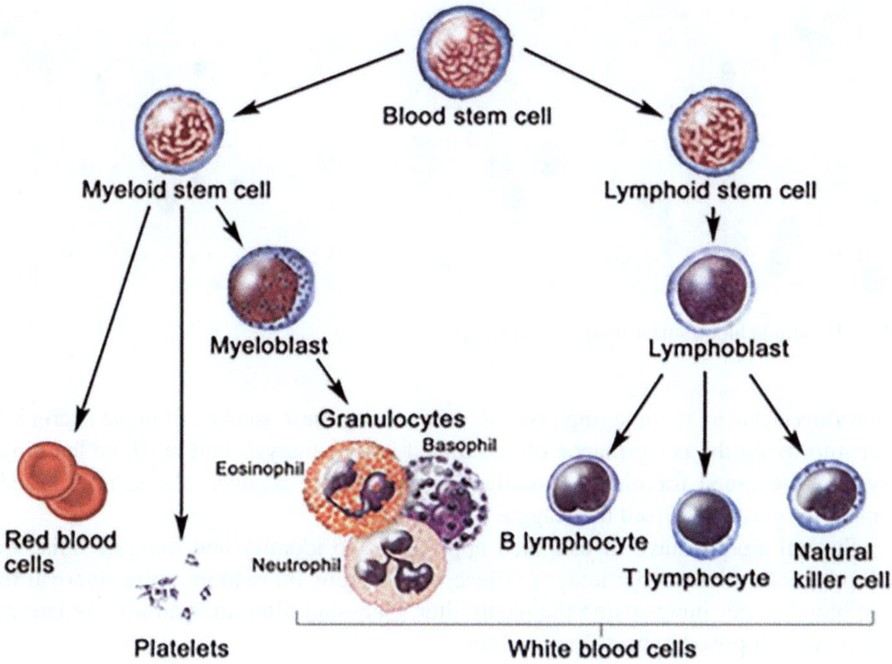

Fig. 2 Stages of blood cell development

development and progression of ALL, and several genetic syndromes and common allelic variants have been associated with an increased susceptibility to the disease [5, 6].

The different states of blood cell development are depicted in Fig. 2. Blood stem cells, or hematopoietic stem cells, may be found in bone marrow and are capable of differentiating into many types of blood cells. These stem cells can differentiate into myeloid and lymphoid progenitor cells. In addition, differentiation generates myeloid stem cells into red blood cells responsible for the circulation of oxygen in tissues, white blood cells particularly the granulocytes for combating infections, and platelets that are required for clotting [7, 8].

1.2 Blood Smear Images

Blood smear testing is an important method for detecting and monitoring ALL. Technical professionals encounter technological obstacles when interpreting images, which might impact the accuracy and dependability of the diagnosis. Noise in blood smear microscopic images can arise from a variety of sources, including variations in sample preparation, staining inconsistencies, and artifacts

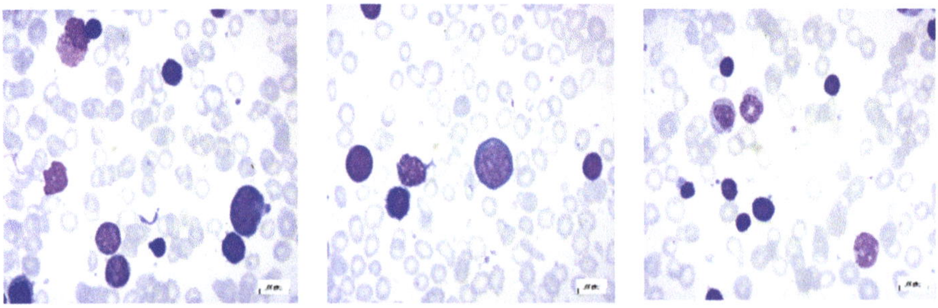

Fig. 3 Sample blood smear images from Kaggle

introduced during the imaging process. Addressing these sources of noise is crucial for improving the performance of automated image analysis and machine learning systems designed for image classification [9]. Figure 3 shows the sample blood smear images taken from the Kaggle depository.

Several studies have investigated approaches to identify and mitigate noise in blood smear microscopic images. These works focus on random noise removal in the blood smear image using the mean filter, Gaussian filter, median filter, bilateral filter, and proposed hybrid Wiener filter.

Comparative analysis is performed by calculating MSE, PSNR, SNR, and SSIM.

2 Related Work

Blood smear images are affected by random noise due to factors like lighting, interference, thermal noise, etc. Various filters are used by the researchers to reduce the noise in the blood smear image. Ahmed et al. [10] proposed an average filter to reduce the noise in the image. A 6*6 factor is set to the average filter, and rotation is performed for each image. To optimize the target pixel, the average value of its neighboring pixels was calculated, and then the Laplacian filter was used to enhance the WBC edges. Devi and Patil [11] compared the median filter and the Weiner filter using PSNR values, with the median filter showing better results compared to the Weiner. To enhance the image histogram, equalization and order statistic filters were used by Jyoti R et al. [12] to remove noise. Nameirakpam D et al. [13] removed noise in segmented images using a median filter. Gaussian blur techniques were proposed by Waghela JA [14] to remove noise in the blood smear image. Mishra et al. [15] proposed a median filter to remove blurriness, and the Weiner filter was used. To find the amount of noise removed from the image, mean solidity was used. Chitra et al. [16] used an adaptive weighted mean filter to reduce noise in the image and used histogram techniques to enhance the edges. Hazlyna HN et al. [17] removed salt and pepper noise by applying a seeded region growing and 7*7 median filter. Ton LB et al. [18] proposed a median filter of size 7*7. Sathiamoorthy and Bennet

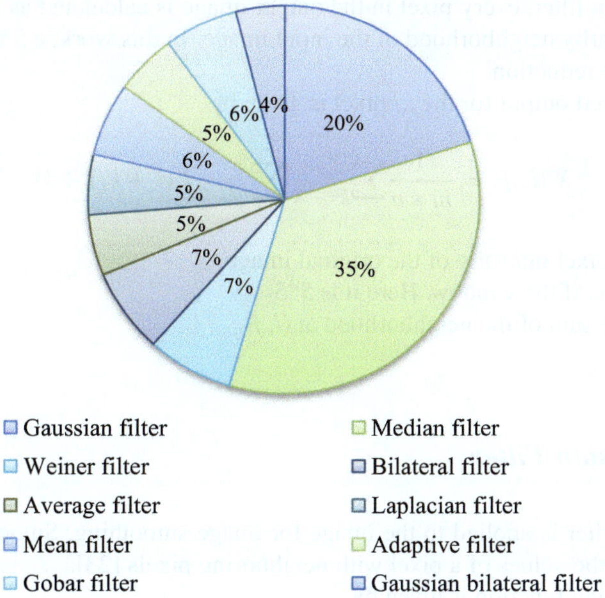

Fig. 4 Filters used for denoising

Rajesh M [19] removed salt and pepper noise using a median filter, and edges are preserved in the images. Karur R and Maini R [20] removed Gaussian noise, salt and pepper noise, speckle noise, and Poisson noise using an alpha-trimmed mean filter, an adaptive filter, a median filter, and contra-harmonic mean filter, with PSNR values calculated for each filter. Azza et al. [21] used a median filter to remove noise from blood smear images. Singh et al. [22] used a Gaussian filter with a better SNR value for the classification of ALL. Figure 4 shows the survey of various filters used for denoising blood smear images.

3 Denoising of Blood Smear Images

3.1 Mean Filter

The mean filter functions by replacing the value of a pixel in an image with the average value of the neighbors or the values around the pixel, and this has the capability of eliminating noise in an image. This approach is more helpful in removing noise that is normally present in the images of the blood smears. This approach can filter and enhance the image quality and help doctors to recognize the various components of the blood sample in a clearer manner.

In the mean filter, every pixel in the output image is calculated as an average of pixels in a nearby neighborhood of the input image. In this work, a 5*5 filter size is used for noise reduction.

Mean filtered output for the *i,j* pixel is given by

$$X\left(i,j\right)=\frac{1}{m*n}\sum\nolimits_{k=-\frac{m}{2}}^{\frac{m}{2}}\sum\nolimits_{l=-\frac{n}{2}}^{\frac{n}{2}}I\left(i+k.j+l\right) \tag{1}$$

$X(i, j)$ is the pixel intensity of the original image.
$m*n$ is the size of the window. Here it is 5*5.
k and l are the sum of the neighborhood at (i, j).

3.2 Gaussian Filter

A Gaussian filter is applied to the image for image smoothing. Smoothing is done by averaging the values of a pixel with neighboring pixels [23].

The Gaussian function is given by

$$G\left(X,Y\right)=\frac{1}{2\pi\sigma^2}e^{-\frac{x^2}{2\sigma^2}} \tag{2}$$

X and Y are the distances from the origin in both axes (vertical and horizontal).

σ = Standard deviation.

By convolving the kernel with the image, the equation is given by

$$I'\left(X,Y\right)=\sum\nolimits_{i=-X}^{X}\sum\nolimits_{J=-X}^{X}I\left(X-i,Y-j\right)*G\left(i,j\right) \tag{3}$$

$I'(X, Y)$ = Filtered output.
K is the kernel.

In this work, the value of σ is 2 and the kernel size is 5*5.

3.3 Bilateral Filter

The bilateral filter removes the noise and prevents the edges of the blood smear image. It performs functions of both spatial and intensity Gaussian filter [24]. The mathematical equation for the bilateral filter is given by

$$I'(X,Y) = \frac{1}{w_p} \sum_{i=-K}^{K} \sum_{j=-K}^{K} I(X-i, Y-j)\, S_G(i,j)$$
$$R_G(I(X,Y), I(X+i, Y+j)) \tag{4}$$

$S_G(i,j)$ is the spatial Gaussian kernel, it is given by

$$S_G(i,j) = e^{\frac{-i^2+j^2}{2\sigma_s^2}} \tag{5}$$

σ_s Smoothing will change depending on σ value.
$R_G(I(X,Y), I(X+i, Y+j)$ is the range Gaussian kernel, it is given by

$$R_G(I(X,Y), I(X+i, Y+j)) = e^{-\frac{\left(I(X,Y)-I(X+i,Y+j)^2\right.}{2\sigma_r^2}} \tag{6}$$

In this work for $\sigma_s = 8$, $\sigma_r = 0.5$ is used for noise removal.

3.4 Hybrid Weiner Filter

Mean and Gaussian filter blur the edges of blood cells in the smear image. The median filter also loses the fine details in the blood smear image. A bilateral filter requires proper tuning of σ_s and σ_r values to get a denoised image. The proposed hybrid Wiener filter presents the fine details and edges of blood cells in the blood smear images. In this filter, a Gaussian function is combined with a Wiener filter for noise removal. The mathematical equation is given by

$$\mathrm{I}(\mathrm{i},\mathrm{j}) = \mu_g(i,j) + \frac{\sigma_g^2(i,j) - \tau^2}{\sigma_g^2(i,j)}$$
$$\left(\sum_{K=-n}^{n} \sum_{L=-n}^{n} g\left(k,l\right) . I(i+K, j+L) - \mu_g(i,j)\right) \tag{7}$$

$\mu_g(i,j) =$ Local mean.
${\sigma_g}^2(i,j) =$ Local variance.
$\tau^2 =$ Noise variance.

4 Evaluation Method

4.1 Sample Output of Filters

Table 1 shows the output of mean, Gaussian, median, bilateral, and hybrid Wiener filters. The hybrid Wiener shows better noise reduction compared to all other filters. Performance of these filters is measured by determining MSE, PSNR, SNR, and SSIM [25–27].

The mean squared error (MSE) calculates the average squared difference between comparable pixels in two images. A lower MSE score suggests a greater similarity between the images, meaning less distortion and noise.

$$\text{MSE} = \frac{1}{XY}\sum_{i=1}^{X}\sum_{j=1}^{Y}(M\,(i,\,j) - K\,(i,\,j))^2 \tag{8}$$

X and Y are image dimensions.

$M(i, j)$ and $K(i, j)$ are the original and filtered pixels value at position(i, j).

PSNR is a metric that quantifies the peak inaccuracy between two images. It is represented in decibels (dB), with greater PSNR values indicating better image quality. PSNR is very useful for evaluating the quality of compressed or denoised images to their originals. It is given by the equation

$$\text{PSNR} = 10.\log_{10}\frac{255^2}{\text{Mean Square error}} \tag{9}$$

SNR is the ratio of the power of the original image to the power of the noise. High SNR readings imply that the image content is stronger than the noise. SNR is commonly used to evaluate overall image quality in terms of clarity and noise.

$$\text{SNR} = 10.\log_{10}\left(\frac{\sum_{i=1}^{X}\sum_{j=1}^{Y}M(i,\,j)^2}{\sum_{i=1}^{X}\sum_{j=1}^{Y}(M\,(i,\,j) - K\,(i,\,j))^2}\right) \tag{10}$$

$M(i, j)$ = Original image pixel

$N(I, j)$ = Filtered image pixel

SSIM evaluates the similarity of two images based on luminance, contrast, and structure. Unlike MSE and PSNR, which analyze pixels alone, SSIM takes into account visual impact and structural information, giving it a more perceptually relevant statistic [28].

$$\text{SSIM} = \frac{\left(2\mu_x\mu_{y+X_1}\right)\,\left(2\sigma_{xy} + X_2\right)}{\left(\mu_x^2 + \mu_y^2 + X_1\right)\,\left(\sigma_x^2 + \sigma_y^2 + X_2\right)} \tag{11}$$

μ_x, μ_y image mean value pixel

Table 1 Denoised output of spatial domain filters

Sample	Mean filter	Gaussian filter	Median filter	Bilateral filter	Hybrid Wiener filter
Sample 1	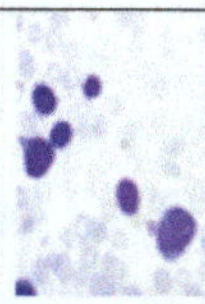	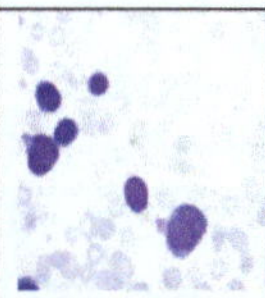	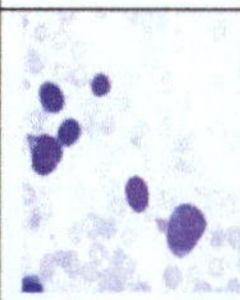	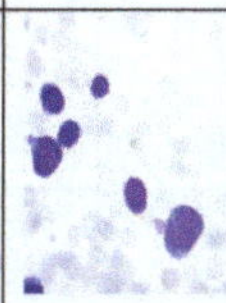	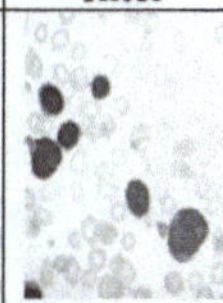
Sample 2	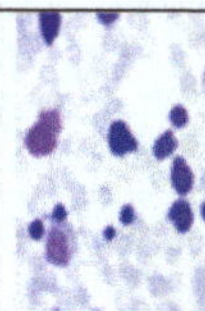	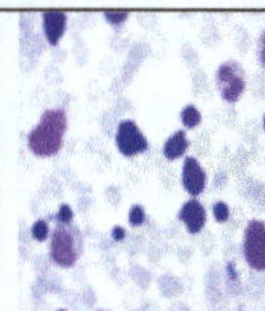	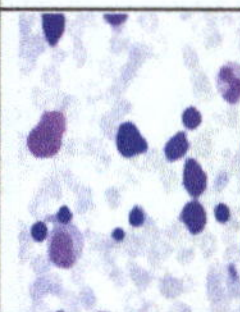	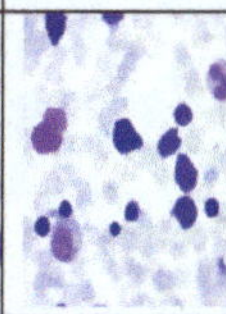	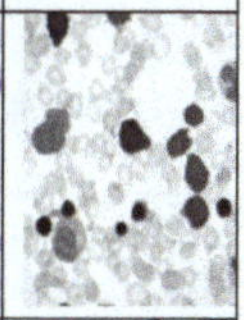
Sample 3	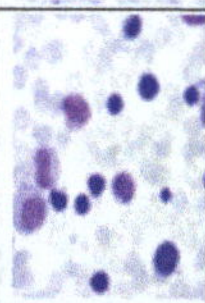	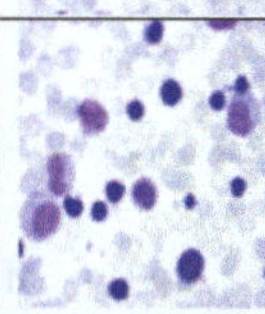	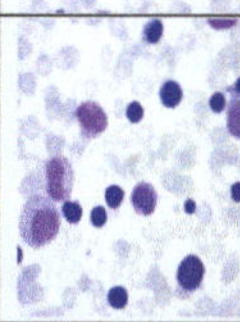	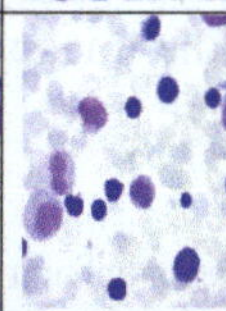	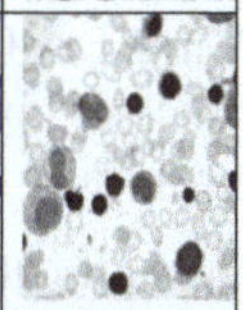
Sample 4	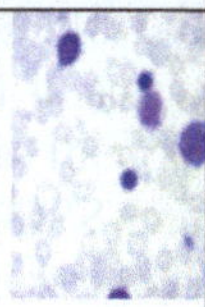	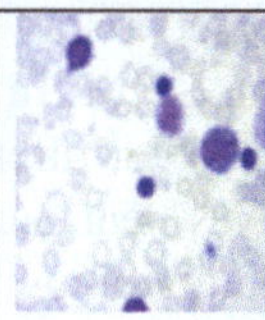	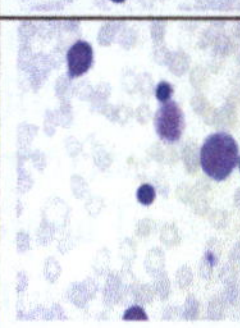	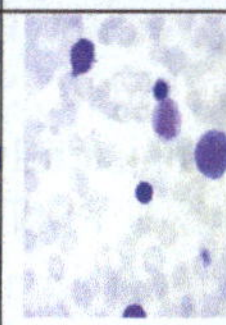	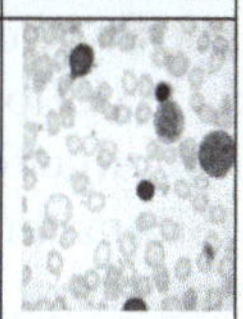
Sample 5	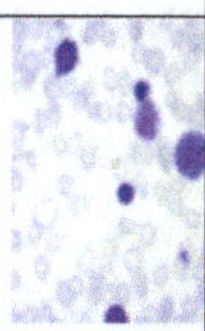	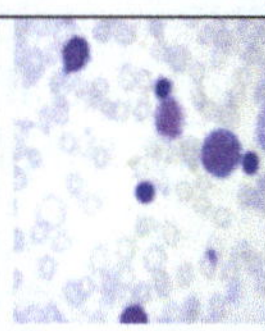	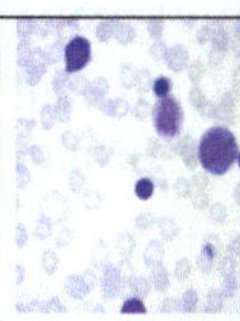	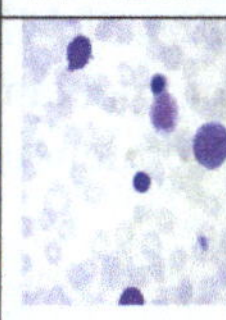	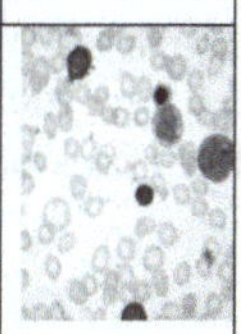
Sample 6	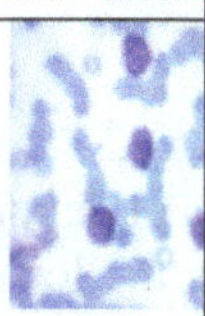	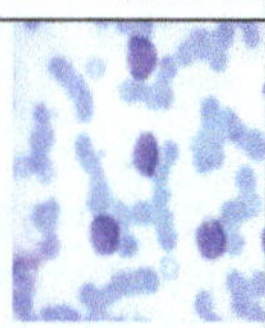	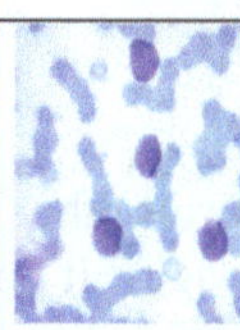	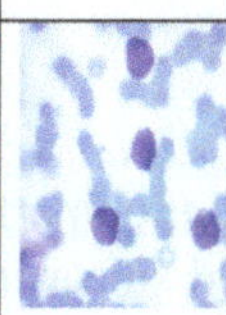	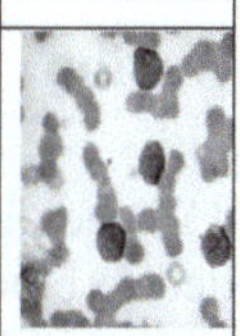
Sample 7	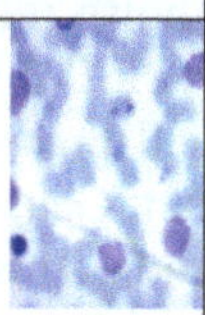	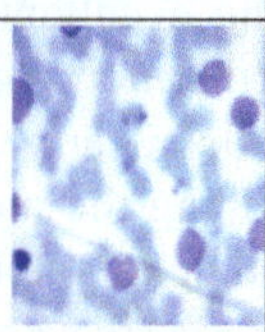	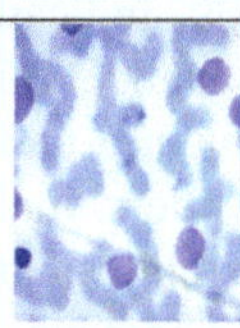	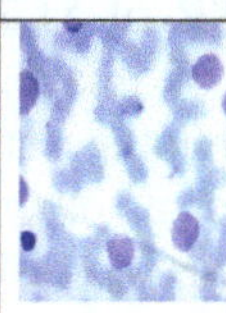	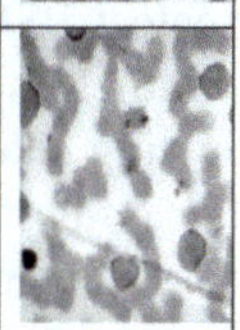
Sample 8	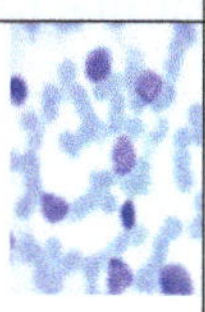	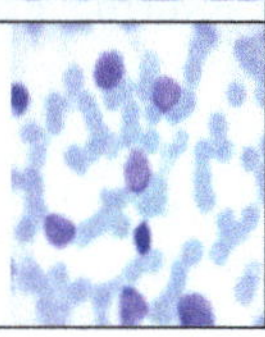	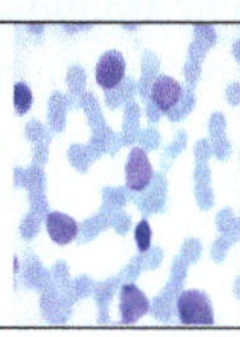	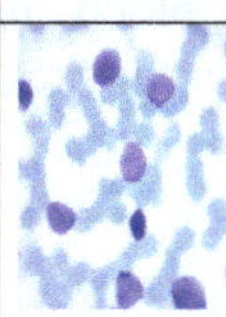	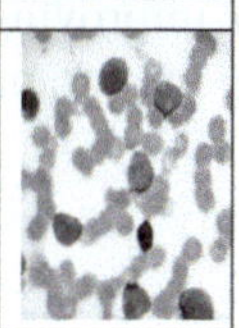

(continued)

Table 1 (continued)

Sample 9	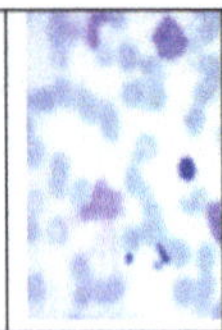	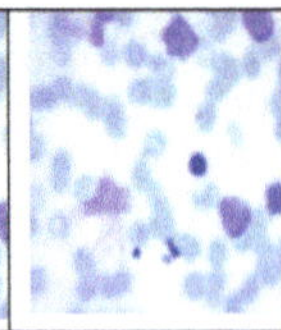	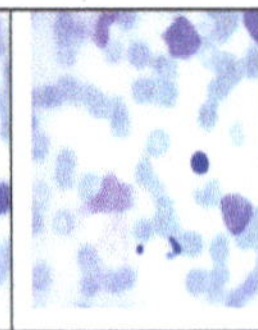	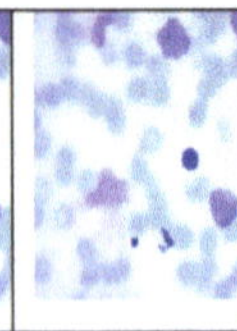	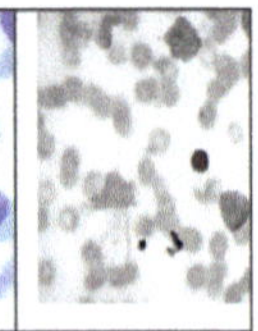
Sample 10	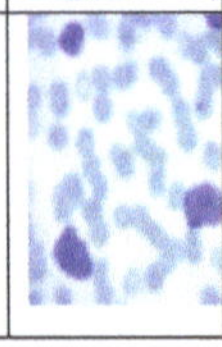	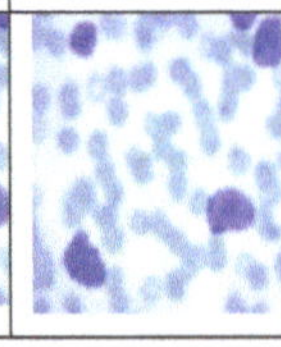	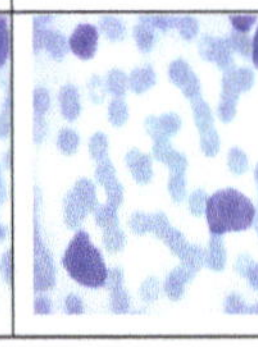	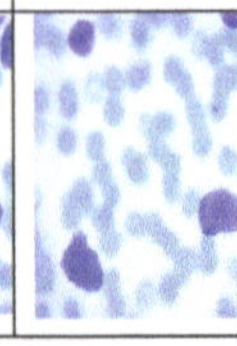	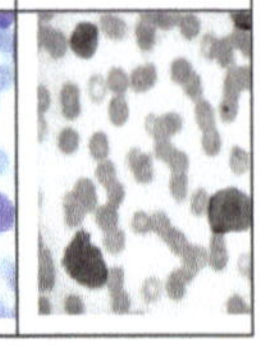

$\sigma^2{}_x$, $\sigma^2{}_x$ Variance of the image pixel
σ_{xy} covariance of image pixel
X_1 and X_2 constant

Table 2 shows the performance metrics values of mean, Gaussian, median, bilateral, and hybrid Wiener filters. Compared to all the filters, the mean filter shows high MSR, low PSNR, SNR, and high SSIM. In terms of image quality, the hybrid Wiener filter consistently outperforms other filters on all samples. This filter reduces MSE, maximizes PSNR, and achieves near-perfect SSIM, making it ideal for denoising blood smear images while maintaining structural integrity. Figure 5 shows the average value of MSE, PSNR, SNR, and SSIM for 10 sample images. Compared to the mean, median, Gaussian, and bilateral filters, the proposed hybrid Wiener filter gives better denoising results for the sample images.

5 Conclusion

The hybrid Wiener filter outperforms other filters in terms of denoising performance when applied to blood smear images. It consistently has the lowest MSE of 0.505 and the greatest PSNR of 51.239, indicating excellent noise reduction of 50.059 with minimum loss of image quality. Furthermore, its SSIM values are nearly ideal, indicating that image structure and detail are well preserved. The hybrid Wiener filter outperforms other filters, including the mean, Gaussian, median, and bilateral filters, in terms of noise removal and detail retention. This makes it the best option for improving blood smear images, assuring great diagnostic clarity while preserving important structural information.

Table 2 Performance metrics of filters

Samples	Parameters	Mean filter	Gaussian filter	Median filter	Bilateral filter	Hybrid Wiener filter
Image 1	MSE	56.74	45.45	34.73	29.14	**0.5**
	PSNR	30.59	31.56	32.72	33.49	**51.11**
	SNR	30.02	30.98	32.15	32.91	**50.53**
	SSIM	0.8866	0.907	0.9322	0.8842	**0.9973**
Image 2	MSE	24.38	49.04	59.71	46.93	**0.51**
	PSNR	27.18	28.17	30.37	31.42	**51.09**
	SNR	25.83	26.82	29.02	30.07	**49.74**
	SSIM	0.8263	0.8574	0.8895	0.8586	**0.9982**
Image 3	MSE	31.07	48.64	63.72	51.78	**0.5**
	PSNR	26.96	27.93	30.09	30.99	**51.12**
	SNR	25.63	26.60	28.76	29.66	**49.76**
	SSIM	1	0.8443	0.8805	0.8511	**0.9982**
Image 4	MSE	69.62	55.68	43.12	40.85	**0.5**
	PSNR	29.70	30.67	31.78	32.02	**51.11**
	SNR	28.38	29.35	30.46	30.70	**49.79**
	SSIM	0.8629	0.8892	0.9092	0.8694	**0.9979**
Image 5	MSE	83.47	67.42	52.17	47.51	**0.51**
	PSNR	28.92	29.84	30.96	31.36	**51.09**
	SNR	27.59	28.52	29.64	30.04	**49.76**
	SSIM	0.8327	0.8622	0.8876	0.8534	**0.998**
Image 6	MSE	83.46	57.42	42.17	43.61	**0.51**
	PSNR	27.82	28.84	32.56	33.06	**51.19**
	SNR	28.59	29.52	29.66	31.04	**49.79**
	SSIM	0.9327	0.8822	0.8889	0.8554	**0.998**
Image 7	MSE	69.62	55.68	43.12	40.85	**0.52**
	PSNR	30.59	31.56	32.72	33.49	**51.11**
	SNR	25.83	26.82	29.02	30.07	**49.74**
	SSIM	1	0.8443	0.8805	0.8511	**0.9982**
Image 8	MSE	82.47	66.42	52.19	48.51	**0.50**
	PSNR	29.92	29.73	32.96	33.36	**51.29**
	SNR	28.59	29.52	28.64	31.04	**49.96**
	SSIM	0.8327	0.8622	0.8876	0.8534	**0.998**
Image 9	MSE	83.49	56.42	44.17	49.61	**0.50**
	PSNR	27.82	28.84	32.56	33.06	**51.17**
	SNR	28.59	29.52	29.66	31.04	**49.99**
	SSIM	0.9327	0.8822	0.8889	0.8554	**0.9986**
Image 10	MSE	56.74	45.45	34.73	29.14	**0.5**
	PSNR	35.9	31.56	32.72	34.49	**52.11**
	SNR	31.12	32.99	33.18	36.91	**51.53**
	SSIM	0.8866	0.907	0.9322	0.8842	**0.973**

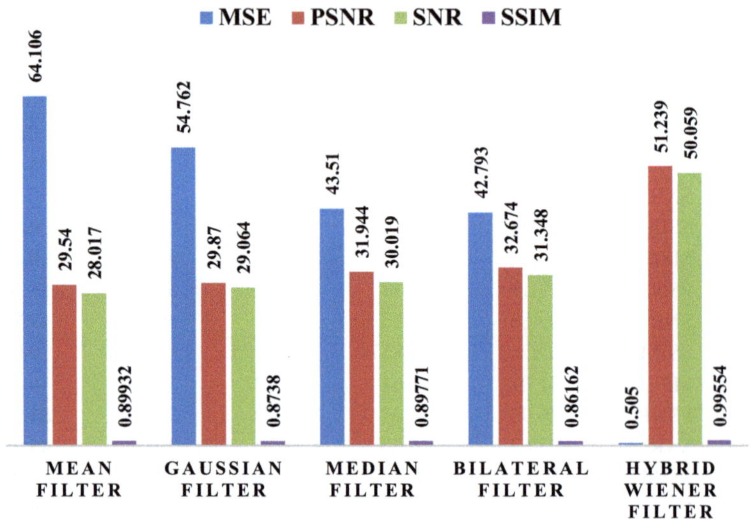

Fig. 5 Average value of MSE, PSNR, SNR, and SSIM

References

1. World Health Organization [Internet], WHO: The Global initiative for childhood cancer; (2018). Available from: https://www.who.int/health-topics/cancer
2. https://www.cancer.gov/types/leukemia/patient/adult-all-treatment-pdq
3. M. Zolfaghari, H. Sajedi, A survey on automated detection and classification of acute leukemia and WBCs in microscopic blood cells. Multimed. Tools Appl. **81**, 6723–6753 (2022). https://doi.org/10.1007/s11042-022-12108-7
4. T.G. Devi, N. Patil, S. Rai, C.S. Philipose, Gaussian blurring technique for detecting and classifying acute lymphoblastic leukemia cancer cells from microscopic biopsy images. Life (Basel) **13**(2), 348 (2023)
5. A. Mittal, S. Dhalla, S. Gupta, A. Gupta, Automated analysis of blood smear images for leukemia detection: A comprehensive review. ACM Comput. Surv. **54**(11s), 1–37 (2022)
6. K. Jędraszek, M. Malczewska, K. Parysek-Wójcik, M. Lejman, Resistance mechanisms in pediatric B-cell acute lymphoblastic leukemia. Int. J. Mol. Sci. **23**(6), 3067 (2022)
7. A. Al-Ghraibah, M. Al-Ayyad, Automated detection of leukemia in blood microscopic images using image processing techniques and unique features: Cell count and area ratio. Cogent. Eng. **11**(1), 2304484 (2024)
8. A.E. Aby et al., A review on leukemia detection and classification using artificial intelligence-based techniques. Comput. Electr. Eng. **118**, 109446 (2024)
9. T.G. Devi, N. Patil, Analysis & evaluation of image filtering noise reduction technique for microscopic images, in *2020 International Conference on Innovative Trends in Information Technology (ICITIIT)*, vol. 13–14, (2020), pp. 1–6
10. I.A. Ahmed, E.M. Senan, H.S.A. Shatnawi, Z.M. Alkhraisha, M.M.A. Al-Azzam, Hybrid techniques for the diagnosis of acute lymphoblastic leukemia based on fusion of CNN features. Diagnostics **13**(6), 1026 (2023)
11. A.P. Patil, M. Hiremath, K. Kavipriya, A study of preprocessing techniques on digital microscopic blood smear images to detect leukemia, in *Data Science and Security. Lecture Notes in Networks and Systems*, ed. by S. Shukla, X.Z. Gao, J.V. Kureethara, D. Mishra, vol. 462, (Springer, Singapore, 2022). https://doi.org/10.1007/978-981-19-2211-4_24

12. R. Jyoti, A. Singh, H.S. Bhadauria, J. Virmani, J.S. Devgun, Computer assisted classification framework for prediction of acute lymphoblastic and acute myeloblastic leukemia. Biocybern. Biomed. Eng. **37**(4), 637–654 (2017)
13. D. Nameirakpam, M. Khumanthem, J.C. Yambem, Image segmentation using K -means clustering algorithm and subtractive clustering algorithm. Procedia Comput. Sci. **54**, 764–771 (2015)
14. J.A. Waghela, *Identification of Acute Lymphocytic Leukemia (Blood Cancer) Through Microscopic Images of Blood samples[Project]* (School of Computing: National College of Ireland, 2021)
15. S. Mishra, B. Majhi, P. Kumar Sa, L. Sharma, Gray level co-occurrence matrix and random forest based acute lymphoblastic leukemia detection. Biomed. Signal Process. Control **33**, 272–280 (2017)
16. P. Chitra, M.R. Ebenezer Jebarani, Kavipriya, K. Srilatha, M. Sumathi, S. Lakshmi, Detection of AML in blood microscopic images using local binary pattern and supervised classifier. Res. J. Pharm. Tech. **12**(4), 1717–1720 (2019)
17. H.N. Hazlyna, H.H. Aini, S. Fadzilah, K. Nadia, M.Y. Mashor, R. Hassan, Fusion 7×7 median filter and seeded region growing area extraction algorithms for effective detection of acute leukemia based on blood images. J. Fundam. Appl. Sci. **10**(3S), 726–737 (2018)
18. L.B. Toh, M. Mashor, P. Ehkan, H. Rosline, A. Junoh, N. Harun, Image segmentation for acute leukemia cells using color thresholding and median filter. J. Telecommun. Electron. Comput. Eng. **10**(1–5), 69–74 (2018)
19. M. Bennet Rajesh, S. Sathiamoorthy, Classification of leukemia image using genetic based K-nearest neighbor. Asian J. Comput. Sci. Technol. **7**(2), 113–117 (2018)
20. R. Kaur, R. Maini, Performance evaluation and comparative analysis of different filters for noise reduction. Graphics and signal processing. Int. J. Image **8**(7), 9–21 (2016)
21. M.B.A. Azza, A.A. Ethar, S.R. Omer, B.A. Ibraheem, Z.A. Mustafa, An innovative leukemia detection system using blood samples via a microscopic accessory. Int. J. Eng. Manuf. **13**(1), 23–32 (2023)
22. G. Singh, P. Swarnalatha, B.K. Tripathy, S. Kakani, Convex Hull based Wbc computation for leukaemia detection. Int. J. Adv. Res. Electr. Electron. Instrument. Eng. **2**(5), 1–5 (2013)
23. M. Bukhari, S. Yasmin, S. Sammad, A.A. Abd El-Latif, A deep learning framework for leukemia cancer detection in microscopic blood samples using squeeze and excitation learning. Math. Probl. Eng. **2022**, 18 (2022)
24. L. Kumari, N. Mittal, M. Modi, Comparative analysis of noise reduction filters for medical images, in *Fourth International Conference on Advances in Electrical, Computing, Communication and Sustainable Technologies (ICAECT)*, (IEEE, 2024)
25. M. Menagadevi, S. Devaraj, N. Madian, D. Thiyagarajan, Machine and deep learning approaches for alzheimer disease detection using magnetic resonance images: An updated review. Measurement **226**, 114100 (2024). https://doi.org/10.1016/j.measurement.2023.114100
26. S. Anitha, L. Kola, P. Sushma, S. Archana, Analysis of filtering and novel technique for noise removal in MRI and CT images, in *2017 International Conference on Electrical, Electronics, Communication, Computer, and Optimization Techniques (ICEECCOT), Mysuru*, (2017), pp. 1–3. https://doi.org/10.1109/ICEECCOT.2017.8284618
27. A. Haldorai, S. Murugan, M. Babitha, An investigation on different approaches for medical imaging, in *Artificial Intelligence for Sustainable Development*, (Springer Nature Switzerland, Cham, 2024), pp. 57–75
28. S. Devaraj, N. Madian, M. Menagadevi, et al., Deep learning approaches for analysing Papsmear images to detect cervical cancer. Wirel. Pers. Commun. **135**, 81–98 (2024). https://doi.org/10.1007/s11277-024-10986-8

A Machine Learning-Based Approach for Parkinson's Disease Diagnosis to Improve Clinical Decision Support Systems

Karina V. Mehta, Maitri N. Shah, Malhar A. Solanki, Dharmesh R. Tank, and Palak J. Parmar

1 Introduction

Early diagnosis of Parkinson's disease (PD) is important for timely interventions, optimization of patient care, and reduction in long-term health-care burdens. The progressive neurodegenerative nature of PD defines its motor symptoms, like tremors, and nonmotor problems, including cognitive impairment. Most of these symptoms are usually very subtle in their presentation and thus tend to be diagnosed late. Traditional diagnostic approaches, however, rely on clinical observation and hence face difficulty in early diagnosis, particularly in the beginning stages. While this is a challenge, recent developments in machine learning provide an excellent alternative through the analysis of a great amount of data and patterns that might not be easily observed by human eyes. More specifically, machine learning models analyze noninvasive biomarkers that include voice recordings because of subtle changes in speech that are evocative of PD. The current study investigates the predictive performance of three different machine learning algorithms, namely random forest (RF), decision tree (DT), and SVM, for the diagnosis of PD based on voice markers. It utilizes publicly available datasets and finds that random forest, with a prediction accuracy of 94.12%, is the best. The better performance of random forest is because of its ability to generalize due to ensemble learning, thereby avoiding overfitting. These results not only beat prior accuracy rates but also demonstrate how much machine learning can contribute to building clinical decision support systems for a possible revolution in early diagnosis of PD, thus allowing for earlier and more effective interventions.

K. V. Mehta (✉) · M. N. Shah · M. A. Solanki · D. R. Tank · P. J. Parmar
LDRP Institute of Technology and Research, KSV, Gandhinagar, India
e-mail: karinavmehta@gmail.com; maitrishah498@gmail.com; malharsolanki2806@gmail.com; dharmesh_ce@ldrp.ac.in; palak_ce@ldrp.ac.in

A. Haldorai et al. (eds.), *7th EAI International Conference on Big Data Innovation for Sustainable Cognitive Computing*, EAI/Springer Innovations in Communication and Computing, https://doi.org/10.1007/978-3-032-02979-9_8

2 Literature Survey

Recent advancements in predictive modeling for heart and Parkinson's diseases have been extensively documented, with a focus on the performance of support vector machines (SVMs), decision trees, and random forests. Studies consistently report that random forests deliver superior accuracy due to their ability to mitigate over-fitting through ensemble learning. Comparative analyses reveal that while SVMs excel in handling high-dimensional data, decision trees offer interpretability that is crucial for clinical applications. Moreover, the incorporation of advanced feature selection strategies has been proven to greatly boost model precision. These outcomes underline the possibility of these approaches in developing reliable and efficient disease prediction systems.

2.1 Parkinson's Disease

Parkinson's disease is a gradual neurological condition largely affecting movement. It occurs from the degeneration of dopamine-producing neurons in the substantia part of the brain. Characterization of the disorder is done through the presence of numerous symptoms such as tremors, muscular rigidity, and postural instability. Although the specific cause of this disease is not fully understood, it is recognized to have both a hereditary and an environmental etiology. This can further develop to Parkinson's disease, resulting in nonmotor indications like cognitive damage, emotional difficulties, and sleep abnormalities. As of now, there is no cure for the illness; however, medications are available to manage its symptoms.

A research paper [1] investigates the early detection of PD using voice markers with machine learning. Based on the open speech database, K-NN showed the highest accuracy at 94.7%, followed by random forest with an accuracy of 88.16%, and a Bayesian neural network with an accuracy of 74.74%.

The authors in Ref. [2] proposed applying the XGBoost algorithm to classify Parkinson's disease based on voice recordings from the UCI Machine Learning Repository, which included 31 subjects. While specific accuracy details were not provided, the study demonstrated the potential of noninvasive diagnosis.

In another research paper [3], the random forest algorithm was used to detect the disease from voice disorder information. The dataset included voice recordings from 20 Parkinson's patients and 20 healthy individuals. The random forest algorithm achieved an accuracy of 79.31%, highlighting the potential of voice analysis in early Parkinson's diagnosis.

Alshammri et al. [4] discuss the application of machine learning for the early detection of Parkinson's disease from voice markers. The model used is a multilayer perceptron, wherein it is reported that, with this model, the highest accuracy was achieved at 98.31%, above the random forest at 92%, and the support

Table 1 Comparative study of several machine learning system in Parkinson's syndrome

Year	Dataset used	Models used	Parameter
2023 [1]	Public database of voice signals from the Parkinson's sufferers	Random forest, decision tree, support vector machine	Precision, F1 score, accuracy
2023 [2]	UCI Machine Learning Repository dataset with 31 subjects	Random forest	Recall, precision, accuracy
2023 [3]	UCI Machine Learning Repository dataset	Random forest	Accuracy
2023 [4]	UCI Machine Learning Repository dataset (195 voice samples from 31 subjects)	Decision tree, random forest	Precision, F1 score, recall, accuracy
2024 [5]	Patient demographics and clinical assessments dataset	Random forest	Accuracy

vector machine at 95%. Achievement of model performance heavily depends on hyperparameter tuning and balancing the data using certain techniques.

The author of this paper [5] evaluates machine learning models, including logistic regression, XGBoost, and SVM, for Parkinson's disease prediction. In their experiments, XGBoost achieved 90%, SVM reached 91%, and logistic regression had an accuracy of 87%. Ultimately, as indicated by the study, feature selection and ensemble methods are also vital to improving the accuracy of predictions (Table 1).

3 Dataset

3.1 Public Database of Speech Signals from Parkinson's Patients (2023)

In Ref. [1], this work is based on the "Public Database of Speech Signals from Parkinson's Patients (2023)." This dataset contains 1000 rows and 30 variables, with a number of vocal measures for each one relevant to the diagnosis of PD, drawn from a public repository or Kaggle. They include some of the basic frequency measures by MDVP, frequency variations by MDVP, jitter and shimmer metrics, that is, noise-to-harmonics ratio, and harmonics-to-noise ratio, among other key vocal features. The target variable was "status," meaning there were cases of Parkinson's and controls. In this chapter, proper data preprocessing was presented to ensure accuracy at full length with normalization of values, validation, and treatment of missing values [6]. Vocal profiles of this dataset represent a large amount of information and allow the highest possible potential for studying acoustic and phonetic markers of Parkinson's. This kind of preprocessing also assured anonymity to guarantee ethical guidelines for all patients [7]. This makes the dataset qualitatively rich and thus appropriate to support vocal biomarkers in the process of diagnostic and therapeutic tool development.

3.2 UCI Machine Learning Repository Dataset with 31 Subjects

In Ref. [8], the analysis is presented utilizing datasets from the UCI machine learning repository, received in 2020, which included recordings of 31 subjects with a total of 5875 rows and considered 22 variables related to different vocal measures for the Parkinson's disease diagnosis. The main features are measures of basic frequency by MDVP, measures of frequency variations by MDVP, jitter and shimmer metrics, the noise-to-harmonics ratio, the harmonics-to-noise ratio, and other crucial vocal features. The target variable is "status," which indicates whether a person has Parkinson's disease. Data quality has been made comprehensive by normalizing the values, certifying accuracy, and dealing with missing values. The dataset contains elaborate voice profiles, in which the detailed analysis of acoustic and phonetic markers for Parkinson's is possible. The highest considerations in following the ethical guidelines included the total anonymization of data to obscure patient identities. It is a rich dataset that will form a firm base in the vocal biomarker study of Parkinson's disease, where this kind of information is needed for the development of diagnostic or therapeutic tools.

3.3 Parkinson's Disease Dataset from UCI Machine Learning Repository: The Word "Data" Is Plural, Not Singular

In Ref. [9], the dataset used was obtained from the UCI machine learning repository in 2021. The dataset comprises 303 observations against 14 variables, each of which represents a different medical measurement and attribute critical to the diagnosis of heart disease. These vectors have been taken as: risk factors, which are: age, sex, chest pain type, resting blood pressure, serum cholesterol, fasting blood sugar, resting electrocardiographic results, maximum heart rate achieved, exercise-induced angina, ST depression induced by exercise relative to the rest, the slope of the peak exercise ST segment, the number of major vessels colored by fluoroscopy, and a categorical variable of the presence or the absence of heart disease. The dataset was preprocessed for quality and integrity by normalizing the values, checking for missing data, and validating the accuracy of information. It was also deidentified to ensure patient confidentiality and adherence to ethical research practices. This dataset provides strong support—with its huge range of medical attributes—to the study of the various factors influencing the diagnosis of heart disease.

3.4 UCI Machine Learning Repository Dataset

The paper [10] is based on the UCI Machine Learning Repository dataset of 195 voice recordings from 31 people. These records include both healthy controls (HC) and patients with Parkinson's disease (PD), and each one has different voice signal features. The data comprise fundamental frequency measures such as MDVP, jitter, and shimmer metrics, and harmonic-to-noise ratios. The key target variable for this dataset is "status," where a 0 value indicates healthy controls (HC) and a 1 value indicates Parkinson's disease (PD) cases.

3.5 Parkinson's Disease Prediction Using Clinical Data

In Ref. [11], the dataset with clinical data from Parkinson's disease patients included demographics, motor symptoms, and all clinical markers. After preprocessing, including dealing with missing values, normalization, and feature selection, the above machine learning models were trained on the dataset, logistic regression, XGBoost, and SVM, for the prediction of Parkinson's disease. The strategy emphasized early detection while maintaining stringent ethics on secrecy toward patient data [12].

The dataset used for our model is detailed in the Experiment section (Sect. 5).

4 Models Used

4.1 Support Vector Machine (SVM)

A support vector machine is a machine learning algorithm, wherein there is direct supervision. It is generally used in classification-type problems [13]. Its main motive aims at providing the best hyperplane which could give good separation between the data points of different classes in high-dimensional space, such that the margin between the closest points—known as support vectors—of each class gets maximized [14]. The core competencies of SVMs include mapping input data through a kernel function in order to deal with both linear and nonlinear classification problems onto a higher-dimensionality space, where a linear separator is more readily found. Due to this, SVMs prove to be a very effective way of dealing with complex datasets that have clear margins of separation and prove to be much more robust against overfitting within high-dimensional spaces. This proves to be a very strong technique in medical diagnostics, including the early detection of diseases like Parkinson's [15].

4.2 Decision Tree (DT)

A decision tree [16] is a machine-learning algorithm used for classification and regression. It classifies and regresses through a recursive partitioning of data over the most significant feature at each node. Each internal node of the tree represents a test on an attribute; each branch represents the outcome of the test, and each leaf node represents a class label or continuous value. Modern decision tree algorithms, like the C4.5 algorithm, were introduced by Ross Quinlan in the early 1990s. In classification, each tree is built on a training set based on features for which the largest amount of gained information, or in other words, the highest decrease in impurity, is gained about the impurity measures of Gini or entropy. Decision trees are intuitive, simple to interpret, and are capable of handling data with both discrete and continuous attributes, but their main drawback is that they probably tend to overfit in the presence of noisy data or if the tree is too deep. Techniques like pruning, ensemble methods, or limiting the depth of the tree overcome this.

4.3 Random Forest (RF)

Random forest is an ensemble learning method used for classification and regression tasks. During training, it creates a multitude of decision trees and outputs the mode of their classes if it is a classification task, or the mean prediction in case of regression tasks. It leverages the power of several trees, randomized, to come out with improved accuracy in prediction, and keeps overfitting in a tab. Each of these trees is built from a random subset of the data and features; this creates diversity and reduces variance, hence making the model robust against noise in the data. Aggregation over several trees creates stability and good generalization. The random forest algorithm is capable of handling high-dimensional datasets with ease, and it also handles complex interactions of variables. Maybe it is because of this that it finds wide applications, from banking to the medical diagnosis of diseases like Parkinson's disease.

5 Experiment

5.1 The Architecture of the Proposed System

The architecture begins with a dataset with raw features, followed by feature extraction and feature engineering to produce optimized features. These features are then passed to various machine learning algorithms, such as random forests, support vector machines, and decision trees, during the training process. The trained

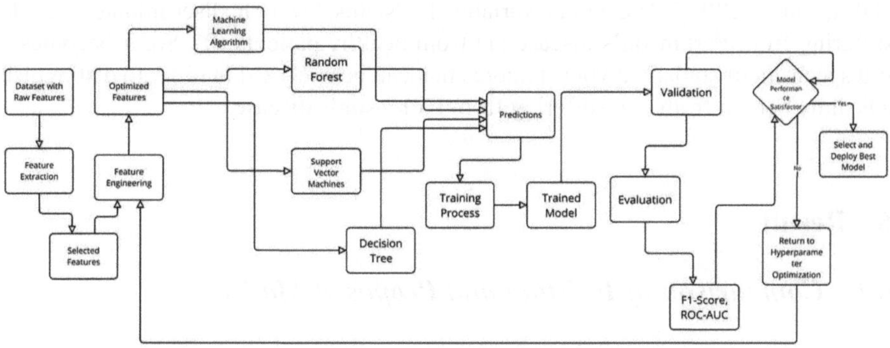

Fig. 1 Proposed system architecture

models generate predictions, which are evaluated using metrics like F1-Score and ROC-AUC.

The evaluation results are validated to determine model performance. If unsatisfactory, the workflow returns to hyperparameter optimization for iterative improvements. Once the performance is deemed satisfactory, the best-performing model is selected and deployed. This architecture highlights the iterative nature of model development, emphasizing feature optimization, algorithm selection, and performance validation (Fig. 1).

5.2 *Configuration System*

All experiments were carried out on a system fitted with an 11th Gen Intel(R) Core(TM) i5-1135G7 processor running at 2.40 GHz and 16 GB RAM (15.6 GB usable). The test environment was a 64-bit version of Windows 11 Pro, Version 23H2. There was no availability of the pen or touch input. In this configuration, effective model training and testing during all the research will be witnessed.

5.3 *Proposed Model*

This dataset comprises 195 observations and 24 attributes that include voice measurements for people, some of whom are suffering from Parkinson's disease. Many measures of the basic frequency include "MDVP (Hz)," etc. Next in line are jitter metrics, such as "MDVP (%)," "MDVP (Abs)," quantifying frequency variation, and shimmer metrics that quantify amplitude variation—"MDVP," "Shimmer." It also contains "NHR" and "HNR," which show the quality of voice. In this dataset, there are included nonlinear dynamical complexity measures such as "RPDE,"

"DFA," and "PPE." The target variable is "status," which discriminates people suffering from Parkinson's disease (1) from healthy patients (0). So, it becomes a full spectrum of statistical voice features that can be analyzed in order to distinguish a healthy person from one suffering from Parkinson's disease.

6 Result

6.1 Comparison of Existing and Proposed Model

The accuracies reported in five studies are pitted against the accuracies realized in a new model. Concretely, for Vishniakou and Yiwei (2023), it reached an accuracy of 88.16% for random forest, 74.74% for the decision tree, and 76.03% for SVM. For the latter, a higher result of 94.12% was realized with a random forest model, and an even greater one with a decision tree: 92.38% (Table 2).

A study by Abdurrahman and Sintawati reports an accuracy of 72.50% using random forest, while the new model realizes an accuracy of 94.12%. On another note, Fahira et al. also reported an accuracy of 79.31% using random forest. Again, in this case, the realization is an accuracy of 94.12% by the new model. All in all, the new model outperforms the results mentioned in the previous studies (Tables 3, 4, 5 and 6).

Table 2 Comparison of reported accuracy and proposed accuracy for our model

Reference	Existing models' accuracy (%)	Proposed model accuracy (%)
Vishniakou, U. A., and YiWei, X. (2023) [1]	RF: 88.16, DT: 74.74 SVM: 76.03	RF: 94.12, DT: 92.38SVM: 87.17
Fahira, et al. (2023) [2]	RF: 79.31	RF: 94.12
Abdurrahman G., and Sintawati M. (2023) [3]	RF: 72.50	RF: 94.12
R. Alshammri and Ghaida Alharbi (2023) [4]	DT: 90, RF: 92	DT: 92.38RF: 94.12
S. Kanakala (2024) [5]	RF: 88	RF: 94.12

Table 3 Comparison of existing accuracy and proposed accuracy using SVM model

Model	Existing accuracy (%)	Proposed accuracy (%)
Vishniakou, et al. [1]	76.03	87.17

Table 4 Comparison of existing accuracy and proposed accuracy using DT model

Model	Existing accuracy (%)	Proposed accuracy (%)
Vishniakou, et al. [1]	74.74	92.38
R. Alshammri, et al. [4]	90	92.38

Table 5 Comparison of existing accuracy and proposed accuracy using random forest (RF) model

Model	Existing accuracy (%)	Proposed accuracy (%)
Vishniakou, et al. [1]	72.50	94.12
Fahira, et al. [2]	79.31	94.12
Abdurrahman G., et al. [3]	88.16	94.12
R. Alshammri, et al. [4]	92	94.12
Srinivas Kanakala [5]	88	94.12

Table 6 Comparison of existing precision value and proposed precision

Reference	Existing precision value (%)	Proposed precision value (%)
Vishniakou, et al. [1]	76.91	94.01
Fahira, et al. [2]	72.28	94.01
Abdurrahman G., [3]	–	94.01
R. Alshammri, et al. [4]	92	94.01
Srinivas Kanakala [5]	–	94.01

6.2 Graphical Representation of the Results (Figs. 2, 3 and 4)

7 Conclusion

In the future, our model for detecting Parkinson's disease will be extended and improved to detect multiple diseases like diabetes, Alzheimer's disease, and heart conditions by adding their respective biomarkers. This makes the model become a multidisease predictor; hence, it can better classify symptoms that may superficially appear similar, thus enhancing its accuracy. This will be achieved by enriching the model with various datasets, including genetic information and cognitive scores that improve the performance of the model, giving it a wider and more comprehensive analysis of patient data. We also want to use advanced machine learning techniques that include hybrid models that use support vector machines for diseases such as diabetes and heart disease, and random forest for Parkinson's. Furthermore, complex patterns and interdependencies in the data shall be captured using deep learning. This will also aid in the identification of common biomarkers across diseases that will allow earlier detection with better insights into disease trajectories.

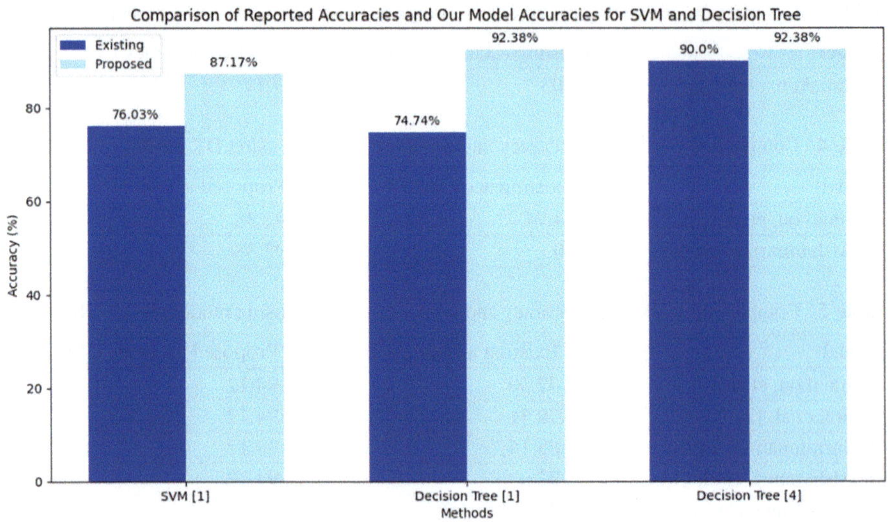

Fig. 2 Comparison of existing accuracies and proposed accuracies for SVM and decision tree

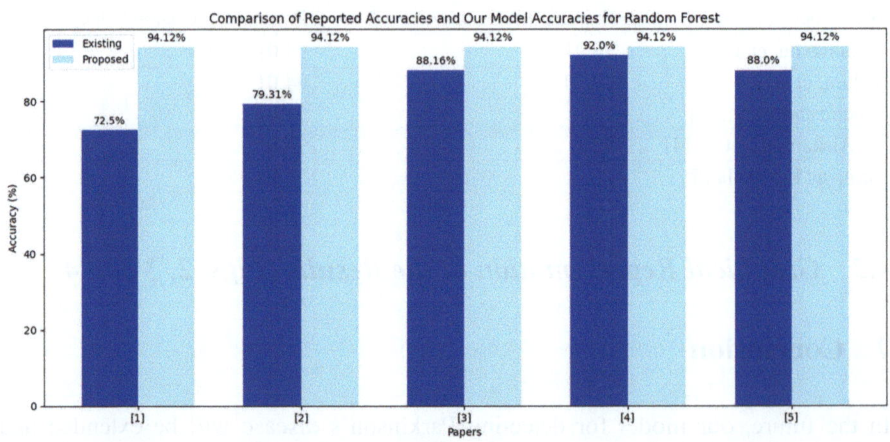

Fig. 3 Comparison of existing accuracies and proposed accuracies for random forest

Furthermore, this shall pave new avenues in developing novel diagnostic tools. Future directions will include real-world clinical testing, continuous model learning, and adaptations to retain accuracy and relevance in dynamic health-care settings.

Acknowledgments We would like to thank our institution, supervisors, and colleagues for their continuous support and guidance throughout the research. Second, we would like to acknowledge the creators of open-source datasets included in this study and those who reviewed the paper for providing valuable feedback. Most importantly, we are grateful to the people who have been diagnosed with Parkinson's disease whose lives motivated this research.

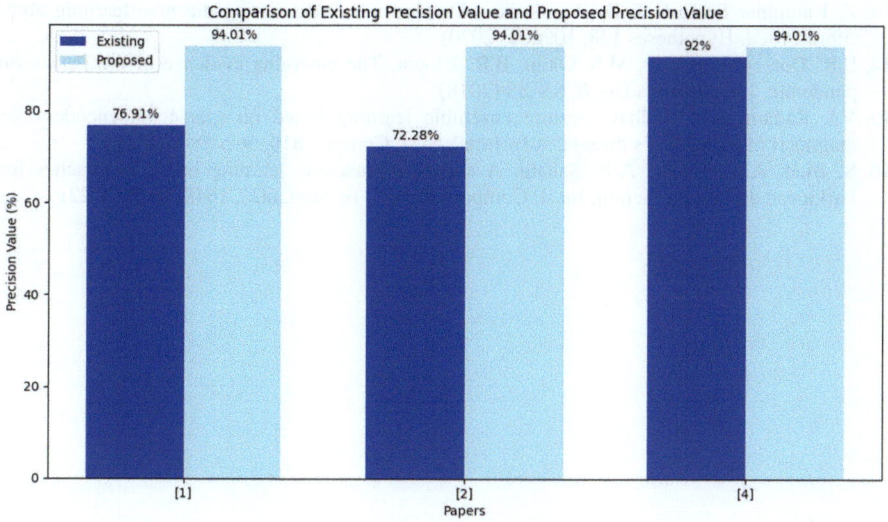

Fig. 4 Comparison of existing precision values and proposed precision values

References

1. U.A. Vishniakou, Y. Xia, IT diagnostics of Parkinson's disease based on the analysis of voice markers and machine learning. Doklady BGUIR **21**(3), 102–110 (2023)
2. N.R. Fahira, A. Lawi, M. Aqsha, Early detection model of Parkinson's disease using random forest method on voice frequency data. J. Nat. Sci. Math. Res. **9**(1), 30–38 (2023)
3. C.D. Anisha, K.G. Dr, Saranya, A machine learning based Parkinson prediction system—A comprehensive review. Recent Res. Rev. J. **2**(1), 27–38 (2023)
4. R. Alshammri, G. Alharbi, E. Alharbi, I. Almubark, Machine learning approaches to identify Parkinson's disease using voice signal features. Front. Artif. Intell. **6**(1084001), 1–17 (2023)
5. S. Kanakala, Parkinson's disease prediction using machine learning models. Int. J. Adv. Comput. Sci. Technol. **13**(7), 78–81 (2024)
6. D. Trabassi, M. Serrao, T. Varrecchia, A. Ranavolo, G. Coppola, R. De Icco, C. Tassorelli, S.F. Castiglia, Machine learning approach to support the detection of Parkinson's disease in IMU-based gait analysis. Sensors **22**(3700), 1–26 (2022)
7. J. Zhang, W. Zhou, H. Yu, T. Wang, X. Wang, L. Liu, Y. Wen, Prediction of Parkinson's disease using machine learning methods. Biomolecules **13**(12) Art. no. 1761, 1–18 (2023)
8. S. Manoja, T. Rajendran, Higher prediction accuracy on Parkinson disease patients using random Forest algorithm over DTA. Adv. Parallel Comput. Algorithms Tools Paradigms **2022**, 569–574 (2022)
9. Y. Ranga Sai Reddy, K.D.S. Prasad, D.A. Ashwin, S. Shaik, Prediction of Parkinson disease using machine learning. Int. J. Sci. Res. Eng. Manag. **7**(1), 1–5 (2023)
10. M.T.H.K. Tusar, M.T. Islam, A.H. Sakil, An experimental study for early diagnosing Parkinson's disease using machine learning. Int. J. Adv. Intell. Inform. **9**(2), 1–5 (2023)
11. N. Thapliyal, R. Nishad, Prediction of Parkinson's disease using machine learning algorithms. Int. J. Sci. Eng. Res. **10**(12), 1405–1411 (2019)
12. V.P. Bhaisare, R.R. Jain, A comprehensive analysis and performance comparison of various machine learning algorithms for Parkinson disease prediction. Int. J. Sci. Res. Sci. Technol. **8**(2), 553–559 (2021)

13. Z. Karapinar Senturk, Early diagnosis of Parkinson's disease using machine learning algorithms. Med. Hypotheses **138**, 109603 (2020)
14. E.R. Dorsey, T. Sherer, M.S. Okun, B.R. Bloem, The emerging evidence of the Parkinson pandemic. J. Parkinsons Dis. **8**, S3–S8 (2018)
15. V.J. Kadam, S.M. Jadhav, Feature ensemble learning based on sparse autoencoders for diagnosis of Parkinson's disease. Adv. Intell. Syst. Comput. **810**, 567–580 (2019)
16. S. Bind, A.K. Tiwari, A.K. Sahani, A survey of machine learning based approaches for Parkinson disease prediction. Int. J. Comput. Sci. Inf. Technol. **6**(2), 1648–1655 (2022)

Part III
Big Data for Real-time Analytics and Artificial Intelligence

Smart Solution for Safer Roads: Identifying and Alerting Accident-Prone Zones Using GIS and Real-Time Data

Preethi Harris, M. S. Adhnan Jeff, G. S. Abhinavu, and M. Harisudhan

1 Introduction

Accidents on roadways represent significant risks to public safety, with Coimbatore, often referred to as the "Manchester of South India," experiencing its own set of road safety challenges, including a notable increase in vehicular accidents in recent years. Identifying accident-prone zones becomes imperative for implementing proactive measures to prevent accidents and mitigate their consequences [1]. Traditional methods of accident detection often rely on manual reporting or historical data analysis, which may lack real-time insights and fail to effectively address emerging risk areas [2].

Consequently, there's a growing demand for innovative approaches to identify and mitigate accident-prone zones. In recent years, advancements in technology have paved the way for more sophisticated strategies in accident detection and prevention. The proposed study focuses on crafting a prototype for pinpointing accident-prone zones, leveraging the capabilities of ArcGIS software. Through ArcGIS, we've devised a data-driven map of accident hotspots by precisely plotting accident coordinates and storing them in an Excel spreadsheet. These coordinates are then read as input in a Python file, and the black spots are identified and stored in a .csv file. The .csv file is used as input dynamically for a HashMap, which is used to calculate the closest accident-prone zone; similarly, this is followed to display the closest hospital. Road safety is a significant concern globally, particularly in India, which ranks second in the world by size of the road network, following the United States. Despite a slight decrease in accident rates, India remains among the most accident-prone countries [3], with 1% of the global vehicle population but 11% of

P. Harris (✉) · M. S. Adhnan Jeff · G. S. Abhinavu · M. Harisudhan
Sri Ramakrishna Engineering College, Coimbatore, India

A. Haldorai et al. (eds.), *7th EAI International Conference on Big Data Innovation for Sustainable Cognitive Computing*, EAI/Springer Innovations in Communication and Computing, https://doi.org/10.1007/978-3-032-02979-9_9

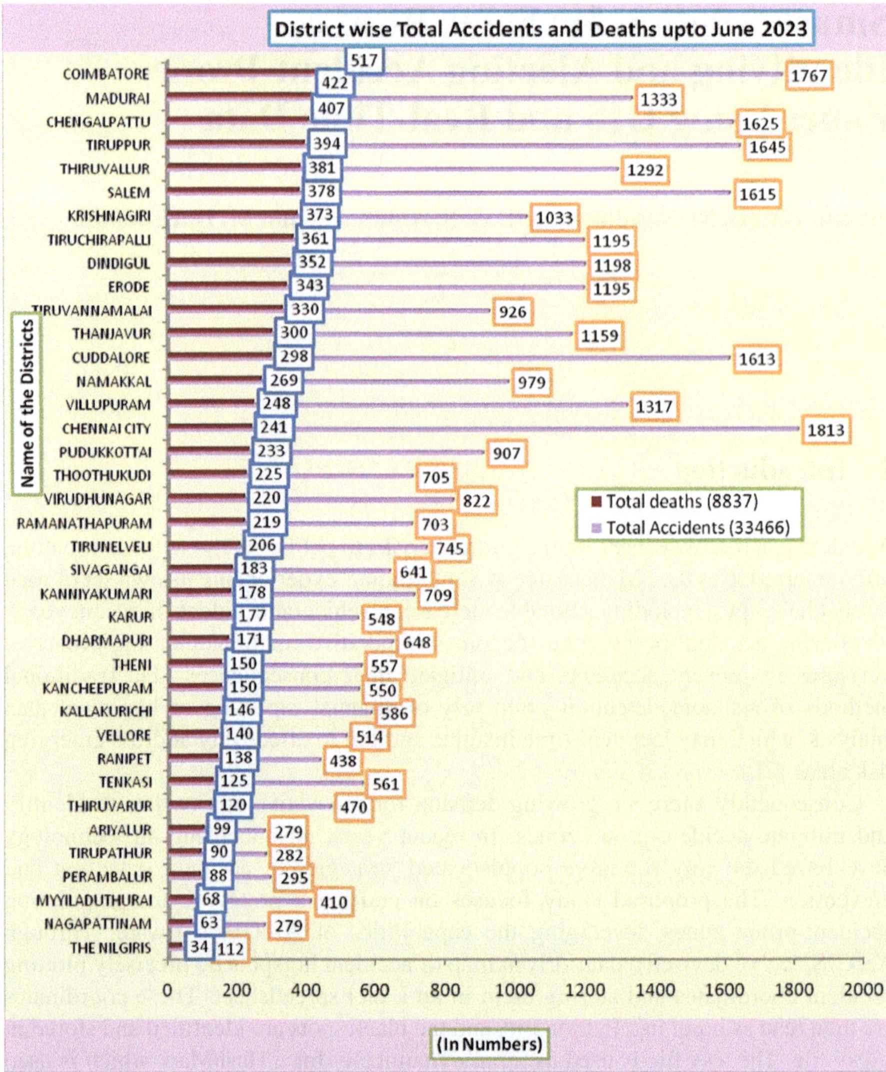

Fig. 1 District-wise total accidents and deaths up to June 2023

two road fatalities in 2020. A recent study on the analysis of accidents reveals that Coimbatore tops the list with the most accidents and deaths, with approximately 29.25% of people who met with an accident having died. The study results are shown in Fig. 1.

There are several factors that lead to road accidents in India, where road type is one of the main reasons. Key findings reveal that 31.8% of accidents occur on national highways (NH), 24.8% on state highways (SH), and 43.4% on other roads. Road safety audit (RSA) is crucial for identifying three safety deficiencies, including

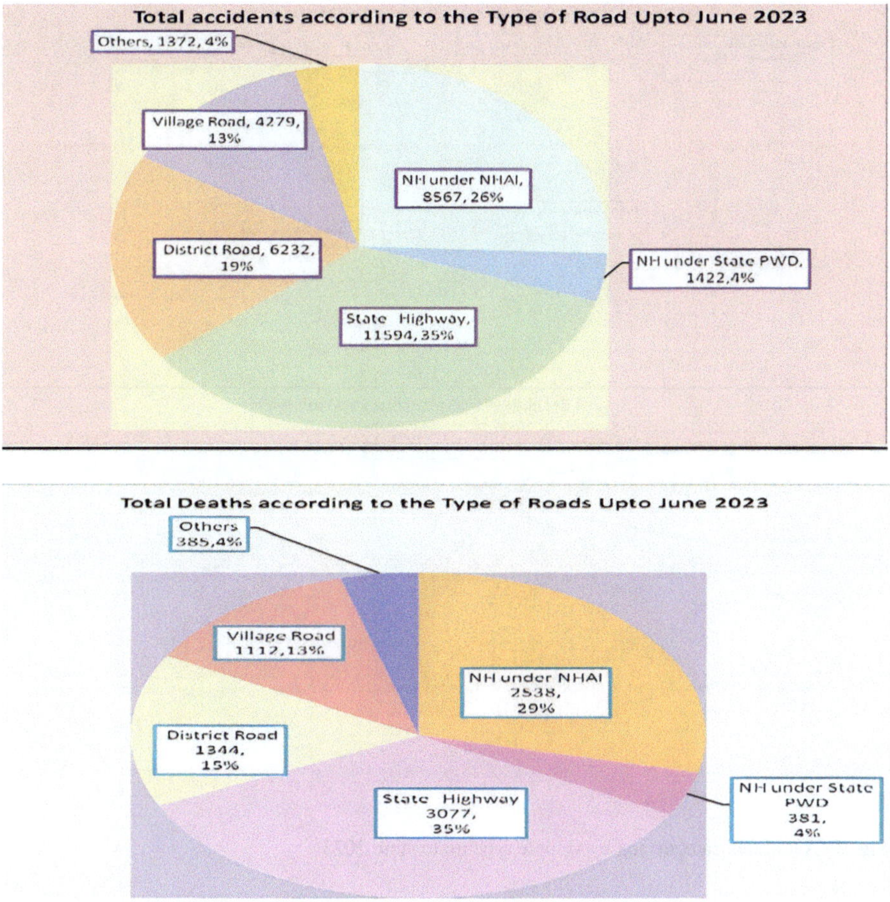

Fig. 2 Accidents categorized by road type up to June 2023

hazardous features such as high embankments and untreated junctions [4]. Figure 2 highlights the distribution of accidents and fatalities by road type up to June 2023 (provisional). The highest share of accidents (35%) and fatalities (35%) occurred on state highways, followed by national highways under NHAI (26% accidents, 29% fatalities), district roads (19% accidents), and village roads (13% accidents) in Tamil Nadu.

Figure 3 displays the overall number of accidents and fatalities broken down by vehicle type. Up until June 2023 (provisional), two-wheelers accounted for 45% of all accidents, while four-wheelers such as cars, jeeps, taxis, vans, and LMVs accounted for 29% of all accidents. Two-wheelers account for 43% of all fatalities, followed by four-wheelers such as cars, jeeps, taxis, vans, Tempo, and LMV (23%), and goods carriages (trucks and lorries) at 15%. Due to helmet nonwearing, two-wheelers are responsible for the greatest number of fatalities [5].

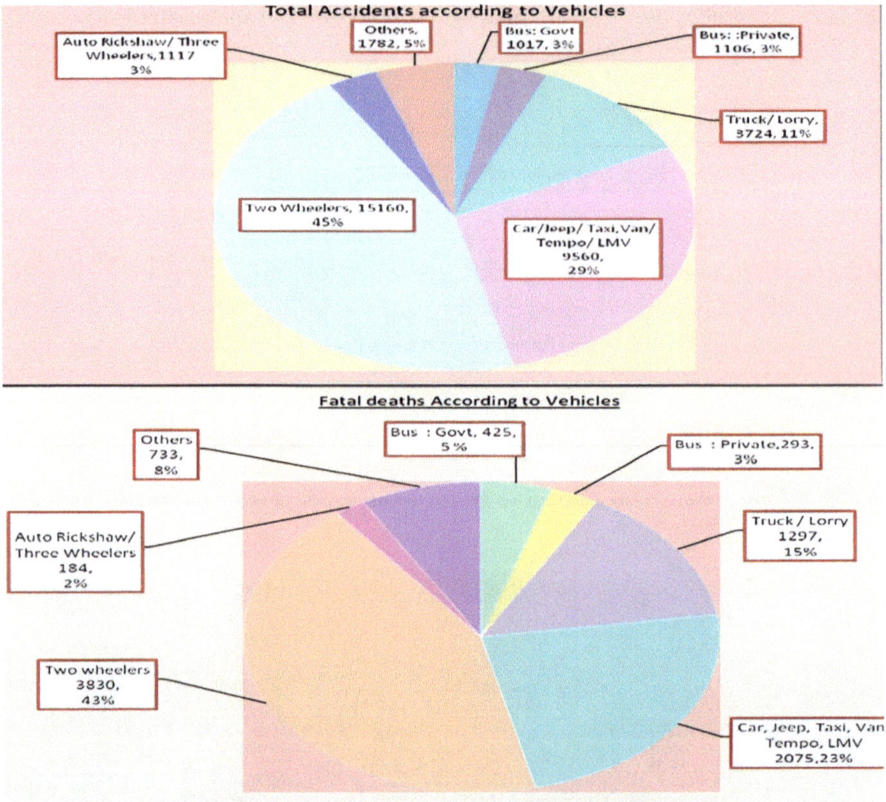

Fig. 3 Accidents categorized by vehicle type up to June 2023

The proposed approach presents several advantages over traditional methods, primarily its capacity to analyze extensive datasets and furnish five real-time insights for proactive decision-making. By seamlessly integrating this prototype, users receive timely alerts when approaching a danger zone or accident-prone area, utilizing the stored coordinates as reference points. This initiative contributes significantly to enhancing road safety and minimizing the impact of accidents on society by furnishing motorists and pedestrians with critical information, thereby fostering the creation of safer roadways for all.

2 Project Description

Geographic information systems (GIS) data sources include local government accident reports, such as Police First Information Report (FIR). The collected data are processed to extract precise geographical coordinates using ArcGIS for plotting on the Coimbatore map. The app ensures timely alerts using optimized latency mechanisms and prioritizes notifications based on proximity to risk zones. Driver distraction is minimized by using auditory signals and simple on-screen cues for critical alerts. Accidents are a major hazard to public safety worldwide, resulting in serious injuries, fatalities, and substantial property damage [6]. For better overall safety and to put efficient safeguards into place, it is essential to comprehend the frequency and trends of incidents. In the specific context of Coimbatore, a bustling district spanning 4723 km^2, the issue of accidents is a matter of concern. Traditional methods of accident zone detection often rely on manual reporting or historical data analysis, resulting in delayed insights and inadequate coverage of emerging risk areas. In recent years, advancements in technology have paved the way for more sophisticated strategies in accident detection and prevention. This study focuses on crafting a robust App for pinpointing accident-prone zones, leveraging the capabilities of ArcGIS software [7]. Through ArcGIS, a data-driven map of accident hotspots has been developed. The plotted accident coordinates are converted into shape files and stored as an Excel spreadsheet. These coordinates are then read as input in a Python file, and the black spots are identified and stored in a .csv file. The .csv file is then used as input dynamically for a hash map, which is used to calculate the closest accident-prone zone [8]; similarly, this is followed for the closest hospital. The proposed work presents several advantages over traditional methods, primarily the capacity to analyze extensive datasets and furnish real-time insights for proactive decision-making. By extending the ARCGIS concept into the App, users can receive timely alerts when approaching a danger zone or accident-prone area, utilizing the stored coordinates as reference points. This initiative contributes significantly to enhancing road safety and minimizing the impact of accidents on society by furnishing motorists and pedestrians with critical information, thereby fostering the creation of safer roadways for all [9].

The proposed work contains three modules in its workflow. The following block diagram in Fig. 4 represents data extraction for RideOSafe, the developed App. The accident data are collected from the official records of the Coimbatore police FIRs, stored in Excel format, then converted to a .csv file and then further converted to a .shp file, to be plotted in a virtual map using ArcGIS software for the visualization and storing of the accident-prone zones [10].

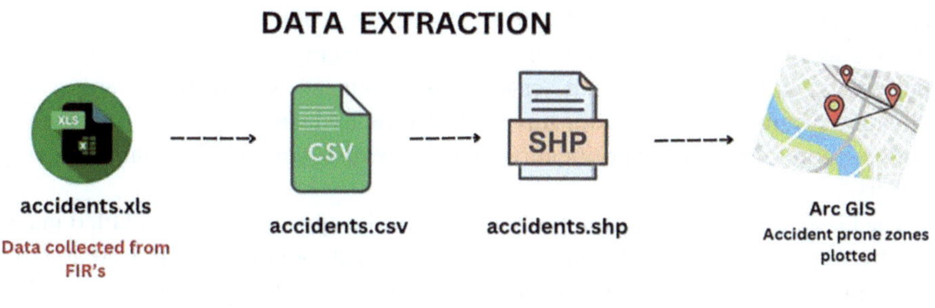

Fig. 4 Data extraction block diagram

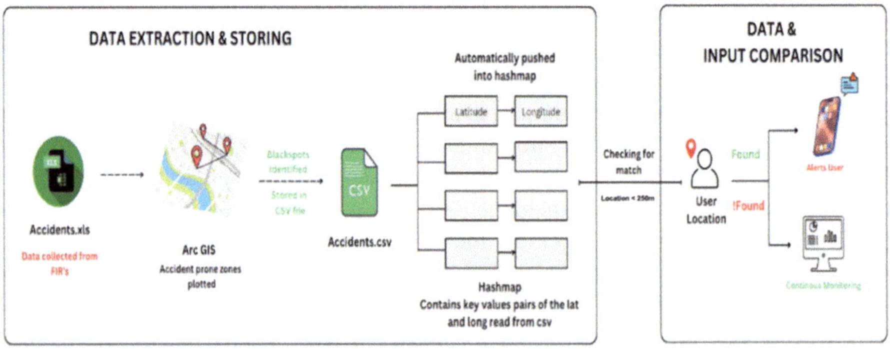

Fig. 5 Data storing block diagram

The conversion and storage of data using a hash map, as illustrated in Fig. 5, is dynamically managed by reading from a .csv file. This approach allows for the efficient updating and retrieval of accident-prone zone information. When a user approaches within approximately 250 m of such a zone, the prototype promptly sends an alert to ensure their safety. Additionally, the RideOSafe app not only warns users of these hazardous areas but also displays the nearest medical facility or infirmary for their reference. This feature ensures that users have immediate access to critical information, enhancing their ability to respond swiftly in emergencies. By combining real-time alerts with the nearest medical facility information, RideOSafe provides a comprehensive safety solution designed to significantly improve user awareness and preparedness while on the road.

The proposed work's principle revolves around providing users with real-time information about accident-prone locations, including latitude, longitude, and the closest accident-prone area and the closest hospitals.

User Interface and Functionalities The Android app integrates spatial datasets taken from the CSV file using geospatial libraries of the accident-prone zones. Alerts are concise and designed to minimize driver distraction, with auditory signals used for critical notifications. The application presents a streamlined and user-friendly interface, offering essential details such as latitude, longitude, and the current location address. Users can easily determine the distance to the nearest accident-prone zone and medical facility. A dedicated button integrates seamlessly with Google Maps for convenient navigation, enhancing the user experience. The app excels in location tracking, calculating distances to hazardous areas, and activating driving mode for safer travel. Real-time updates keep users informed of changing conditions, while notification alerts provide timely warnings about nearby dangers.

The Google Maps integration further enriches the app by offering detailed maps, step-by-step navigation, and route options to medical facilities or safer areas. This integration ensures that users can take quick action in case of emergencies. The platform is designed for Android, taking full advantage of built-in location services,

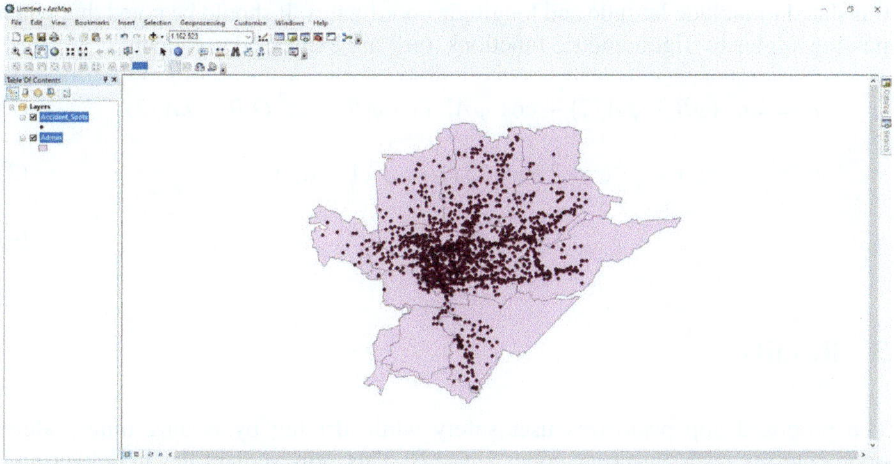

Fig. 6 Accident zones plotted in the ArcGIS tool

notifications, and Google Maps for enhanced functionality. Accident-prone area coordinates are securely stored in a HashMap to ensure both data integrity and user privacy. The app follows strict security protocols to safeguard sensitive information, protecting it from unauthorized access and ensuring confidentiality. For precise distance measurements, the calculations in Fig. 6 are utilized to determine the separation between two locations on the surface of the Earth and to convert degrees to radians. By combining real-time tracking, alerts, and precise navigation, the app provides a comprehensive solution for safer driving and hazard awareness.

Haversine Formula This formula accounts for the spherical shape of the Earth, providing a more accurate distance than simple planar methods, especially for longer distances. In practical applications, the Haversine formula (1) is used in global positioning systems (GPS), mapping applications, and location-based services to calculate distances between geographic coordinates accurately.

$$\text{haversine}\,(\theta) = \sin^2(\theta/2) \tag{1}$$

The calculation of geographic coordinates is as follows:

- Convert degrees to radians: Since trigonometric functions in the formula use radians, convert the latitude and longitude from degrees to radians.
- Calculate differences: Compute the differences in latitudes and longitudes.
- Apply the Haversine formula: Compute the value of "*a*," then compute "*c*" using the arctangent function, and finally calculate the distance "*d*" by multiplying "*c*" by the Earth's radius.

The value φ (2) represents the latitude and λ the longitude. *R* represents the Earth's radius, where the mean radius is considered as 6371 km (3) and (4) and are translated to include latitude and longitude coordinates. It should be noted that when passing angles to trigonometric functions, they are expressed in radians.

$$a = \sin^2(\varphi B - \varphi A/2) + \cos\,\varphi A^* \cos\,\varphi B * \sin^2(\lambda B - \lambda A/2) \tag{2}$$

$$c = 2 * \text{atan2}\left(\surd a, \surd(1-a)\right) \tag{3}$$

$$d = R \cdot c \tag{4}$$

3 Results

The proposed app prioritizes user safety while driving by issuing timely alerts upon entering accident-prone zones [11, 12]. This achievement lies in delivering a robust solution that empowers individuals to navigate with heightened awareness, significantly promoting road safety and providing peace of mind for all users. The application offers real-time data and alerts to help users avoid dangerous areas, ultimately contributing to safer driving practices and enhanced situational awareness. Figure 6 illustrates the plotted map of all identified accident zones, meticulously compiled from first information reports (FIRs) over the past 3 years. This comprehensive dataset allows the application to offer precise and up-to-date information about high-risk areas, further strengthening its effectiveness as a safety tool. Direct analysis of accident data is used to identify hotspots based on predefined criteria and thresholds. Historical data validation ensures the reliability of alerts by cross-referencing with past patterns and trends. The prototype ensures users are well-informed and can take proactive measures to avoid potential hazards, making every journey safer and more secure.

The FIR points plotted were sent as data into a Python program, which facilitated the identification of potential hotspots—areas with more than eight recorded accidents. These hotspots, also referred to as blackspots [13, 14], were then stored in a hash map. This hash map is crucial for determining the closest distance between the user's current location and the nearest accident-prone zone, as well as the closest medical facility [15, 16].

The integration of Google Maps plays a vital role in this process, as it provides real-time navigation and location services essential for guiding users away from hazardous areas and toward medical assistance. The seamless access to Google Maps from the app, as shown in Fig. 7, ensures that users receive accurate and timely information, thereby improving the ability to respond effectively in emergencies and ensuring a higher level of road safety.

The proposed application is designed to significantly enhance road safety by identifying accident hotspots and pinpointing the nearest medical facilities. A comparative analysis presented in Table 1 illustrates the differences between traditional methods and the cutting-edge RideOSafe app, providing valuable insights into areas that are prone to accidents. This initiative effectively leverages advanced

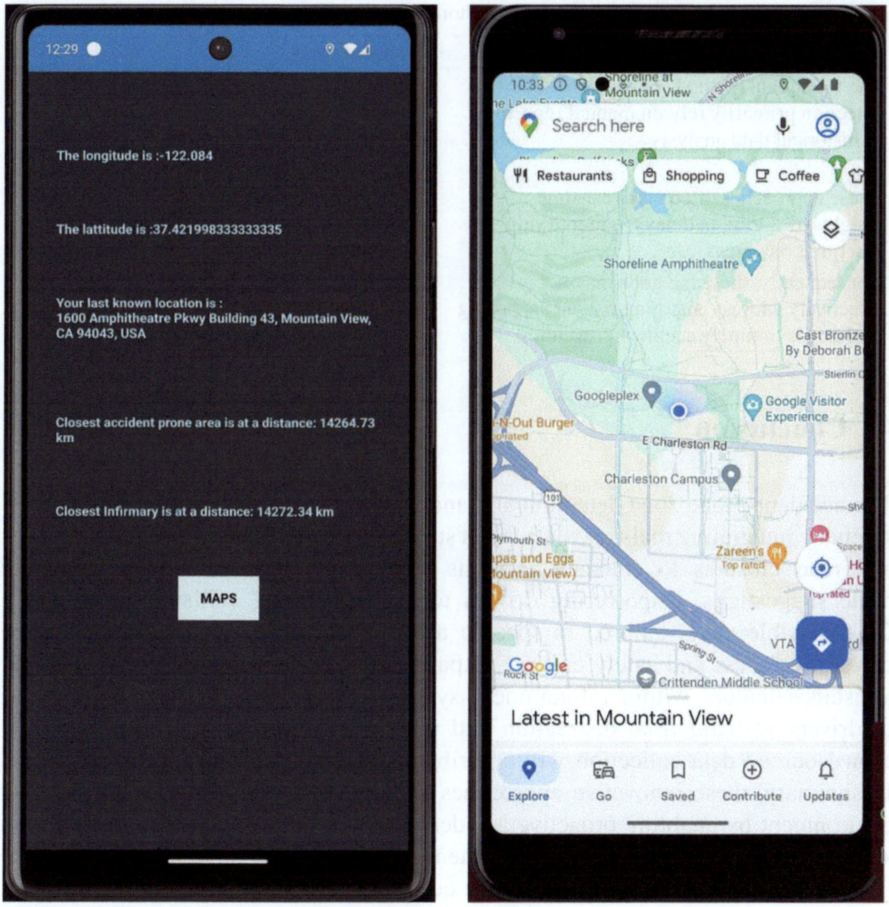

Fig. 7 Application interface and Google Maps for navigation

technologies such as ARCGIS, accident FIRs, and Google Maps to make critical information readily available at users' fingertips. By integrating these powerful tools, the RideOSafe app aspires to improve response times and increase awareness of hazardous zones. Ultimately, this contributes to the creation of safer roadways for all users, empowering them to navigate more confidently and make informed decisions when on the road.

Thus, the proposed App RideOSafe introduces a revolutionary method for identifying accident-prone zones by autonomously detecting them without user intervention. This innovation eliminates manual reporting and greatly reduces the time needed to identify these hazardous areas. With this prototype, users can proactively avoid dangerous zones, enhancing safety and significantly lowering the risk of accidents.

Table 1 Comparison between traditional methods and RideOSafe

Traditional method	RideOSafe app
The traditional methods of accident zone detection primarily rely on manual reporting or historical data analysis	In contrast, our app offers a revolutionary solution by autonomously detecting accident-prone zones without any manual intervention
These approaches often lack real-time insights, resulting in delays in identifying emerging risk areas	This not only eliminates the need for manual reporting but also significantly reduces the time required for accident zone detection
Consequently, these methods may not effectively address emerging risk areas, posing challenges in timely accident prevention	Users can now proactively avoid hazardous areas, thereby enhancing overall safety and minimizing the risk of accidents

4 Conclusion

The accident-prone zone detection and analysis app is poised to revolutionize road safety by integrating real-time data feeds such as traffic updates, weather conditions, and road closures to enhance accident predictions. This will enable dynamic route suggestions, empowering drivers to avoid potential collisions proactively. Customizable alerts tailored to specific areas or conditions will further enhance driver awareness and safety culture. Expansion to iOS and implementing a driver assistance feature within infotainment systems will deliver relevant information to drivers with minimal distraction and will ensure broader accessibility, while crowdsourced data collection will contribute to a comprehensive safety database. In summary, these innovative approaches promise to foster a safer transportation environment by enabling proactive accident avoidance and enhancing overall road safety awareness. With these advancements, the app is set to significantly reduce the likelihood of accidents and create a culture of safety on the road for all users, making it an essential tool for modern transportation systems. Future enhancements will explore the integration of clustering techniques to improve the precision of accident-prone zone identification to complement the direct data analysis approach. Additionally, dynamic risk factors like traffic and weather updates may be integrated, with efforts directed toward reducing alert response time for a more proactive and effective safety system.

References

1. S. Basu, P. Saha, Evaluation of risk factors for road accidents under mixed traffic: Case study on Indian highways. IATSS Res. **46**(4), 559–573 (2022). https://doi.org/10.1016/j.iatssr.2022.09.004
2. N. Tamascelli, R. Solini, N. Paltrinieri, V. Cozzani, Learning from major accidents: A machine learning approach. Comput. Chem. Eng. **162**, 1–15 (2022). https://doi.org/10.1016/j.compchemeng.2022.107786

3. Road accidents in India 2020. https://morth.nic.in/sites/default/files/RA_2020.pdf
4. Ashish, S.Y. Mhaske, Road safety audit (RSA) guidelines of selected nations—a comparative review. J. Road Saf. **34**(4), 42–50 (2023). https://doi.org/10.33492/JRS-D-23-209291
5. K. Adhikari, *A Project Report on Identification of Accident-Prone Areas and Development of Prediction Model Using Multiple Linear Regression Model* (Thesis Kathmandu University, 2024). https://doi.org/10.13140/RG.2.2.19505.15208
6. S.K. Ahmed, M.G. Mohammed, S.O. Abdulqadir, R.G. Abd, N.A. El-Kader, D.C. El-Shall, M.E. Ur Rehman, K. Dhama, Road traffic accidental injuries and deaths: A neglected Global Health issue. Health Sci. Rep. **6**, 5 (2022). https://doi.org/10.1002/hsr2.1240
7. L. Srikanth, I. Srikanth, Identification of traffic accident hotspots using geographical information system (GIS). Int. J. Eng. Adv. Technol. **9**(2), 4429–4438 (2019). https://doi.org/10.35940/ijeat.B3848.129219
8. A.V. Vitianingsih, D. Cahyono, Geographical information system for map-ping accident-prone roads and development of new road using multi-attribute utility method, in *The Proceedings of the 2ndInternational Conference on Science and Technology-Computer (ICST)*, (Yogyakarta, Indonesia, 2016), pp. 66–70. https://doi.org/10.1109/ICSTC.2016.7877349
9. M.N. Khan, S. Das, Advancing traffic safety through the safe system approach: A systematic review. Accid. Anal. Prev. **199**(3) (2024). https://doi.org/10.1016/j.aap.2024.107518
10. Ashish, S.Y. Mhaske, Road safety audit (RSA) guidelines of selected nations—a comparative review. J. Road Saf. **34**(4), 42–50 (2022). https://doi.org/10.33492/JRS-D-23-209291
11. M. Ahmadi, A. Valinejadi, A. Goodarzi, A. Safari, M. Hemmat, H.A. Majdabadi, A. Mohammadi, Geographic information system (GIS) capabilities in traffic accident information management: A qualitative approach. Electron. Physician **9**(6), 4533–4540 (2017). https://doi.org/10.19082/4533
12. A. Mateen, M.Z. Hanif, N. Khatri, S. Lee, S.Y. Nam, Smart roads for autonomous accident detection and warnings. Sensors **22**(6), 2077 (2022). https://doi.org/10.3390/s22062077
13. K.T. Rajesh, S. Athithan, S. Sarumathi, M. Aruna, Blackspot alert and accident prevention system, in *The Proceedings of the 10th International Conference on Computing, Communication and Networking Technologies*, (2019), pp. 1–4
14. S.R. Saagar, S. Chaturvedim, Road safety audit SH-27 analysis. Int. J. Res. Anal. Rev. **6**(2), 615–621 (2019)
15. J. Chaudary, V. Mishra, B. Jain, S. Akshay, D.S. Reddy, P.G. Kaduluri, To find accident prone zones and nearest health facility by GIS based spatial and network analysis in Amritsar, Punjab. Natl J. Community Med. **11**(04), 156–159 (2022). https://doi.org/10.5455/njcm.20200405092812
16. S. Yunus, I.A. Abdulkarim, Road traffic crashes and emergency response optimization: A geospatial analysis using closest facility and location-allocation methods. Geomat. Nat. Haz. Risk **13**(1), 1535–1555 (2022). https://doi.org/10.1080/19475705.2022.2086829.erences

Road Accident Risk Assessment to Develop Strategic Recommendations for Road Safety of the NorthWestern Province of Sri Lanka: A Systematic Literature Review

J. H. K. I. Jayamaha and Thilini V. Mahanama

1 Introduction

Road safety remains a critical issue in Sri Lanka, characterized by a persistently high rate of road traffic accidents (RTAs), resulting in substantial human and economic losses. With an average of 38,000 crashes annually, leading to approximately 3000 deaths and 8000 serious injuries, the country faces a severe public health crisis that extends beyond mere statistics. The Northwestern Province, in particular, has emerged as one of the most affected regions, exhibiting some of the highest fatality rates despite its lower vehicle density compared to other provinces [1, 2]. This alarming trend is compounded by a general lack of public awareness and an underestimation of the broader impacts of RTAs on societal well-being. While issues such as poverty, malnutrition, and terrorism dominate public discourse, the significance of road safety is often overshadowed, resulting in insufficient attention from both the public and policymakers [1]. The economic implications are also profound, with the cost of road accidents in Sri Lanka estimated to exceed Rs 10,000 million annually, encompassing direct costs like medical care and vehicle repairs, as well as indirect costs including police resources and insurance claims [1, 2]. In the past, road safety in Sri Lanka has been a concern for a limited group of professionals, with the broader public largely indifferent to its gravity. However, recent media coverage has heightened public awareness, emphasizing the urgent need for proactive measures to address the escalating road safety challenges [3, 4]. In this context, the need for comprehensive research and targeted interventions is critical, particularly in regions like the Northwestern Province, where the road safety situation is dire (see Tables 1 and 2).

J. H. K. I. Jayamaha · T. V. Mahanama (✉)
Faculty of Science, University of Kelaniya, Kelaniya, Sri Lanka
e-mail: thilinim@kln.ac.lk

A. Haldorai et al. (eds.), *7th EAI International Conference on Big Data Innovation for Sustainable Cognitive Computing*, EAI/Springer Innovations in Communication and Computing, https://doi.org/10.1007/978-3-032-02979-9_10

Table 1 Descriptive statistics of accidents by type and province 2002 [5]

	Type of accident				Total
Province	Fatal	Grievous	Non-grievous	Damaged only	
Western Province	796	1288	5488	25,106	32,678
	39%	39%	43%	68%	60%
Northern Province	58	119	387	414	978
	3%	4%	3%	1%	2%
Southern Province	239	351	1370	1562	3522
	12%	11%	11%	4%	6%
Central Province	149	258	1361	2965	4733
	7%	8%	11%	8%	9%
Northwestern Province	424	578	1595	2502	5099
	21%	18%	13%	7%	9%

Table 2 Fatality rates in provinces according to police reports of road accidents in Sri Lanka [5]

Province	Vehicle, km	Rank	Per 10^5 population	Rank	Composite rank
Central	0.22	6	7.41	5	4
North Central	0.15	7	3.8	7	5
Northwestern	0.38	1	19.14	1	1
Sabaragamuwa	0.27	4	11.97	3	3
Southern	0.32	2	10.23	4	2
Uva	0.3	3	5.89	6	5
Western	0.22	5	17.01	2	3

This systematic literature review aims to examine the existing body of knowledge related to road traffic accidents in Sri Lanka, focusing on identifying the key risk factors, the effectiveness of current safety measures, and the gaps in research that need to be addressed. By synthesizing findings from previous studies, this review seeks to inform the development of strategic recommendations for improving road safety management in the Northwestern Province and beyond. This paper is structured as follows. Section 2 provides the significance of this study based on a review of related studies on RTAs. Section 3 discusses the factors affecting RTAs, Section 4 presents the strategic recommendations to minimize RTAs, and Sect. 5 summarizes with concluding remarks.

2 Methodology

This systematic literature review followed the preferred reporting items for systematic reviews and meta-analyses (PRISMA) approach to ensure a comprehensive and methodologically sound process (see Fig. 1). The review focused on gathering relevant literature related to road accident risk, factors contributing to road safety issues, and strategic recommendations for improving road safety in the

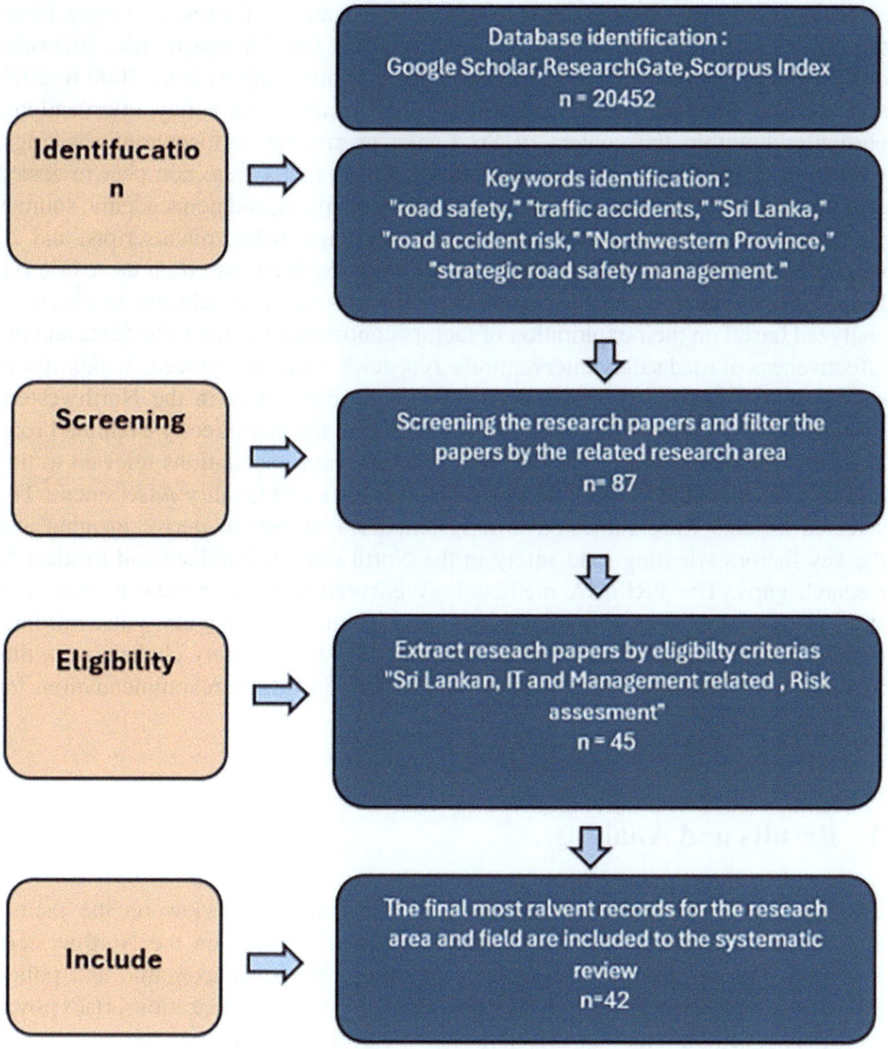

Fig. 1 Literature retrieval process using the PRISMA methodology

Northwestern Province of Sri Lanka. Data Sources and Search Strategy: To collect pertinent literature, academic databases including IEEE Xplore, Google Scholar, and Scopus were systematically searched. The search strategy was constructed using targeted keywords such as "road safety," "traffic accidents," "Sri Lanka," "road accident risk," "Northwestern Province," and "strategic road safety management." Boolean operators "AND" and "OR" were used to refine the search results and connect different aspects of the research focus. Study Selection and Screening: The initial search yielded 20,452 documents. These documents were then screened by reviewing the titles and abstracts to assess their relevance to the research topic. From this initial screening, 87 documents were selected for full text review. Inclusion criteria for this review included English-language publications from 2000 to 2023 that focused on road safety, accident severity factors, and safety interventions, particularly within the context of Sri Lanka or similar socioeconomic settings. Exclusion criteria involved studies unrelated to the focus area, non-peer-reviewed articles, and gray literature such as reports, book chapters, and nonacademic sources (excluded: 14 health-related articles of RTA, 11 unpublished manuscripts, and 20 nonacademic sources). A total of 45 articles were excluded based on these criteria. Data Extraction and Synthesis: During the full-text review, the selected articles were analyzed based on their exploration of factors contributing to road accidents and the effectiveness of road safety interventions. A specific focus was placed on identifying studies that addressed the unique challenges of road safety in the Northwestern Province of Sri Lanka. Ultimately, 42 research articles that directly examined road accident risks, contributing factors, and strategic recommendations relevant to this region were included in the final synthesis. Analysis and Quality Assessment: The selected articles were subjected to a systematic synthesis to derive insights into the key factors affecting road safety in the Northwestern Province and to identify research gaps. The PRISMA methodology ensured that this review process was standardized, unbiased, and transparent. The synthesized insights will contribute to developing a comprehensive understanding of the road safety challenges in this region and inform the formulation of evidence-based strategic recommendations for improving road safety.

3 Results and Analysis

This section synthesizes the findings from the literature review on the factors contributing to RTAs in Sri Lanka, with a particular focus on the Northwestern Province. The results are presented by analyzing the socioeconomic and policy influences, long-term trends in RTAs, the impact of vehicle registrations, risks posed by different vehicles, and the effectiveness of current road safety measures.

3.1 The Impact of Socioeconomic Changes and Policy Shifts on RTAs

The literature indicates that road traffic crashes in Sri Lanka have been significantly influenced by socioeconomic changes and policy shifts over the decades (see Fig. 2). Notably, the establishment of the traffic police division and the introduction of mandatory vehicle registration improved the reporting and management of traffic incidents. However, the introduction of an open economy in the late 1970s,

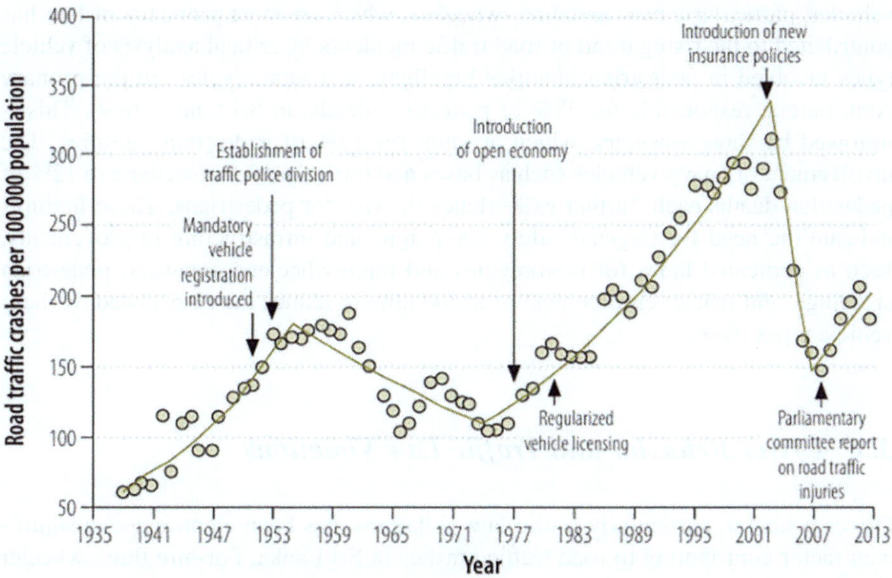

Fig. 2 Road traffic crashes in Sri Lanka, 1938–2013 [6]

while contributing to economic growth, also led to a rise in vehicle ownership and a corresponding increase in traffic crashes [6–8]. The 1980s to the early 2000s reported a peak in road traffic crashes, attributed to the rapid growth in vehicle numbers and changes in traffic policies. A decline in crashes was observed post-2007, following enhanced road safety awareness and legislative measures introduced after a parliamentary report on road traffic injuries [6–8].

Long-term trends in road traffic accidents reveal significant increases in crashes, injuries, and fatalities in Sri Lanka. From 1938 to 2013, the number of crashes per 100,000 people rose from 61.2 to 183.6, while injury and fatality rates also showed substantial increases [1]. These trends are closely linked to Sri Lanka's political, economic, and infrastructural developments, particularly following the introduction of an open economic policy in 1977, which led to a surge in vehicle numbers, especially two- and three-wheeled vehicles, contributing to higher accident rates [1].

3.2 *Impact of Vehicle Registrations on RTAs*

The rapid increase in vehicle registrations in Sri Lanka since 2003, reflecting the country's economic growth, has had significant implications for road safety. The total number of registered vehicles nearly doubled between 2003 and 2012, with two-wheeled vehicles representing the largest category [6–8]. This surge in vehicles, particularly two- and three-wheelers, which are more prone to crashes, has contributed to the rising trend of road traffic incidents. A critical analysis of vehicle types involved in pedestrian fatalities highlights that motorcycles are the primary contributors, responsible for 38% of pedestrian deaths in Sri Lanka [6–9]. This is followed by three-wheelers, which account for 17% of pedestrian fatalities. The involvement of heavy vehicles such as buses and lorries, which contribute to 12% of pedestrian deaths each, further exacerbates the risk for pedestrians. These findings indicate the need for targeted safety campaigns and infrastructure improvements, such as dedicated lanes for motorcycles and three-wheelers, enhanced pedestrian crossings, and stricter enforcement of traffic laws to reduce the risks posed by these vehicle types [6–8].

3.3 Driver Behavior and Traffic Law Violations

Driver behavior, particularly traffic law violations, has been identified as a significant factor contributing to road traffic crashes in Sri Lanka. For-hire three-wheeler drivers, especially in urban areas, exhibit a higher likelihood of being involved in crashes due to behaviors such as overloading and operating old vehicles [6–9]. A study [7] highlighted that 89.8% of drivers involved in crashes had allowed more than three passengers, which significantly increased the risk of accidents. This finding emphasizes the importance of stricter enforcement of passenger limits and regular vehicle maintenance to ensure roadworthiness and reduce the incidence of crashes. The effectiveness of increased traffic fines as a deterrent for traffic violations remains inconclusive. While higher penalties have been shown to reduce violations, their overall impact on road safety is limited without consistent enforcement and comprehensive public awareness initiatives [10]. The literature suggests that increased fines must be complemented by other measures, such as improved enforcement and public education, to achieve significant improvements in road safety.

3.4 Limitations in Policy Implementation Due to Underreporting

This review identified significant underreporting of road traffic injuries in Sri Lanka, with studies suggesting that actual figures may be higher than reported. Factors contributing to high accident rates include human errors, vehicle defects, poor road conditions, and environmental influences [11, 12]. A study on blood alcohol concentration levels in fatal accidents revealed that 51% of cases were above the legal limit, underscoring the need for stricter enforcement of alcohol-related traffic laws and better identification of accident-prone areas or 'black spots' to

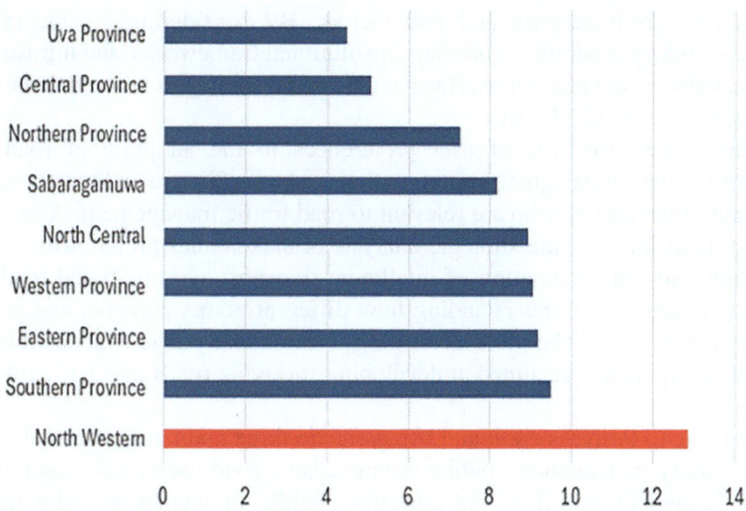

Fig. 3 Severity index by province in Sri Lanka [5]

implement specific safety measures [13–15]. The implementation of road safety policies in Sri Lanka faces significant challenges due to incomplete and low-quality data, which undermines the accuracy of road safety audits and hotspot analyses. Another study [10] demonstrated that police data, while frequently used, often lacks comprehensive geolocation details, leading to difficulties in accurately identifying accident hotspots. The use of survey data as an alternative showed a high level of accuracy in identifying road traffic crash hotspots, suggesting that more reliable data sources are needed to improve road safety measures [16].

3.5 The Significance of Designing Road Safety Assessment

There are significant regional disparities in road traffic accidents in Sri Lanka, with the Northwestern Province showing a particularly high fatality rate and severity index (see Fig. 3). The Northwestern Province recorded a fatality rate of 19.14 per 100,000 population and a severity index of 12.86%, indicating that accidents in this region are more likely to result in fatalities compared to other provinces [17]. The high percentage of motorcycle-related fatalities in this region highlights the need for focused safety campaigns and stricter enforcement of road safety regulations aimed at motorcycle riders [17].

The literature emphasizes the importance of a holistic approach to road safety assessment, integrating both qualitative and quantitative factors to develop more effective and targeted safety strategies. Advanced analytical methods, such as machine learning and statistical techniques, have been suggested to provide better insights into crash severity and risk factors. By considering a wide range of factors, including economic conditions, institutional frameworks, and infrastructure, policymakers can develop more effective road safety strategies tailored to the unique challenges in Sri Lanka [2, 6].

Understanding the role of user preferences in the adoption of road safety measures is vital for designing strategies that are both effective and widely accepted by various stakeholders who are relevant to road traffic management. According to Periyasamy et al. [18], the strategic analysis of stakeholder preferences regarding the design and implementation of intelligent transport systems-based road safety measures is critical in understanding how different actors perceive and prioritize various interventions. The multi-actor multi-criteria analysis to evaluate the potential road safety measures aimed at developing forgiving roads and self-explanatory roads.

According to Periyasamy et al. [18], there are three main groups of stakeholders for road safety management: public policymakers, road users, and manufacturers, each with distinct priorities and concerns. Public policymakers, who represent societal interests, place a high emphasis on road safety interventions that can significantly reduce traffic accidents. They prioritize speed-related measures such as variable message signs that provide real-time speed warnings, as well as speed alert systems that function by recognizing traffic signs. These measures are seen as having a high potential to improve road safety by directly addressing one of the leading causes of speeding accidents. These interventions are rated highly by policymakers because they align with broader public safety objectives and have a positive impact on reducing traffic fatalities and injuries [18].

4 Discussion

The evaluation of road safety across different provinces in Sri Lanka highlights that the Northwestern Province stands out as the most vulnerable, primarily due to its high fatality rates (see Table 2). This region's top ranking in fatality rates per vehicle kilometer traveled and per 100,000 population underscores the severity of road safety challenges faced in the area (see Table 1). While the Western Province records the highest number of road accident fatalities, the adjusted fatality rates in the Northwestern Province indicate a more critical issue, as accidents in this region are more likely to result in deaths. This disparity between the number of accidents and the severity of outcomes highlights a critical research gap that requires urgent attention. Despite the high incidence and severity of road accidents in the Northwestern Province, there has been a lack of focused research addressing the specific causes and potential solutions for this region. Most existing studies have concentrated on more urbanized regions like the Western Province, leaving a significant gap in understanding the unique challenges faced by rural and semi-urban areas. The Northwestern Province, with its lower accident rate but significantly higher fatality rate presents a compelling case for targeted road safety interventions. The high severity index, indicating that nearly 13% of all road accident injuries in the province result in fatalities, further emphasizes the urgent need for dedicated research and policy action [17].

A key finding from the literature is the prevalence of motorbikes in the Northwestern and Southern provinces, which significantly contributes to the high fatality rates in these areas. Motorbikes, due to their inherent vulnerability, exhibit substantially higher fatality rates compared to other vehicle types. This underscores the heightened risk posed by motorbike usage, necessitating targeted measures such as enhanced rider safety programs, stricter enforcement of helmet laws, and the construction of dedicated lanes for motorbikes to mitigate the associated road safety hazards. Given these findings, a comprehensive approach is necessary to improve road safety in the Northwestern Province. Addressing the specific challenges in this region requires a detailed analysis of road accident data to identify key risk factors and develop targeted solutions.

Strategic recommendations for enhancing road safety should be based on thorough investigations into the factors causing these accidents, including road infrastructure, vehicle conditions, driver behavior, and environmental influences. Prioritizing high-risk areas and vulnerable road user groups, such as motorbike riders, through comprehensive risk assessments is essential. Evaluating the effectiveness of current road safety measures and policies in the Northwestern Province is also crucial for developing evidence-based recommendations that can enhance road safety management practices at the local police station level.

The implications of this research for policymakers and stakeholders involved in road safety management in Sri Lanka are significant. The high fatality rates and severity index in the Northwestern Province indicate a critical need for policy actions tailored to the unique challenges of this region. Developing methodologies for assessing and predicting future accident risks, along with formulating strategic recommendations, will provide valuable insights for evidence-based decision-making. Future research should focus on addressing the identified gaps, particularly the need for more localized studies that consider the specific socioeconomic and environmental factors influencing road safety in the Northwestern Province. By exploring the complex dynamics of road accidents in this region, future studies can contribute to developing more targeted and effective road safety interventions, ultimately leading to safer transportation systems across Sri Lanka.

5 Conclusion

The analysis of road safety in Sri Lanka, with a particular focus on the Northwestern Province, underscores the critical need for targeted interventions and policy actions to address the unique challenges faced by this region. The findings reveal that despite a lower overall accident rate, the Northwestern Province exhibits significantly higher fatality rates, indicating a severe road safety issue that requires urgent attention. The prevalence of motorbikes, which contributes disproportionately to high fatality rates, highlights the need for specific safety measures such as improved rider safety programs and infrastructure adjustments. This research emphasizes the importance of a comprehensive approach to road safety, one that integrates detailed analysis of accident data, considers the socioeconomic and environmental factors at play, and prioritizes high-risk areas and vulnerable road user groups. The recommendations derived from this study aim to provide a foundation for enhancing road safety management practices, particularly at the local level, through evidence-based strategies. Ultimately, the insights gained from this research have significant implications for policymakers and stakeholders. Addressing the identified gaps, particularly in the Northwestern Province, will be crucial in reducing road traffic fatalities and injuries. Future research should continue to explore the complex dynamics of road safety in Sri Lanka, contributing to the development of more effective and targeted interventions that can lead to safer roads and a reduction in the socioeconomic impacts of road accidents across the country.

References

1. https://www.roadsafetyfacility.org/country/sri-lanka, accessed on 28/04/2024
2. https://www.themorning.lk/articles/XdBrr2QxsVhnfEH9qQgE, accessed on 28/04/2024 https://www.who.int/publications/i/item/9789241565684, accessed on 02/05/2024
3. https://road-safety.transport.ec.europa.eu/index_en, accessed on 02/05/2024
4. S.D. Dharmaratne et al., Road traffic crashes, injury and fatality trends in Sri Lanka: 1938–2013 (2015)
5. A.S. Kumarage, S.M. Wickremasinghe, M.D.R.P. Jayaratne, Where do road accidents occur, in *Analysis of Road Accidents in Sri Lanka*, ed. by P. Fisher, (Supreme Group of Companies, 2003), pp. 2–11
6. M. Ben Laoula et al., Evaluating Expressway Traffic Crash Severity by using Logistic Regression and Explainable & Supervised Machine Learning Classifiers. Updated literature review
7. A.U. Jayatilleke et al., Increased traffic fines and road traffic crashes in Sri Lanka (2012)
8. S.D. Dharmaratne, A.U. Jayatilleke, Risk posed by different vehicles to pedestrian safety in Sri Lanka (2015)
9. B. Perera et al., Enhancing the effectiveness of risk management practices in Sri Lankan road construction projects: A Delphi approach (2014)
10. K. Devasurendra et al., An insight into motorized two- and three-wheel crashes in developing countries: A case study in Sri Lanka (2017)
11. S.A.R. Shah et al., Road safety risk assessment: an analysis of transport policy and management for low-, middle-, and high-income Asian countries (2018)
12. P.W. Liyanage, K.P.G.C.D. Sucharitharathna, Short-term road traffic forecast using LSTM-based deep learning model, in *Proceedings of the 2023 International Research Conference of Sri Lanka Technology Campus, Colombo, Sri Lanka*, (2023), pp. 1–8. December 2023
13. K. De Brucker et al., Strategic analysis of stakeholder preferences regarding the design of ITS-based road safety measures (2014)
14. P.G. Amarasinghe, S.D. Dharmaratne, Epidemiology of road traffic crashes reported in the Kurunegala police division in Sri Lanka. Sri Lanka J. Med. **28**(1), 10–19 (2019)
15. M. Eddleston, N. Udayakumara, S. Adhikari, D. de Silva, M.H.R. Sheriff, D.L. Waidyaratne, The importance of poisoning vs. road traffic injuries as a cause of death in rural Sri Lanka. PLoS One **2**(7), e599 (2007)

16. A.U. Jayatilleke et al., Factors associated with RTCs among for-hire three-wheeler drivers in Sri Lanka: a case–control study (2015)
17. A.S. Edirisinghe, I.D. Kitulwatte, U.D. Senarathne, A study into blood alcohol concentration in fatal accidents among vulnerable road users in a tertiary care hospital, Sri Lanka. Int. J. Inj. Contr. Saf. Promot. **22**(2), 158–164 (2015)
18. N. Periyasamy, C.A. Lynch, S.D. Dharmaratne, D.B. Nugegoda, Underreporting of road traffic injuries in the district of Kandy, Sri Lanka. BMJ Open **3**, e003640 (2013)

Revolutionizing Supply Chains: A Systematic Review of Integration of Artificial Intelligence in Supply Chain Management

Archana Sasi, S. Sreenath, Shreehari Manu, Aadith Nidi, C. Shamdan, and Nitesh Yadav

1 Introduction

Supply chain management (SCM) is an intricate and multifaceted field that orchestrates the seamless interaction among various sectors such as marketing, logistics, and production. Achieving excellence in SCM is equivalent to achieving overall business success. Modern corporate practices like lean management and the just-in-time philosophy, which significantly impact logistics and production, necessitate continuous improvements in SCM. Moreover, the frequent occurrence of adverse events like natural disasters, political instability, and other disruptions further underscores the need for robust and adaptive SCM practices [1, 2]. The supply chain (SC) idea spans all stages of manufacturing and distribution, from vendors and producers to distributors and, finally, to the end consumer.

The rise of artificial intelligence (AI) has been transformative in the realm of SCM. AI, which can be described as the capability of computers to autonomously resolve issues that they were not specifically built to handle [3, 4], has witnessed remarkable progress since its start in 1956 by John McCarthy, who organized the pioneering Dartmouth workshop [5, 6]. Rapid improvements in mobile computing, secure online data preservation, cloud-based machine learning (ML), as well as sophisticated data processing algorithms, have dramatically broadened the applicability and usability of AI technology [7]. AI-driven systems are now capable of performing difficult tasks such as pattern recognition, data collection, inference, and forecasting functions that previously required human insight [8, 9]. The advent

A. Sasi (✉) · S. Sreenath · C. Shamdan
Department of Computer Science and Engineering, Jain University, Bangalore, India

S. Manu · A. Nidi · N. Yadav
Department of Information Science and Engineering, Jain University, Bangalore, India

A. Haldorai et al. (eds.), *7th EAI International Conference on Big Data Innovation for Sustainable Cognitive Computing*, EAI/Springer Innovations in Communication and Computing, https://doi.org/10.1007/978-3-032-02979-9_11

of deep learning (DL) algorithms has enabled machines to surpass human ability in specific jobs, including playing complex games like Atari and Go [10, 11], and excelling in visual tasks such as object detection [12].

AI empowers machines with human-like intelligence, facilitating the ubiquitous use of digital data to enhance the performance of digital services, products, and operations [13]. The deployment of AI applications has provided businesses with a competitive edge by delivering near-perfect accuracy in forecasting, improving research and development processes, enhancing product efficiency and quality, and significantly improving customer targeting and overall experience. In the past few years, the use of AI technology has gained significant traction across various industries [14]. AI is described as the potential of machines to gather information from experience and to arrive at judgements that are similar to human intelligence [15]. Deep neural networks (DNN), convolutional neural networks (CNN), and optimization approaches employed in operations analysis have advanced rapidly, allowing computers to accomplish activities that were previously only performed by humans. According to Russell and Norvig (2016), the aim of AI is to build rational agents proficient in perceiving and acting to optimize some objective function [16].

Many leading corporations, such as Amazon, Walmart, Philips, and eBay, have successfully integrated AI into their SCM processes [15]. Another noteworthy technology in SCM is radio-frequency identification (RFID). RFID technology, which offers flow control capabilities, is widely used in food SCM to monitor products at various stages of production [17]. Its appeal lies in its relative affordability and widespread adoption. However, RFID's limited memory capacity can be a drawback [18]. To improve comprehension of AI's function in SCM, this study tackles two main research inquiries: What possible uses for AI exist in SCM and operations? What are the potential effects of implementing AI on business? By exploring these inquiries, the paper aims to identify the core technologies driving AI in the business planning phase of SCM, the value AI adds to the SC field, the infrastructure required for AI implementation, the process of training AI models to achieve desired outcomes, and the types of data that facilitate AI's effectiveness. The key objective is to provide a comprehensive overview of AI's role in SCM. This study investigates the revolutionary effects of AI on SCM through a comprehensive review of existing literature. It delves into AI concepts and their application in SCM, including demand forecasting, inventory management, and logistics optimization, from various business perspectives, and discusses the necessary infrastructure for AI deployment.

2 Theoretical Background

2.1 Conceptualization of AI

AI, initially conceptualized by John McCarthy in the 1950s during the era of the Turing test, is a broadly defined field. A widely accepted definition of AI is "the capacity of a computer or robot under computer control to carry out actions often linked with intelligent beings". Since the early 2000s, AI has garnered substantial support and interest, combining scientific and engineering goals to create intelligent machines that address challenges in both domains. Consequently, AI has become a focal point for numerous research projects [19]. Russell and Norvig (2016) state AI as "the scientific investigation and creation of intelligent agents designed to comprehend their surroundings" [16].

The scientific aspect of AI seeks to understand human thought processes, while the engineering aspect focuses on creating systems that exhibit intelligent behaviour. Min (2010) argues that AI involves computers learning from large datasets, identifying patterns, and arriving at appropriate solutions [20]. AI primarily concerns data processing, environmental perception, and intelligent action through algorithms [16]. Modern AI relies on cutting-edge technology, vast amounts of data, and advanced statistical methods to achieve optimal results. AI encompasses several subfields, including ML, machine vision, natural language processing (NLP), and robotics, which collectively aim to achieve different aspects of intelligence and often overlap and support one another [15]. ML is widely used for predictive analytics, helping with demand forecasting, inventory management, and logistics optimization by learning and adapting over time. Despite its popularity and accessibility, AI faces significant barriers to adoption in businesses [21–23]. Table 1 delves into the subfields of AI and their contributions to achieving intelligent computing.

2.2 *AI-Driven SCM*

An SC frequently involves production, transformation, and logistics companies all over the world [41]. The SC is a comprehensive system that includes products, organizations, people, resources, and services [39, 42, 43]. SCM solutions facilitate critical flows within and between organizations. Several AI technologies are highly effective in optimizing inventory management and logistics within SCs. AI-driven SCM systems can autonomously make decisions to achieve SCM objectives and operate in partially unknown SC environments [31, 44]. Soleimani (2018) [45] highlights optimization, forecasting, modelling, and simulation as the four core attributes through which AI enhances SC performance. Roden et al. [46] highlight six areas where big data and ML are applied in operations management [47, 48], relevant to AI including Forecasting, Inventory Control, Sales Management and Marketing, Transportation, SCM, and Risk Assessment [49, 50].

2.3 *AI in Demand Forecasting*

AI enhances demand forecasting in SCM by utilizing advanced machine learning algorithms that analyse vast amounts of historical data, market trends, and external

Table 1 Subset of AI

Technology used	Definition	Features	Reference
Machine learning (ML)	The procedure of finding relevant groupings within data or fitting models for predictions to data	Substantial improvement in performance with minimal human input	[24–26]
Artificial neural network (ANN)	ANN is a computing approach that creates many processing units using interlinked connections	Helps in consumer behaviour prediction, and understanding sophisticated buyer segments are applications of it	[27, 28]
Machine vision (MV)	They help in simulating human visual functions through a computer, to examine and comprehend the objective world	MV technology is commonly utilized for object recognition and image interpretation They are suitable for acquiring images, image processing, and pattern identification	[29, 30]
Expert system	Computer programs that simulate the knowledge and experience of domain experts to solve complex problems	It has both a rule system, where a set of defined rules are used to govern as well as a decision system	[21, 31–33]
Natural language processing (NLP)	The goal of the AI and linguistics tract is to enable systems to understand words or sentences recorded in human languages	Two types of NLP: Statistical NLP Semantic NLP	[21, 34, 35]
Speech processing	Digital signal processing (DSP) methods to deliver speech-into-speech digital signals	It can provide an audio prompt to users for directing a task, while the user can respond to it verbally to reply	[36–38]
Robotics	Interdisciplinary field of science and technology that encompasses computer science, engineering in electronic devices	Has two parts: presentation layer and technical layer	[21, 39]
Evolutionary computing	The techniques of EC typically comprise genetic algorithm, ant colony optimization (ACO), swarm intelligence, and neuro-evolution	The tendency of the rule of natural selection is displayed where the strongest survive and reproduce	[40]

factors such as weather, holidays, or economic shifts. AI also continuously refines its models as new data becomes available, ensuring forecasts remain up to date and adaptable to changing market conditions. This dynamic approach improves inventory management, production planning, and customer satisfaction by minimizing stock-outs and overstocking. AI advances the accuracy of demand forecasts in SCM through sophisticated methods in ML, DL, and predictive analytics. These systems look at vast, complex datasets and analyse them by looking for patterns that cannot be seen with ordinary forecasting methods. AI systems can integrate and analyse data from diverse sources, including market trends, historical sales, and external factors like weather or economic shifts. This capability leads to forecasts that dynamically adjust to real-time market conditions, ensuring relevance and precision [51]. Table 2 shows the role of AI in business.

3 Creating and Employing an AI Model in SCM

Deploying AI in SCM requires a robust technological infrastructure. This includes having sufficient data storage capacity, high-speed processing power, and reliable networks to handle the vast amounts of data generated by SC operations. First, a robust data infrastructure is essential. This includes centralized data lakes and analytics platforms that can store and process large volumes of real-time data. These platforms enable the use of AI models for predictive analytics and decision-making. Cloud computing is another vital aspect of AI deployment, as it provides the necessary scalability to support computationally intensive AI tasks, such as training models or running algorithms at scale [60]. Lastly, integration platforms are required to ensure that AI systems work seamlessly with existing enterprise resource planning (ERP) and SCM software, facilitating smooth operations across the entire ecosystem [61].

3.1 AI Training

Recent research works in the domain of ML and forecasting on the training of NN models have revealed that AI can execute a different type of intelligent business task: currently, there are several different implemented forecasting methods able to identify credit risk [52], reduce the bullwhip effect, predict the inventory level [62], and forecast customer demand [38]. Cavalcante et al. (2019) looked at supplier selection through supplier risk profiles [62]. Both Goli et al. (2019) and Baryannis Dani et al. (2019) showed how predictions could be made and risks managed in product portfolios and SCs [3, 31].

Table 2 Roles of AI in business

Business partner	Business purpose	Use of AI in supply chain	AI method	Reference
Supplier	Supplier selection	By forecasting demand, AI mitigates major risks and helps optimize businesses	Machine learning (ML) expert systems.	[50–52]
Manufacturer	Order intake	Helping in order intake Chatbots help in customer support areas, which help improve relationships with customers.	Expert system natural language processing	[33, 53]
	Quality checking	Inventory control automated inspection	Expert system; ANN, ML	[33, 54]
	Production planning and control Digital-enabled products	Production optimization and improvement in production capacity Observe and forecast the state of the bearing.	ML neural networks (NN); NLP DL, reinforcement learning	[30] [54]
Logistics	Transportation Warehouse	Precise demand prediction (lessen the bullwhip effect) Postprocessing surveillance Automobile navigation and planning Automatic container unloading AI helpers	NN MV Evolutionary computation Speech processing, NN ML, Robotics, speech processing	[37, 55] [32, 56, 57]
Retailer	Sales process	Sentiment analysis Sales process optimization	ML ML	[58]
Customer	Customer interfaces	Chatbots, which help in customer support areas.	NLP	[59]

3.2 Data Gathering and Training

There are two major activities performed in this phase. The first task is to collect data from commercial sensors, real-time IoT devices, and economic transaction-driven networks for model training [61]. Also gathered are unstructured data such as text and documents [61]. Preprocessing the raw information is the second task. It is essential to map the data according to business model knowledge.

3.3 Model Training

It is at this phase that actual learning takes place. Machines can extract knowledge through repetitive learning, hence making accurate forecasts from data, no matter how vast it is [50–52]. Several different training algorithms are then used for a training dataset, which refers to a portion of the whole data on which learning takes place. The training goal is to create a reliable mapping relationship using information from the data and current algorithms. The training datasets ought to include all of the characteristics required by the model and display low noise [62].

3.4 Runtime-Model

The last step is to put the framework into practice in an actual company context. The test dataset is where the model's accuracy is tested on its performance [53]. However, it is entirely distinct from the training set, and simultaneously, the test set maintains an identical probability distribution. Revalidation of the model should be done in an updated environment, and accordingly, modifications to the model must be made using the outcomes and feedback [61]. Properly trained models can make predictions on their own and arrive at decisions during actual business activities. All steps from steps 1 to 3 are iterative. Initially, substantial manual labour is required for the set-up and awareness of the business. In successive cycles, the computers automatically perform the tasks, interfacing with the physical data sources of IoT and the cyber network [63]. The frequency of these cycles is estimated by company demands as well as the performance of the machine. Table 3 illustrates this:

4 Technological Infrastructure

We can note that the Technological Infrastructure is mainly comprised of two parts, namely Data Collection Infrastructure and Model Training Infrastructure. The Model Training Infrastructure is the other component of the technological framework, which focuses on CC, big data analytics (BDA), and ML. These are interfaced with the Data Collection Infrastructure for decentralized and protected extensive data examination of the gathered data. The AI-enabled data centres have to work on cloud-based servers. These are sensor and computational systems that store and handle massive and huge amounts of heterogeneous and diversified information at high-speed processing. This requirement has, for the most part, been the reason

Table 3 AI in value creation

Sl no.	Value	Contribution	Example	Reference
1.	Demand forecast	AI can automatically process and analyse data, enabling accurate prediction and reliable forecasting of demand	"DeepMind" was developed by Google in collaboration with National Grid in the UK to predict the stock while considering facts such as even the weather	[52, 64]
2.	Production	AI can help in production phases by: (a) Assets optimization and streamlining of processes (b) Better team selection (c) Error-free (d) Maintenance times are lowered	Semiconductor companies are utilizing AI engines to discover the root reasons of production losses, which can be eliminated by altering their manufacturing procedures	[9, 39, 65]
3.	Delivery	AI algorithms analyse real-time data such as traffic circumstances, weather predictions, and road closures	Google collaborated with Chipotle to serve burritos at Virginia Tech	[64]

why BDA has proliferated. Besides, more often than not, for successful machine learning, there is a demand for training data. Continual learning through the training data improves ML techniques and becomes a source of competitive advantage. Davenport (2018) argues that knowledge about the relevance of the training data and incremental learning in AI is much more beneficial in AI than mere mastery of knowledge of AI technology [19]. The application of AI in SCM presents several challenges that organizations need to address for successful implementation.

5 Conclusion and Future Scope

Advances in processing power have considerably surpassed the expansion and complexity of AI applications in recent times. Building on this advancement, our study sought to elucidate AI's contributions to SCM through an extensive literature review. Our research indicates that among the various AI practices, some are more prevalently utilized than others. The most frequently employed AI method is ANNs, which are adept at identifying complex patterns often beyond human capability. These can be applied to numerous problem-solving categories such as pattern recognition, calculation, optimization, clustering, feature forecasting, content fetching, and process management. The next popular approach is fuzzy logic (FL), a type of logic that deals with degrees of truth rather than the binary true or false. FL is widely utilized in the development of hybrid intelligent systems and is a prominent modelling tool. Our findings also underscore several AI techniques that require further exploration and commercial implementation, such as NLP for human–machine interactions, task scheduling (TS) for optimization and robotic dynamics, and Markov decision processes (MDP) for modelling decision-making strategies. The networked characteristics of SCM and logistics present an intuitive structure for AI application. An extensive system of suppliers creates vast quantities of data and necessitates agile decision-making. Hence, employing AI methods for BDA and decision support systems (DSSs) is strongly suggested. SCM organizations depend on physical and digital networks that must operate seamlessly amid huge volumes, lean asset allocation, low margins, and tight deadlines. AI optimizes and enhances network orchestration effectively, often beyond human capability.

Investigation on dynamic decision-making systems can improve the capabilities of AI solutions and expand our knowledge of them. These systems enable AI to shift SCM processes from reactive to proactive, support procedures to become autonomous, transform standardized services into customized ones, and evolve production planning to prediction. In conclusion, AI plays a crucial role in advancing SCM by leveraging cutting-edge computing power and innovative techniques. The ongoing development and integration of AI technologies in SCM promise to revolutionize business procedures, rendering them more intelligent, responsive, and effective. Future research should focus on exploring underutilized AI methods and their potential applications, further propelling the field of SCM.

References

1. Z. Seyedghorban, H. Tahernejad, R. Meriton, G. Graham, Supply chain digitalization: Past, present and future. Prod. Plan. Control **31**(2–3), 96–114 (2020). https://doi.org/10.1080/09537287.2019.1631461
2. I. Tammela, A.G. Canen, P. Helo, Time-based competition and multiculturalism: A comparative approach to the Brazilian, Danish and Finnish furniture industries. Manag. Decis. **46**(3), 349–364 (2008). https://doi.org/10.1108/00251740810863834
3. A. Haas, Logistics and supply chain intelligence, in *Integration of Information Flow for Greening Supply Chain Management*, ed. by A. Kolinski, D. Dujak, P. Golinska-Dawson, (Springer, Cham, 2020), pp. 111–129
4. R. Dash, M. McMurtrey, C. Rebman, U.K. Kar, Application of artificial intelligence in automation of supply chain management. J. Strateg. Innov. Sustain. **14**, 3 (2019) Retrieved from https://articlearchives.co/index.php/JSIS/article/view/4867
5. J. McCarthy, M.L. Minsky, N. Rochester, C.E. Shannon, A proposal for the Dartmouth summer research project on artificial intelligence. AI Mag. **27**, 4 (2006)
6. R.J. Solomonoff, The time scale of artificial intelligence: Reflections on social effects. Hum. Syst. Manag. **5**(2), 149–153 (1985)
7. U.K. Kar, R. Dash, The future of health and healthcare in a world of artificial intelligence. Arch. Biomed. Eng. Biotechnol. (2018)
8. G. Zhang, M.Y. Hu, B.E. Patuwo, D.C. Indro, Artificial neural networks in bankruptcy prediction: General framework and cross-validation analysis. Eur. J. Oper. Res. **116**(1), 16–32 (1999)
9. J. Bughin, E. Hzan, S. Ramaswamy, M. Chui, et al., *Artificial Intelligence: The Next Digital Frontier?* (McKinsey Global Institute, 2017)

10. D. Perez, S. Samothrakis, S. Lucas, Knowledge-based fast evolutionary MCTS for general video game playing, in *IEEE Conference*, (2014) 978-1-4799-3547-5
11. D. Silver, A. Huang, C.J. Maddison, A. Guez, et al., Mastering the game of go with deep neural networks and tree search. Nature **529**(7587), 484–489 (2016)
12. A. Esteva, B. Kuprel, R.A. Novoa, J. Ko, S.M. Swetter, et al., Dermatologist-level classification of skin cancer with deep neural networks. Nature **542**(7639), 115–118 (2017, February 2)
13. D.W. Hall, J. Pesenti, Growing the artificial intelligence industry in the UK. Department for Digital, Culture, Media & Sport and Department for Business, Energy & Industrial Strategy. Part of the Industrial Strategy UK and the Commonwealth (2017)
14. R. Dubey, A. Gunasekaran, S.J. Childe, D.J. Bryde, M. Giannakis, C. Foropon, B.T. Hazen, Big data analytics and artificial intelligence pathway to operational performance under the effects of entrepreneurial orientation and environmental dynamism: A study of manufacturing Organisations. Int. J. Prod. Econ. **226**, 107599 (2019)
15. Y. Duan, J.S. Edwards, Y.K. Dwivedi, Artificial Intelligence for Decision Making in the Era of Big Data–Evolution, Challenges and Research Agenda. Int. J. Inf. Manag. **48**, 63–71 (2019). https://doi.org/10.1016/j.ijinfomgt.2019.01.021
16. S.J. Russell, P. Norvig, *Artificial Intelligence: A Modern Approach* (Pearson Education Limited, Malaysia, 2016)
17. C. Costa, F. Antonucci, F. Pallottino, J. Aguzzi, D. Sarriá, P. Menesatti, A review on agri-food supply chain traceability by means of RFID technology. Food Bioprocess Technol. **6**, 353–366 (2013)
18. T.P. da Costa, J. Gillespie, X. Cama-Moncunill, S. Ward, J. Condell, R. Ramanathan, F. Murphy, A systematic review of real-time monitoring technologies and its potential application to reduce food loss and waste: Key elements of food supply chains and IoT technologies. Sustainability **15**, 614 (2023). https://doi.org/10.3390/su15010614
19. Petri Helo, Yuqiuge Hao, Artificial intelligence in operations management and supply chain management: an exploratory case study. Prod. Plan. Control. **33**(16), 1573–1590 (2022). https://doi.org/10.1080/09537287.2021.1882690
20. H. Min, Artificial intelligence in supply chain management: theory and applications. Int. J. Log. Res. Appl. **13**(1), 13–39 (2010). https://doi.org/10.1080/13675560902736537
21. T.H. Davenport, R. Ronanki, Artificial intelligence for the real world. Harv. Bus. Rev. **96**(1), 108–116 (2018)
22. S. Ransbotham, P. Gerbert, M. Reeves, D. Kiron, M. Spira, Artificial intelligence in business gets real. MIT Sloan Manag. Rev. (2018) https://sloanreview.mit.edu/projects/artificial-intelligence-inbusiness-gets-real/
23. M. Chui, S. Malhotra, *AI Adoption Advances, but Foundational Barriers Remain* (Mckinsey & Company, New York, 2018)
24. J.G. Greener, S.M. Kandathil, L. Moffat, et al., A guide to machine learning for biologists. Nat. Rev. Mol. Cell Biol. **23**, 40–55 (2022). https://doi.org/10.1038/s41580-021-00407-0
25. T. Paul, R. Bommu, Strategic employee performance analysis in the USA: Leveraging intelligent machine learning algorithms. Unique Endeavor Bus. Soc. Sci. **3**(1), 113–124 (2024)
26. J. Alanya-Arango, J. Alanya-Beltran, S.K. Ravichandran, B. Singh, A. Sasi, D. Gangodkar, Internet of Things and Machine Learning Based Intelligent Irrigation System for Agriculture, in *2022 5th International Conference on Contemporary Computing and Informatics (IC3I), Uttar Pradesh, India*, (2022), pp. 67–72. https://doi.org/10.1109/IC3I56241.2022.10073254
27. R. Dastres, M. Soori, Artificial neural network systems. Int. J. Imaging Robot. **21**(2), 13–25 (2021) ⟨hal-03349542⟩
28. S.M. Ali, A.U. Rahman, G. Kabir, S.K. Paul, Artificial intelligence approach to predict supply chain performance: Implications for sustainability. Sustainability **16**, 2373 (2024). https://doi.org/10.3390/su16062373
29. P. Wu et al., Next-generation machine vision systems incorporating two-dimensional materials: progress and perspectives. InfoMat **4**(1), e12275 (2022)
30. M. Tzampazaki, C. Zografos, E. Vrochidou, G.A. Papakostas, Machine vision—Moving from industry 4.0 to industry 5.0. Appl. Sci. **14**, 1471 (2024). https://doi.org/10.3390/app14041471

31. G. Baryannis, S. Validi, S. Dani, G. Antoniou, Supply chain risk management and artificial intelligence: state of the art and future research directions. Int. J. Prod. Res. **57**(7), 2179–2202 (2019). https://doi.org/10.1080/00207543.2018.1530476
32. M.H. Jarrahi, Artificial intelligence and the future of work: Human-AI symbiosis in organizational decision making. Bus. Horiz. **61**(4), 577–586 (2018). https://doi.org/10.1016/j.bushor.2018.03.007
33. X. Yang, C. Zhu, Industrial expert systems review: A comprehensive analysis of typical applications. IEEE Access **12**, 88558–88584 (2024). https://doi.org/10.1109/ACCESS.2024.3419047
34. D. Khurana, A. Koli, K. Khatter, et al., Natural language processing: state of the art, current trends and challenges. Multimed. Tools Appl. **82**, 3713–3744 (2023). https://doi.org/10.1007/s11042-022-13428-4
35. A. Manoharan, G. Nagar, Maximizing learning trajectories: an investigation into AI-driven natural language processing integration in online educational platforms. Int. Res. J. Mod. Eng. Technol. Sci. **3**, 1–10 (2021)
36. T. Fu, B. Sun, Application of speech recognition Technology in Logistics Selection System, in *International Conference on Human Centered Computing*, (Springer, Cham, 2017), pp. 654–659
37. F. Levy, Computers and populism: Artificial intelligence, jobs, and politics in the near term. Oxf. Rev. Econ. Policy **34**(3), 393–417 (2018). https://doi.org/10.1093/oxrep/gry004
38. B.O. Kang, H.-B. Jeon, Y.K. Lee, AI-based language tutoring systems with end-to-end automatic speech recognition and proficiency evaluation. ETRI J. **46**(1), 48–58 (2024)
39. M.R. Mia, J. Shuford, Exploring the synergy of artificial intelligence and robotics in industry 4.0 applications. J. Artif. Intell. General Sci.. ISSN:3006-4023 **1**, 1 (2024). https://doi.org/10.60087/jaigs.v1i1.31
40. X. Wu et al., Evolutionary computation in the era of large language model: Survey and roadmap. *arXiv preprint arXiv:2401.10034* (2024)
41. H. Zijm, M. Klumpp, Logistics and supply chain management: Developments and trends, in *Logistics and Supply Chain Innovation*, ed. by H. Zijm, M. Klumpp, U. Clausen, (Springer, Cham, 2016), pp. 1–20
42. Q. Li, A. Liu, Big data driven supply chain management. Procedia CIRP **81**, 1089–1094 (2019). https://doi.org/10.1016/j.procir.2019.03.258
43. N. Stefanovic, D. Stefanovic, Supply chain business intelligence: technologies, issues and trends, in *Artificial Intelligence: An International Perspective*, (Springer, Berlin, Heidelberg, 2009), pp. 217–245
44. E.E. Kosasih, E. Papadakis, G. Baryannis, A. Brintrup, A review of explainable artificial intelligence in supply chain management using neurosymbolic approaches. Int. J. Prod. Res. **62**(4), 1510–1540 (2023). https://doi.org/10.1080/00207543.2023.2281663
45. S. Soleimani, A perfect triangle with: artificial intelligence, supply chain management, and financial technology. Arch. Bus. Res. **6**(11), 5681 (2018). https://doi.org/10.14738/abr.611.5681
46. S. Roden, A. Nucciarelli, F. Li, G. Graham, Big data and the transformation of operations models: A framework and a new research agenda. Prod. Plan. Control **28**(11–12), 929–944 (2017). https://doi.org/10.1080/09537287.2017.1336792
47. K. Lamba, S.P. Singh, Big data in operations and supply chain management: Current trends and future perspectives. Prod. Plan. Control **28**(11–12), 877–890 (2017). https://doi.org/10.1080/09537287.2017.1336787
48. A. Kolinski, A. Horzela, M. Cudzilo, R. Domanski, Reference model of information flow in business relations with 4pl operator, in *Integration of Information Flow for Greening Supply Chain Management*, ed. by A. Kolinski, D. Dujak, P. Golinska-Dawson, (Springer, Cham, 2020), pp. 19–45
49. M. Korolov, AI in the Supply Chain: Logistics Gets Smart, CIO. *Supply Chain Management* (2018). Available at: https://www.cio.com/article/3269513/ai-in-the-supply-chain-logistics-get-smart.html

50. Y. Zhu, L. Zhou, C. Xie, G.J. Wang, T.V. Nguyen, Forecasting SMEs' credit risk in supply chain finance with an enhanced hybrid ensemble machine learning approach. Int. J. Prod. Econ. **211**, 22–33 (2019). https://doi.org/10.1016/j.ijpe.2019.01.032
51. C. Wu, D. Barnes, Partner selection in agile supply Chains: A fuzzy intelligent approach. Prod. Plan. Control **25**(10), 821–839 (2014). https://doi.org/10.1080/09537287.2013.766037
52. O.A. Elufioye, C.U. Ike, O. Odeyemi, F.O. Usman, N.Z. Mhlongo, AI-driven predictive analytics in agricultural supply chains: A review: Assessing the benefits and challenges of AI in forecasting demand and optimizing supply in agriculture. Comp. Sci. IT Res. J. **5**(2), 473–497 (2024). https://doi.org/10.51594/csitrj.v5i2.817
53. W.J. Lee, H. Wu, H. Yun, H. Kim, M.B. Jun, J.W. Sutherland, Predictive maintenance of machine tool systems using artificial intelligence techniques applied to machine condition data. Procedia CIRP **80**, 506–511 (2019). https://doi.org/10.1016/j.procir.2018.12.019
54. N. Rezki, M. Mansouri, Improving supply chain risk assessment with artificial neural network predictions. Acta Logist. **10**(4), 645–658 (2023)
55. L.P. Singh, R.T. Challa, Integrated forecasting using the discrete wavelet theory and artificial intelligence techniques to reduce the bullwhip effect in a supply chain. Glob. J. Flex. Syst. Manag. **17**(2), 157–169 (2016). https://doi.org/10.1007/s40171-0150115-z
56. K. De Backer, T. DeStefano, Robotics and the global organisation of production, in *Robotics, AI, and Humanity: Science, Ethics, and Policy*, (2021), pp. 71–84
57. S.K. Ravichandran, A. Sasi, Effective storage of goods in a warehouse using farm optimisation algorithm. International Journal of Cloud Computing (IJCC) **9**, 2/3 (2020)
58. A. Muchairi, *Business Process Reengineering for Process Optimization: a Case Study* (University of Johannesburg, Diss, 2022)
59. T. Lalwani, S. Bhalotia, A. Pal, V. Rathod, S. Bisen, Implementation of a Chatbot System using AI and NLP (May 31, 2018). Int. J. Innov. Res. Comput. Sci. Technol. **6**, 3 (2018), Available at SSRN: https://ssrn.com/abstract=3531782 or). https://doi.org/10.2139/ssrn.3531782
60. E. Brynjolfsson, A. McAfee, Artificial intelligence, for real. Harv. Bus. Rev. (2017) https://hbr.org/2018/01/artificial-intelligencefor-the-real-world
61. Q. Min, Y. Lu, Z. Liu, C. Su, B. Wang, Machine learning based digital twin framework for production optimization in petrochemical industry. Int. J. Inf. Manag. **49**, 502–519 (2019). https://doi.org/10.1016/j.ijinfomgt.2019.05.020
62. I.M. Cavalcante, E.M. Frazzon, F.A. Forcellini, D. Ivanov, A supervised machine learning approach to data-driven simulation of resilient supplier selection in digital manufacturing. Int. J. Inf. Manag. **49**, 86–97 (2019). https://doi.org/10.1016/j.ijinfomgt.2019.03.004
63. S.K. Singh, S. Rathore, J.H. Park, Block IoT intelligence: A Blockchain-enabled intelligent IoT architecture with artificial intelligence. Futur. Gener. Comput. Syst. **110**, 721–743 (2020). https://doi.org/10.1016/j.future.2019.09.002
64. A. Sasi, T. Subramanian, Forecasting stochastic consumer portability visitation pattern in fair price shops of India. J. Inf. Optimiz. Sci. **44**(3), 439–454 (2023)
65. A. Sasi, T. Subramanian, S.K. Ravichandran, Systematic literature review on industry revolution 4.0 to enhance supply chain operation performance, in *Proceedings of the 2022 5th International Conference on Computers in Management and Business*, (2022), pp. 173–179

Enhancing Data Warehousing Efficiency and Decision-Making Through Fuzzy Dimensional Hierarchies

P. P. G. Dinesh Asanka and Amal S. Perera

1 Introduction

Dimensions in data warehousing are essential for performing detailed analysis across various industries, including telecommunications, education [1], logistics and transportation, financial [2], retail, health care, insurance, and e-commerce [3]. These dimensions often incorporate hierarchies such as Product Category → Product Subcategory → Product, enabling more sophisticated descriptive and diagnostic analytics. However, these hierarchies are typically crisp, which diverges from complex, real-world scenarios where a single attribute may be associated with multiple higher-level categories, each with varying weightages. For instance, a single product might belong to multiple subcategories. When analyzing with crisp values, descriptive and dignostics analysis remains far from the real-world context.

To accommodate this complexity, the implementation of fuzzy hierarchies is necessary. Fuzzy logic allows for more flexible relationships within the data, reflecting the nuances of real-world classifications. However, defining fuzzy membership functions, which are critical to this process, requires more data-driven and scientific techniques.

This research paper introduces a novel approach for implementing fuzzy hierarchies within data warehouses. By incorporating fuzzy logic into these dimensions, it becomes possible to create a more accurate representation of the relationships

P. P. G. Dinesh Asanka (✉)
Department of Industrial Management, University of Kelaniya, Kelaniya, Sri Lanka
e-mail: dasanka@kln.ac.lk

A. S. Perera
Department of Computer Science and Engineering, University of Moratuwa, Moratuwa, Sri Lanka
e-mail: amal.perera@uconn.edu

A. Haldorai et al. (eds.), *7th EAI International Conference on Big Data Innovation for Sustainable Cognitive Computing*, EAI/Springer Innovations in Communication and Computing, https://doi.org/10.1007/978-3-032-02979-9_12

between different levels of data, thereby enhancing the capability for detailed analysis.

2 Identifying the Research Objectives

This research covers the areas of data warehousing, dimensional hierarchies, and fuzzy logic. Hence, a literature review was carried out in those domains.

A data warehouse is essential for modern businesses because it serves as a centralized repository that aggregates data from various operational databases and other sources, transforming it into structures optimized for business analysis and decision support [4]. The evolution of data collection and database systems, from hierarchical and relational databases to the development of SQL, has led to the creation of data warehouses to manage the increasing volume and complexity of data [5]. Data warehouses facilitate the extraction, transformation, and loading (ETL) of data, enabling the integration of disparate datasets into a cohesive, multidimensional data store that supports comprehensive analysis [6]. This integration is crucial for companies with branches in different locations, as it consolidates historical data, which is vital for making informed decisions [7]. The primary role of a data warehouse is to enhance the efficiency of data querying and support long-term analysis and prediction, overcoming the limitations of traditional database systems that struggle with large volumes of data and complex queries [8]. By utilizing technologies such as online analytical processing (OLAP), data warehouses enable businesses to perform in-depth analyses, identify trends, and make strategic decisions to improve their operations and profitability. Overall, the implementation of a data warehouse transforms raw data into valuable insights, thereby expanding the scope, accuracy, and accessibility of data for decision-making processes [5, 6].

Dimensional hierarchies significantly enhance data analysis by providing structured, multilevel perspectives that facilitate the understanding and interpretation of complex datasets. These hierarchies allow for data organization into nested levels, which can be crucial for identifying patterns and relationships that might not be apparent in flat data structures.

In the context of temporal event data, dynamic hierarchical dimension aggregation allows for the interactive exploration of event groupings, enabling analysts to select the most informative grouping levels for their specific analysis needs [9]. Additionally, generating organized hierarchies from related datasets using probabilistic models can streamline the process of building and understanding complex data structures [10]. Hierarchically ordered groups of variables are also essential in high-dimensional linear models, where they help in detecting significant groups of correlated variables, thus enhancing the explanatory power of the models [11]. User-defined hierarchical relationships in multidimensional data models allow for customizable and client-specific data analysis, which can be stored and modified as part of report metadata [12]. Furthermore, the use of tree tables and pivot grids to replicate hierarchical properties facilitates the dynamic updating and transformation of displayed data, making the analysis more flexible and responsive to new information [13]. Microsoft SQL Server Analysis Service [14] has Natural, Unbalanced, Range, and Parent–Child hierarchies to improve the data analysis [15].

A fuzzy data warehouse is needed to handle imprecise, uncertain, or ambiguous data, enabling more flexible and nuanced analysis that can better reflect real-world complexities [16]. Implementing a fuzzy data warehouse (FDW) offers numerous benefits, particularly in enhancing the accuracy and interpretability of data analysis across various domains. One of the primary advantages is the ability to handle imprecise and ambiguous data, which is often encountered in real-world scenarios. Traditional data warehouses rely on crisp analysis, which assumes that every situation belongs to a single state, thereby limiting the scope of analysis and potentially leading to less accurate outcomes [17]. In contrast, a fuzzy data warehouse incorporates fuzzy logic, allowing for the storage and operation of fuzzy measures, relationships, and levels, which can provide a more nuanced and accurate representation of data [18]. This is particularly beneficial in educational organizations, where integrating large amounts of diverse data from different sources can be challenging [18]. Moreover, fuzzy data warehouses facilitate linguistic analytics, enabling the interpretation of numeric values in meaningful, nonnumeric terms. This is crucial for business users who may find numeric data difficult to interpret correctly [19]. By using fuzzy classification-based approaches, FDWs ensure a smooth transition between classes, thereby providing a more accurate interpretation of measures [20]. This capability is exemplified in a case study involving a movie rental company, where the application of fuzzy concepts significantly improved data analysis and performance measurement. Additionally, the integration of fuzzy logic does not affect the core structure of the existing classical data warehouse, allowing for simultaneous fuzzy and crisp analysis [21]. The flexibility of fuzzy data warehouses is further enhanced by the use of various fuzzy membership functions, such as arbitrary, data-driven, linguistic, derived, and survey-based functions, which can be tailored to different cases and requirements [22]. This adaptability makes FDWs suitable for a wide range of applications, from web analytics systems to business decision-making processes [23]. Furthermore, the development of specialized tools like the FDW Browser, which adheres to the Fuzzy-OLAP class of data exploration tools, underscores the practical utility and functional features of FDWs in modern analytical environments [24]. In summary, the implementation of a fuzzy data warehouse offers significant benefits, including improved accuracy and interpretability of data, the ability to handle imprecise and ambiguous information, and enhanced flexibility through various fuzzy membership functions. These advantages make FDWs a valuable tool for organizations seeking to gain a competitive edge through more precise and meaningful data analysis [25].

Fuzzy hierarchies in data warehouses represent an advanced approach to managing and analyzing data that incorporates the inherent imprecision and uncertainty of real-world information. The concept of a fuzzy data warehouse involves several key tasks, including the identification of target attributes, linguistic terms, and the definition of membership functions, which collectively enable the integration of fuzzy logic into the data warehouse without altering its base structure [25]. This approach is particularly useful in handling inaccuracy and uncertainties in digital data, as it allows for the design of a fuzzy data warehouse from a traditional one by defining fuzzy classification tables and membership degrees [26]. Moreover, the introduction of the if-cube model in an OLAP-based environment demonstrates how fuzzy hierarchies can operate over data with imprecision in both facts and dimensional hierarchies, utilizing the closure of an intuitionistic fuzzy set over a hierarchical structure to guide decision-making processes [27]. Additionally, data generalization on fuzzy information can be achieved through attribute induction using a fuzzy hierarchy, which further enhances the ability to effectively manage and analyze fuzzy data [28]. Collectively, these methodologies underscore the importance of fuzzy hierarchies in data warehouses, as they provide a robust framework for dealing with the complexities of imprecise data, thereby facilitating more accurate and insightful data analysis and decision-making.

A comprehensive review of existing literature on fuzzy data warehousing and hierarchical structures reveals a significant gap in integrating fuzzy hierarchies within data warehouse environments. This research aims to bridge this gap by proposing a novel framework for incorporating fuzzy hierarchies into data warehouses. To achieve the above objective, the following research questions will be addressed:

1. How can fuzzy logic be effectively integrated into traditional dimensional hierarchies to handle data ambiguity and uncertainty in data warehouses?
2. What are the specific challenges in applying fuzzy logic to dimensional hierarchies, and how can these challenges be addressed?
3. How can fuzzy dimensional hierarchies improve the accuracy and efficiency of data analysis compared to traditional crisp hierarchies?

3 Defining Fuzzy Hierarchies

In a fuzzy data warehouse, one level will be mapped to one or many hierarchal sublevels. Due to the many-to-many relationships that will be introduced to the fuzzy nature of the data warehouse dimension, a bridge table will be introduced as per the database normalization theory.

A critical aspect of this research involves the development of appropriate fuzzy membership functions to assign weights to distinct product categories accurately. To exemplify this challenge, coconut oil, a versatile product with multiple applications in Sri Lankan supermarkets, has been selected as a case study. Data on coconut oil purchases were collected through interviews with 130 customers who bought coconut oil. In this data collection process, the quantity of purchases is ignored, and only the reason for the purchases were recorded. The diverse utilization of coconut oil, encompassing culinary, cosmetic, religious, lubricant, and medicinal purposes, underscores the complexity of defining clear category boundaries and necessitates the application of fuzzy logic for nuanced classification. Suppose any descriptive analytics, such as product category-wise sales, or diagnostic analytics, such as the reason for sales differences, are performed using the crisp analysis. In that case, the analysis will be limited to the assumption that the customer has bought the coconut oil for one reason, which is incorrect.

In most research, weightages were considered subjectively, which is not data-driven. This research introduced subjective and data-driven methods such as the horizontal method, rank-based horizontal method, and pairwise comparison method [29].

3.1 *Subjective Method*

A straightforward approach to defining the fuzzy membership function for fuzzy categorization is through a subjective method. In this method, a domain expert assigns weightages to each category based on their judgment. For example, consider the categorization of coconut oil, where the weightages for the categories "Cooking," "Cosmetic," and "Other" are assigned as 0.6, 0.3, and 0.1, respectively, as shown in Eq. 1.

$$\text{Coconut Oil Categorization} = \frac{0.6}{\text{Cooking}} + \frac{0.3}{\text{Cosmetic}} + \frac{0.1}{\text{Other}} \tag{1}$$

In this method, the domain expert frequently adjusts the weightages, as these values tend to change over time.

3.2 *Horizontal Method*

The horizontal method is a key technique for defining fuzzy membership functions. A survey was conducted to implement the Horizontal method. Coconut oil was chosen for this example, as preliminary research indicates that it falls into multiple product categories such as cooking, medicine, cosmetics, lubricant, and religious use within the Sri Lankan context. A total of 130 customers who purchased coconut oil in a semi-urban area of Sri Lanka during the given month were interviewed, and the results are presented in Table 1.

Table 1 Purchase intentions of customers on coconut oil

Category	Number of customers	Percentage
Cooking	75	57.69
Cosmetic	35	26.92
Religious	12	9.23
Lubricant	4	3.08
Medicine	4	3.08

Using the horizontal method, the fuzzy categorization membership function can be defined by rounding the above percentages to two decimal points, as illustrated in Eq. (2).

$$\text{Coconut Oil Categorization} = \frac{0.58}{\text{Cooking}} + \frac{0.27}{\text{Cosmetic}} + \frac{0.09}{\text{Religious}} + \frac{0.03}{\text{Lubricant}} + \frac{0.03}{\text{Medicine}} \tag{2}$$

The primary disadvantage of this technique is the potential presence of numerous categories with minimal contributions. To address the issue, a threshold can be established whereby all minor contributions are aggregated into a single category labeled as "Other." This scenario is illustrated in the Eq. 3.

$$\text{Coconut Oil Categorization} = \frac{0.58}{\text{Cooking}} + \frac{0.27}{\text{Cosmetic}} + \frac{0.09}{\text{Religious}} + \frac{0.06}{\text{Other}} \tag{3}$$

The main drawback of this technique is that it is time-consuming. Additionally, the method operates under the significant assumption that each customer purchases the product for a single purpose, which does not accurately reflect real-world behavior.

3.3 Rank-Based Horizontal Method

As discussed in Section A, a significant disadvantage of the previously outlined horizontal method is its reliance on the assumption that each customer purchases the product for a single purpose, which does not reflect actual consumer behavior. To address the above limitation, the Rank-Based Horizontal method is introduced.

In this method, customers can rank up to three categories in order of preference. The weightage assigned is as follows: a rank of one receives a weightage of three, a rank of two receives a weightage of two, and a rank of three receives a weightage of one. Customers are required to select at least one category while selecting the second and third options, which are optional. The ranked weightage is computed using Eq. 4:

$$\text{Ranked Weightage} = (3 \times \text{Rank1}) + (2 \times \text{Rank2}) + (1 \times \text{Rank3}) \tag{4}$$

The calculated ranked weightages for various product categories are presented in Table 2.

For instance, the cooking category receives a ranked weightage of 307, calculated as follows: $(3 \times 75) + (2 \times 35) + (1 \times 12)$. The percentage for this category is determined by dividing 307 by the total ranked weightage of 728.

Table 2 Ranked weightages for different product categories

Category	R1	R2	R3	Total	Ranked Total	Ratio
Cooking	75	35	12	122	307	0.42
Cosmetic	35	48	20	103	221	0.30
Religious	12	30	40	82	136	0.19
Lubricant	4	12	4	20	40	0.05
Medicine	4	3	6	13	24	0.03
Total	130	128	82	340	728	1.00

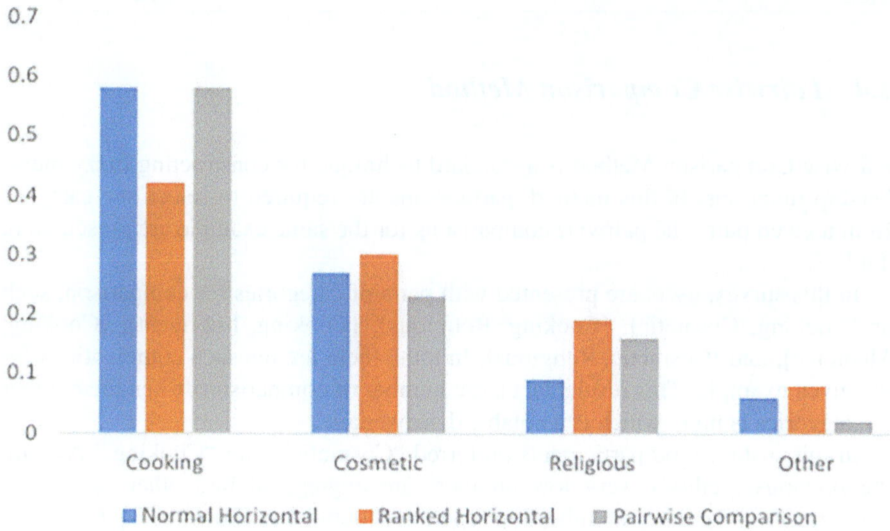

Fig. 1 Weightages for each category by different techniques

Using a similar approach to the standard horizontal method, a fuzzy categorization function is constructed for the Rank-Based Horizontal Method, as detailed below. This technique is then compared with the previous method using the example of coconut oil, as illustrated in Fig. 1.

Figure 1 demonstrates that the introduction of the Rank-Based Horizontal Method has not significantly altered the contribution numbers.

The fuzzy membership function derived from the Rank-Based Horizontal Method is presented in Eq. 5.

$$\text{Coconut Oil Categorization} = \frac{0.42}{\text{Cooking}} + \frac{0.30}{\text{Cosmetic}} + \frac{0.19}{\text{Religious}} + \frac{0.08}{\text{Other}} \tag{5}$$

Table 3 Pairwise comparison with each category

Category	Cooking	Cosmetic	Religious	Lubricant	Medicine	Total	Ratio
Cooking		101	102	112	116	431	0.58
Cosmetic	60		85	16	12	173	0.23
Religious	24	17		60	19	120	0.16
Lubricant	2	1	0		1	4	0.01
Medicine	3	7	2	5		17	0.02
Total						245	1.00

3.4 Pairwise Comparison Method

Pairwise Comparison Method is a standard technique for constructing fuzzy membership functions. In this method, participants are required to select one category from a given pair. The pairwise comparisons for the same example are presented in Table 3.

In this survey, users are presented with pairs of categories for comparison, such as {Cooking, Cosmetic}, {Cooking, Religious}, {Cooking, Lubricant}, {Cooking, Medicine}, and {Cosmetic, Religious}. In total, there are ten such combinations for the given example. This results in a large number of comparisons when the number of categories is high, which is a notable disadvantage.

In this instance, 60 participants preferred "Cosmetic" over "Cooking." As with the previous methods, very low numbers are aggregated into other categories. Consequently, the fuzzy membership function is defined as shown in Eq. 6.

$$\text{Coconut Oil Categorization} = \frac{0.58}{\text{Cooking}} + \frac{0.23}{\text{Cosmetic}} + \frac{0.16}{\text{Religious}} + \frac{0.03}{\text{Other}} \tag{6}$$

The primary disadvantage of this method is the need to gather a large amount of data, which can be impractical to collect from customers. Figure 1 illustrates the differences between the various methods used to determine the fuzzy membership function.

The above Fig. 1. indicates that the horizontal and pairwise methods are similar to the rank-based horizontal method for the given case.

4 Evaluating Fuzzy Hierarchies

The proposed method was evaluated at a supermarket in a semi-urban area of Sri Lanka using coconut oil as the experimental product. In the current operational practice, the organization categorizes coconut oil into three groups: Cooking, Cosmetic, and Other. However, the data-driven approach proposes four categories:

Table 4 Revenue comparison between the horizontal method and actual revenue

Category	Actual revenue	Revenue calculated from the horizontal method	Difference
Cooking	105,533	102,015	(3518)
Cosmetic	52,766	47,490	(5276)
Other	17,589	26,383	8794

Table 5 Revenue comparison between ranked method and actual revenue

Category	Actual revenue	Revenue calculated from ranked based	Difference
Cooking	105,533	73,873	(31,660)
Cosmetic	52,766	52,766	0
Other	17,589	49,248	31,660

Table 6 Revenue comparison between ranked method and actual revenue

Category	Actual revenue	Revenue calculated from pairwise	Difference
Cooking	105,533	102,015	(3518)
Cosmetic	52,766	40,454	(12,312)
Other	17,589	33,419	15,830

Cooking, Cosmetic, Religious, and Other. For comparison, the "Religious" category is aggregated into the "Other" category.

For the given month, total sales of coconut oil amounted to Rs. 175,888. The actual categorized weightages were 0.6 for Cooking, 0.3 for Cosmetic, and 0.1 for Other. The same sales amount was distributed among these categories using three methods: Normal Horizontal, Rank-Based, and Pairwise Comparison. Each method's results were compared with the current actuals, and the standard deviation was calculated to determine which method most closely aligns with the actual data.

Table 4 presents a revenue comparison between the Horizontal Method and the actual categorized data.

Table 5 presents the revenue comparison between the Rank-Based Method and the actual categorized data for the different categories.

Table 6 presents the revenue comparison between the Pairwise Method and the actual categorized data for the different categories.

As illustrated in Fig. 1, the Normal Horizontal Method is closer to the actual categorization of coconut oil for the given month. Similarly, this method can be employed to evaluate the most accurate categorization approach for each product. The pairwise comparison of these four techniques has computational complexity and needs more resources than other methods.

5 Concluding Remarks

This research has introduced several methods for defining membership functions to construct fuzzy hierarchies. These fuzzy hierarchies can be integrated into fuzzy data warehouses, enhancing data analysis capabilities. The research findings suggest that incorporating these detailed fuzzy hierarchies enables more precise and effective data analytics within data warehouses. This research introduced four fuzzy hierarchy membership function-defining methods: Subjective, horizontal, rank-based, and pairwise comparison. Depending on the situation, any data warehouse implementation can incorporate fuzzy hierarchies to improve the effectiveness of the data analysis.

Since this research focused on designing hierarchies for coconut products, the methodology should be extended to various other products to validate its effectiveness. Additionally, the use of interval-valued fuzzy sets can further refine the fuzzy membership functions. Given that data warehouse dimensions often involve slowly changing dimensions (SCD), it is crucial to explore how these fuzzy hierarchies can be adapted to accommodate Type 2 SCD.

References

1. C.U. Serasinghe, D.C.R. Jayakody, K.T.M.N. Dayananda, D. Asanka, Design and implementation of data warehouse for a higher educational Institute in Sri Lanka, in *2021 6th International Conference for Convergence in Technology (I2CT), Maharashtra, India*, (2021), pp. 1–5. https://doi.org/10.1109/I2CT51068.2021.9417820
2. W.R.A. Fonseka et al., Use of data warehousing to analyze customer complaint data of consumer financial protection Bureau of United States of America, in *2016 IEEE International Conference on Information and Automation for Sustainability (ICIAfS)*, (2016), pp. 1–6. https://doi.org/10.1109/ICIAFS.2016.7946520
3. R. Kimball, M. Ross, W. Thornthwaite, J. Mundy, B. Becker, *The Data Warehouse Lifecycle Toolkit: Practical Techniques for Building Data Warehouse and Business Intelligence Systems*, 2nd edn. (Wiley, 2008)
4. Data Warehouse Concepts (2022), pp. 45–74. https://doi.org/10.1007/978-3-662-65167-4_3
5. V. Haxhiu, Decision making based on data analyses using data warehouses. **6**(3), 1–6 (2017). https://doi.org/10.33107/IJBTE.2018.6.3.04
6. W.S. Fana, R. Sovia, R. Permana, A. Islam, Data warehouse design with ETL method (extract, transform, and load) for company information Centre. Int. J. Artif. Intell. **5**(2), 132–137 (2021). https://doi.org/10.29099/IJAIR.V5I2.215
7. A.F. Alasta, M.A. Enaba, Data warehouse on manpower employment for decision support system. *arXiv: Databases* (2019), [Online]. Available: https://arxiv.org/pdf/1904.01134
8. G. Ma, Research on the data warehouse in the design process of the database application system (2015), pp. 50–54. doi: https://doi.org/10.2991/ICCMCEE-15.2015.11
9. D. Gotz, J. Zhang, W. Wang, J. Shrestha, D. Borland, Visual analysis of high-dimensional event sequence data via dynamic hierarchical aggregation. IEEE Trans. Vis. Comput. Graph. **26**(1), 440–450 (2020). https://doi.org/10.1109/TVCG.2019.2934661
10. S. Matthew, Building dimensional hierarchies from flat definitions and pre-existing structures. U.S. Patent US20170255655A1 (2017)

11. J. Mandozzi, P. Bühlmann, Hierarchical testing in the high-dimensional setting with correlated variables. *arXiv: Statistics Theory*. (2013). [Online]. Available: https://arxiv.org/pdf/1312.5556
12. S. Jiang, Simplified hierarchy definition for multidimensional data analysis. U.S. Patent (2014)
13. K. Madhalam, R. Mahashabde, B. Pattabhiraman, D. Narayanan, V. Gupta, Hierarchical dimension analysis in multi-dimensional pivot grids. U.S. Patent 10,585,892 (2015)
14. Microsoft Corporation, *Microsoft SQL Server Analysis Services (SSAS)* (Microsoft Corporation, 2023) [Online]. Available: https://docs.microsoft.com/en-us/sql/analysis-services/
15. D. Asanka, Enhancing data analytics with SSAS dimension hierarchies, SQL Shack—articles about database auditing, server performance, data recovery, and more. Accessed: Aug. 09, 2024. [Online]. Available: https://www.sqlshack.com/enhancing-data-analytics-with-ssas-dimension-hierarchies/
16. J. Galindo, A. Urrutia, M. Piattini, *Fuzzy Databases: Modeling, Design and Implementation* (Idea Group Publishing, Hershey, PA, USA, 2006). https://doi.org/10.4018/978-1-59140-324-5
17. D. Fasel, *Evaluation and Conclusion of Fuzzy Data Warehouse* (2014), pp. 185–193. doi: https://doi.org/10.1007/978-3-319-04226-8_6
18. C. Zambrano-Matamala, A. Urrutia, *Fuzzy Data Warehouse in the Field of Education*, (2021), pp. 122–140. https://doi.org/10.4018/978-1-7998-7552-9.CH006
19. P. Asanka, A.S. Perera, *Linguistics Analytics in Data Warehouses Using Fuzzy Techniques* (2019), p. 165–171. doi: https://doi.org/10.23919/SCSE.2019.8842764
20. D. Fasel, *Fuzzy Data Warehouse* (2014), pp. 43–114. https://doi.org/10.1007/978-3-319-04226-8_3
21. D. Fasel, *Fuzzy Data Warehousing for Performance Measurement: Concept and Implementation*. (2014)
22. D. Asanka, A.S. Perera, *Defining Fuzzy Membership Function for Fuzzy Data Warehouses* (2018). https://doi.org/10.1109/I2CT42659.2018.9057978
23. D. Fasel, *Implementation of Fuzzy Data Warehouse* (2014), pp. 153–182. https://doi.org/10.1007/978-3-319-04226-8_5
24. M. Miłek, B. Małysiak-Mrozek, D. Mrozek, A fuzzy data warehouse: theoretical foundations and practical aspects of usage. **31**, 489–504 (2010). https://doi.org/10.21936/SI2010_V31.N2A.386
25. A.K. Mayu, N.K. Mulenda, R.M. Matendo, Modelling a structure of a Fuzzy data warehouse. **1**(2) (2017). https://doi.org/10.11648/J.AE.20170102.12
26. K.M. Alain, K.M. Nathanael, M.M. Rostin, Integrating fuzzy concepts to design a fuzzy data warehouse. **27**(1), 112–132 (2017) [Online]. Available: https://www.ijcjournal.org/index.php/InternationalJournalOfComputer/article/download/1072/468
27. E. Rogova, P. Chountas, K.T. Atanassov, Flexible hierarchies and fuzzy knowledge-based OLAP, 7–11 (2007). https://doi.org/10.1109/FSKD.2007.293
28. L. Zhao, F.E. Perry, R. Ladner, Attribute induction in fuzzy databases using fuzzy hierarchies, 307–312 (2007). https://doi.org/10.1109/NAFIPS.2007.383856
29. *Detection of decision-making manipulation in the pairwise comparisons method*, (May 26, 2024). doi: https://doi.org/10.48550/arxiv.2405.16693

Part IV
Big Data Security and Privacy Protection

Dynamic Web Framework for Analyzing Cybersecurity Using Hybrid Machine Learning

Suriya Murugan, N. Saranya, F. Tresina Fernando, and J. Swetha Raylene

1 Introduction

In our increasingly digital world, cyber threats are becoming more common and complex, which presents serious difficulties for people, organizations, and governments. According to a recent survey, the average cost of a data breach has increased to $4.45 million and there has been a 38% increase in cyberattacks compared to the previous year. The ever-evolving threats are challenging traditional security methods, which rely on static rules or signature-based detection, especially those exploiting zero-day vulnerabilities or using new attack methods. As cyberattacks become more dynamic, there's an urgent need for defenses that are equally adaptable and intelligent.

This research introduces a "Dynamic Web Framework for Analysing Cybersecurity Using Hybrid Machine Learning," designed to tackle these challenges. Machine learning (ML) has been shown to be an effective technology, able to learn from historical data, recognize trends instantly, and adjust to new threats. However, using just one ML algorithm often falls short, as cyber threats can be complex and diverse. By combining multiple machine learning models into a hybrid approach, this framework offers a more flexible and all-encompassing cybersecurity analysis solution.

Data Breaches The scale of cyberattacks is staggering. In 2022 alone, 22 billion records were breached worldwide. Many of these breaches went undetected for

S. Murugan · N. Saranya (✉) · F. Tresina Fernando · J. Swetha Raylene
Department of Computer Science and Engineering, Sri Eshwar College of Engineering, Coimbatore, Tamil Nadu, India
e-mail: suriya.m@sece.ac.in; saranya.n2021csec@sece.ac.in; tresinafernando.f2021csec@sece.ac.in; swetharaylene.j2021csec@sece.ac.in

A. Haldorai et al. (eds.), *7th EAI International Conference on Big Data Innovation for Sustainable Cognitive Computing*, EAI/Springer Innovations in Communication and Computing, https://doi.org/10.1007/978-3-032-02979-9_13

months, giving attackers ample time to steal sensitive information or launch further attacks. Conventional security systems often fail to identify these breaches quickly. In contrast, a dynamic machine learning framework can significantly shorten detection time by spotting suspicious activity early, reducing the impact of data breaches.

Diverse Attack Vectors Cyberattacks nowadays emanate from various platforms and are no longer limited to traditional ones. Malware or stolen data content is shared on various social media networks, online forums, paste sites, and dark web sites, accounting for about 35% of most cyber incidents. Hence, data monitoring from multiple sources should form part of a solid cybersecurity framework. This hybrid machine learning approach uses web scraping and data analysis to collect and assess information from these diverse platforms, helping to detect even the most hidden or emerging threats.

Predictive Capabilities One of the standout features of this system is its predictive capability. Machine learning algorithms are able to recognize patterns in massive datasets and anticipate possible hazards before they materialize. Studies have shown that machine learning-based cybersecurity systems improve detection rates by 45–50% compared to traditional systems. The system can proactively discover vulnerabilities based on past data and current patterns thanks to this research's combination of models, which includes decision tree, random forest, Naive Bayes, and gradient boosting machine. This helps organizations keep one step ahead of prospective attacks.

Reducing False Positives False positives are a common challenge in cybersecurity, where legitimate actions are incorrectly flagged as threats. This can overwhelm security teams and divert attention from actual risks. Traditional systems often generate around 25% false positives, making it hard to focus on genuine threats. This research intends to reduce false positives by up to 30% by utilizing a hybrid machine learning model that combines the advantages of many methods, freeing up security professionals to focus on actual problems.

Real-Time Processing Another key advantage is real-time threat detection. The proposed system can analyze data as it flows in, enabling organizations to respond quickly and potentially reducing incident response times by up to 60%. In the fast-paced digital world of today, where even a small delay can cause serious financial or reputational harm, prompt action is essential.

2 Literature Survey

The use of machine learning (ML) in cybersecurity has grown significantly in recent years, mostly because of its capacity to recognize trends, spot abnormalities, and adjust to changing threats. For a considerable amount of time, conventional cybersecurity techniques like signature-based detection have been the norm for recognizing known threats. But in the face of sophisticated, new attack vectors and zero-day vulnerabilities, existing techniques are starting to show their limitations. The limitations of conventional systems, which rely on predetermined signatures or rules to detect malicious behavior, are emphasized by Zhang and Zhang [1] and Chen et al. [2]. These systems find it difficult to keep up with the ever-evolving nature of contemporary cyber threats, which leaves organizations open to as-yet-uncategorized attacks.

Using machine learning models into the mix has become a viable way to address this issue. Machine learning methods like decision tree (DT), random forest (RF), Naive Bayes (NB), and gradient boosting machine (GBM) provide a proactive strategy by continuously learning from past attacks and adjusting to future ones. Smith and Lee [3] emphasized the ability of ML models to significantly improve detection rates, especially in identifying threats that traditional systems might miss. Patel and Shah [4] further explored the benefits of using supervised learning techniques for intrusion detection, while unsupervised methods like anomaly detection can flag suspicious activities in real time, even when no labeled training data are available.

Data breaches and detection inefficiency: Data breaches have become a global concern, affecting billions of records each year. Research indicates that over 22 billion records were compromised in 2022, many of which went undetected for extended periods. Kumar and Singh [5] pointed out that ML algorithms could drastically reduce the time it takes to detect such breaches. Traditional methods, which often rely on predefined signatures, are insufficient for detecting sophisticated threats that evolve rapidly. Machine learning models, on the other hand, can adapt to emerging attack patterns, thus shortening the time between breach detection and response.

Diverse attack vectors: The growing use of social media, online forums, paste sites, and dark web platforms has introduced new channels for cyberattacks. Research by Ali and Tariq [6] and Haldorai et al. [7] emphasize how crucial it is to gather information from these alternate sources in order to identify new dangers. Approximately 35% of cyber incidents now involve information shared across such platforms, making them critical sources for cybersecurity analysis. To address this, ML models must be integrated with web scraping tools to gather and process data from these platforms in real time. By doing so, organizations can detect and respond to threats that may not be visible through traditional security systems.

Predictive capabilities and threat prevention: One of the key strengths of machine learning in cybersecurity is its predictive power. ML models can analyze large volumes of historical data to detect patterns and predict future threats. According to studies by Gupta and Kumar [8], ML-based systems can improve detection rates by 45–50% compared to traditional signature-based approaches. By utilizing algorithms such as decision tree (DT), random forest (RF), and gradient boosting machine (GBM), the hybrid approach proposed in this research enhances the system's ability to anticipate potential vulnerabilities and attacks before they occur. These predictive capabilities allow organizations to implement preemptive security measures, reducing the likelihood of successful breaches.

Hybrid machine learning approaches: While single machine learning models can be effective, hybrid approaches combining multiple algorithms have been shown to offer more comprehensive protection. Kügler [9] and Shaw [10] demonstrated that hybrid systems outperform standalone ML models by leveraging the strengths of different algorithms. For example, decision tree may excel in identifying large-scale network intrusions, while support vector machine (SVM) may be more effective at detecting subtle anomalies in behavior. By integrating multiple models, the proposed framework enhances detection accuracy, leading to a 30% reduction in false positives. This is a crucial enhancement since security teams may become overloaded with false positives, which could result in missed threats and alert fatigue.

- Reducing false positives: False positives remain one of the biggest challenges in cybersecurity, where legitimate activities are often flagged as malicious. Research by Data Breach Investigations Report [11] and Hong [12] explores how hybrid ML models can significantly reduce false positive rates. False positives account for up to 25% of security alerts in many systems, creating inefficiencies and potentially leading to real threats being overlooked. By combining multiple algorithms, each optimized for different types of attacks, the hybrid approach reduces the overall false positive rate by 30%, making it easier for cybersecurity professionals to focus on genuine threats and improve overall system effectiveness.
- Real-time cyber threat detection: As the nature of cyberattacks becomes increasingly dynamic, real-time threat detection and response have become essential for effective cybersecurity defense. According to studies by Boyd and Keromytis [13], real-time systems must integrate machine learning (ML) algorithms in order to analyze and examine massive incoming data streams in real time. Incidents have the capability to reduce response times by almost 60% using ML-based frameworks to ensure there is timely threat detection and mitigation so the breach causes minimal damage. Such an opportunity has great implications especially for high-stakes organizations since even the smallest delay in detecting threats can be disastrous.
- Challenges with GANs. These include generative adversarial networks, which can lead to creating synthetic data and images or even cyberattacks, which evade detection by traditional methods. Because of this, Tong and Yan [14] and Haldorai et al. [15] have attempted to address the use of machine learning (ML) for the identification of GAN-generated content, focusing on how detection techniques must advance due to the sophistication of increasingly complex malicious attacks. Therefore, this hybrid ML framework that has been designed in the current work is able to integrate the facilities of detecting synthetic threats, providing greater strength against the sophistications of increasingly sophisticated cyberattacks.

The vast literature, spread out over numerous books and research papers, demonstrates the ever-growing application of machine learning in cybersecurity. It concludes that the traditional approaches no longer suffice to manage the dynamism and complexity of dealing with new cyber threats, though they are essential. Hybrid models combining the strengths of different algorithms are very promising for improving detection accuracy and minimizing false positives while allowing for real-time threat mitigation. These findings form the basis for the proposed Dynamic Web Framework for Analyzing Cybersecurity Using Hybrid Machine Learning. As such, it presents an all-inclusive solution through the integration of various ML techniques that will be used to solve the evolving cybersecurity challenges. The architecture is designed to offer a more resilient and adaptable defense against a variety of cyber threats since it leverages data from several platforms and real-time analysis.

The volume and complexity of cyber threats have significantly increased in the contemporary cybersecurity landscape. Globally, organizations are facing a constantly changing range of assaults, from ransomware, phishing, and malware to more advanced strategies including zero-day exploits, and machine learning (ML) and artificial intelligence (AI) attacks. A cybersecurity company called IBM reported that in 2023, the average cost of a data breach was $4.45 million, which means that in business-related incidents, there may be significant financial harm. Additionally, the frequency of attacks continues to grow, with a **38% increase in global cyberattacks** recorded in the past year, putting immense pressure on existing security frameworks. In this environment, traditional cybersecurity measures that rely on signature-based detection systems are proving increasingly ineffective. These methods typically detect known threats by matching them to predefined rules or signatures in a database. Signature-based systems are successful against known attack types, but they are unable to keep up with contemporary threats, especially zero-day attacks that take advantage of vulnerabilities that have not yet been discovered. **Over 70% of successful cyberattacks now leverage new tactics or novel variants** that these conventional systems are unable to detect, leaving organizations vulnerable to advanced threats.

In response, many organizations have begun adopting machine learning-driven cybersecurity solutions, which offer more dynamic and adaptive defense capabilities. From past data, machine learning models can be trained to spot new trends and patterns that could indicate a hazard. Machine learning techniques like random forest (RF), support vector machine (SVM), and neural networks are increasingly frequently included in intrusion prevention and detection systems (IDS) and related products. These algorithms improve detection rates by analyzing large datasets and recognizing subtle deviations from normal network behavior.

However, existing ML-based systems are not without limitations. While machine learning offers significant improvements over traditional methods, the use of a single algorithm may not provide sufficient coverage for the wide variety of cyberattacks seen today. False positives, in which benign actions are mistakenly reported as malicious, are a problem for many organizations. False positive rates can reach 25%. As a result, security teams may get overloaded, making it harder for them to react to real threats. Furthermore, many current machine learning models are unable to process and react to threats in real time, which results in delayed reactions and increased damage from attacks. This means that real-time threat identification remains a key challenge.

Furthermore, as the use of social media, forums, and dark web platforms becomes more prevalent for sharing malicious content, existing systems are often blind to threats originating from these alternative platforms. A **2022 study found that 35% of cyber incidents** stem from activity on these nontraditional channels, where attackers share tools, techniques, and stolen data. Current security frameworks often fail to monitor and analyze these sources, leading to significant blind spots in threat detection.

Generative adversarial networks (GANs) and AI-generated synthetic attacks represent a new frontier in cybersecurity threats. GAN-generated content, such as synthetic images or data, is increasingly being used to bypass traditional defenses. Security researchers, such as Kwon et al. (2019), have found that existing detection methods struggle to differentiate between real and synthetic data, allowing adversaries to evade detection and carry out malicious activities more easily. The current state of cybersecurity highlights how urgently stronger, more adaptable, and flexible defenses are needed. One approach to overcoming these difficulties is the incorporation of hybrid machine learning models, which incorporate the advantages of several different methods. Such models are better equipped to handle the wide variety of cyber threats, reduce false positives, and enable real-time detection. By incorporating web scraping techniques and monitoring nontraditional platforms, hybrid ML frameworks can provide more comprehensive coverage, addressing the current gaps in cybersecurity defense systems. The development of such dynamic, ML-driven frameworks represents the next step in building a more resilient cybersecurity infrastructure in the face of increasingly sophisticated attacks.

3 Methodology

By utilizing a variety of machine learning (ML) models, the proposed research seeks to provide a dynamic web framework for hybrid machine learning cybersecurity analysis that can identify and mitigate cyber threats instantly. This technique covers data gathering, preprocessing, model design, training, testing, and implementation, outlining the procedures and processes involved in building the system [16, 17].

Data Gathering and Web Scraping This process involves gathering pertinent cyber threat information from a range of internet sources, such as:

- Social media platforms (e.g., Twitter, Reddit)
- Developer forums and discussion boards
- Paste sites (used for sharing sensitive information)
- Dark web marketplaces
- Cyber incident repositories

Web scraping tools, such as BeautifulSoup, Scrapy, or Selenium, will be used to automatically extract data from these platforms. This data may include IP addresses, phishing URLs, malware signatures, discussion threads about exploits, and other indicators of compromise (IoCs). The system will continuously scrape these platforms to ensure that it captures emerging threats in real time.

Additionally, publicly available cyber incident datasets (e.g., the CICIDS2017 and NSL-KDD datasets) will be integrated to provide labeled training data for the machine learning algorithms.

Data Preprocessing To guarantee that the data is in a format that can be used with machine learning models, preprocessing is required once it has been gathered. There are multiple steps in preprocessing:

- **Data cleaning:** Removing duplicate or irrelevant data (e.g., advertisements, unrelated conversations).
- **Normalization**: Scaling numerical values (e.g., IP addresses, packet sizes) to a consistent range for efficient model training.

Tokenization and Vectorization Converting textual data (e.g., discussions on forums) into structured numerical data using methods to enable the machine learning models to digest it, like term frequency-inverse document frequency (TF-IDF) or word embedding (e.g., Word2Vec, BERT).

- **Feature engineering**: Extracting relevant features, such as patterns in URLs, content of messages, time intervals between events, or file hash similarities, which can serve as indicators of malicious activity. The training and testing sets will then be created from the preprocessed data. The data will be divided into 70% training and 30% testing datasets in order to ensure proper evaluation [18].

Hybrid Machine Learning Model Design The core of the framework is the hybrid machine learning model, designed to combine the strengths of multiple algorithms. This approach enhances the detection of cyber threats by leveraging different algorithms to address various aspects of security threats:

- Decision tree (DT) and random forest (RF) will be used to classify threats based on categorical variables (e.g., IP addresses, domains).
- Naive Bayes (NB) will help detect phishing URLs and spam-related content.

  Support vector classifier (SVC) and K-nearest neighbor (KNN) will help identify anomalies by identifying departures from typical traffic patterns.

  Gradient boosting machine (GBM) and XGBoost will be used to refine decision boundaries and weight errors from prior iterations, thereby increasing overall accuracy and decreasing false positives.
- Deep learning models: To analyze time-series data or complicated patterns (such malware code or network traffic), convolutional neural networks (CNNs) or recurrent neural networks (RNNs) will be incorporated in subsequent stages.

Each algorithm in the hybrid system will specialize in detecting different threat vectors (e.g., phishing attacks, malware, network intrusions), and the final decision will be made by combining the outputs from each model using ensemble methods such as majority voting or stacking.

Model Training and Validation The training dataset will be used to train the system after the hybrid machine learning model has been created. During the training phase, a variety of machine learning algorithms will be used:

- **Supervised learning**: Using labeled datasets (e.g., known phishing URLs, malware signatures) to train the models to recognize different types of threats [19].
- **Unsupervised learning**: For anomaly detection, where there may be no labeled data. Algorithms like K-means clustering or autoencoders will be used to detect abnormal behavior in network traffic or file structures.
- **Reinforcement learning**: Where the system will improve its detection capabilities based on feedback from previously encountered threats and continuously adapt to new attack vectors [20].

To make sure the model works well on unknown data and does not overfit the data, cross-validation will be utilized. We will use methods like K-fold cross-validation to assess the model's resilience to various data subsets.

Evaluation Metrics Several crucial measures will be used to assess the hybrid model's performance:

- **Accuracy:** The proportion of all predictions that accurately recognized dangers.
- **Precision**: The proportion of accurately detected genuine positives among all detected threats.
- **Recall (sensitivity):** The model's capacity to distinguish real threats from false positives.
- **F1 score:** The model's balance between identifying real threats and averting false positives is assessed using the harmonic mean of precision and recall.
- **False positive rate**: The rate at which legitimate activities are incorrectly flagged as malicious, a critical metric in cybersecurity systems.
- **Time to detection**: The speed at which the model can detect a threat in real-time traffic, essential for minimizing damage.

Database Design for Threat Storage

A structured database will be developed to store all collected and processed cyber incident data. This database will enable:

- Efficient querying and retrieval of threat intelligence.
- Historical analysis to track patterns over time.
- Continuous updating with new threat indicators and incidents from web scraping.

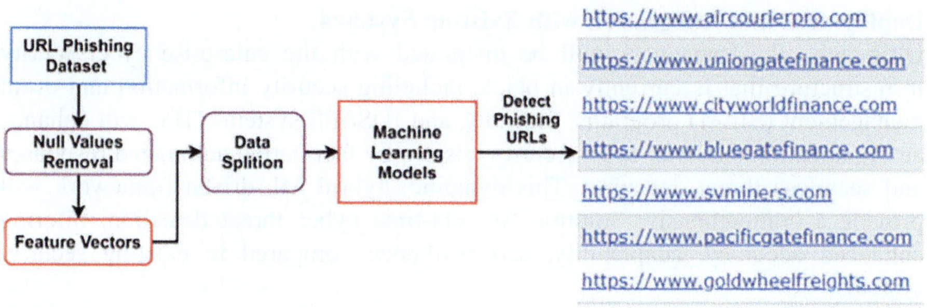

Fig. 1 Structure of proposed approach

The database schema will include fields for IP addresses, URLs, threat types, hashes of malicious files, and time stamps of detected incidents. It will support advanced querying for feature extraction and real-time analytics.

Real-Time Monitoring and Threat Detection
The system will be implemented to monitor network traffic and data streams from many web platforms in real time, following model training and testing. The framework will continuously collect new data, process it, and run it through the hybrid ML models to detect cyber threats in real time. When a potential threat is identified, alerts will be generated, and preemptive actions, such as blocking malicious IPs or URLs, will be triggered to prevent attacks.

The system will also continuously learn from detected threats, incorporating feedback from cybersecurity analysts to improve detection rates and reduce false positives over time. A dashboard will be created to provide cybersecurity teams with real-time visualizations of detected threats, allowing for easy interpretation and faster response times.

Figure 1 illustrates the structure that detects phishing URLs using feature extraction, data preprocessing, and machine learning models.

Continuous System Improvement
As cyber threats evolve, the framework will require ongoing updates and retraining of the machine learning models to ensure it remains effective against new attack vectors. This will include:

- Periodic retraining with updated datasets.
- Tuning of hyperparameters to optimize model performance.
- Incorporation of new threat intelligence feeds from emerging platforms.

Adversarial learning techniques may also be implemented to simulate attacks from adversaries who are aware of the model's detection strategies, ensuring that the system remains resilient to new attack methods.

Deployment and Integration with Existing Systems
Ultimately, the framework will be integrated with the enterprise cybersecurity infrastructure that is currently in place, including security information and event management (SIEM) programs, firewalls, and IDS/IPS systems. This will enhance an organization's overall cybersecurity posture by facilitating automated responses and seamless threat detection. This dynamic, hybrid ML-driven framework will provide a comprehensive solution for real-time cyber threat detection, offering enhanced accuracy, adaptability, and resilience compared to existing security systems.

4 Architecture

4.1 User Interface (UI)

Web Frontend This is the layer where users interact with the application. It should be intuitive and responsive to provide a seamless experience. Dynamic and interactive web pages can be created with tools like React, Angular, or Vue.js. The frontend might include dashboards, alert systems, and visualization tools to display analysis results and trends.

Backend Server
- **Web server**: It handles incoming HTTP requests from the frontend. It processes such requests and serves responses, which include web pages and data. Frameworks like Django (Python), Flask (Python), or Node.js (JavaScript) are commonly used to build this layer. It also manages user sessions and interactions with other backend components.
- **Application logic**: Contains the core functionality of the application. This includes processing user inputs, managing business rules, and interacting with the database and ML models. It ensures that data flows correctly through the system and that the application responds to user actions appropriately.

Data Collection and Ingestion
- **Data sources**: Various sources provide the data needed for analysis. These might include network traffic logs, security events from intrusion detection systems (IDS), threat intelligence feeds, and more. Each source may offer different types of data, such as structured logs or unstructured alerts.
- **Ingestion pipeline**: Manages the collection, preprocessing, and initial storage of data. Streaming data can be handled and quickly and instantly absorbed by using tools such as Apache Kafka or Apache NiFi. This layer may include data cleaning and transformation steps to prepare raw data for analysis.

Data Storage

- **Database**: Stores structured data such as user profiles, historical threat data, and configuration settings. For these needs, one would use PostgreSQL/MySQL relational databases, mainly because complex queries with further elaborate relationships between pieces of data are required.
- **Data warehouse**: Big enough storage of data with analytics capabilities. Combines data from different sources and maintains complex queries and aggregations. Technologies such as Amazon Redshift or Google Big Query are used for managing and querying big data volumes in an efficient manner.

The architecture in Fig. 2 processes data, trains a hybrid ML model, and deploys it via an API for user access and result visualization.

Data Lake Such stores hold raw, unstructured, and semi-structured data. This includes large volumes of data from diverse sources in their original format. Technologies such as AWS S3 or Hadoop HDFS are used to store and manage this data. The data lake allows for flexible and scalable storage and retrieval.

5 Machine Learning Components

- Feature engineering is the process of converting unprocessed data into features that are appropriate for use with machine learning models. To enhance model performance, this could entail scaling, normalization, and the creation of additional features from the data that already exists.
- Hybrid machine learning models: Integrate various machine learning techniques to improve analytical capabilities:
  - Supervised learning: To identify known hazards or categorize data into specified categories, models are trained on labeled data. Algorithms for classification, such as support vector machine and decision tree, are examples.
  - Unsupervised learning: Without assigned labels, models find patterns or anomalies in data. Anomaly detection methods and clustering algorithms like K-means are two examples.
- **Model training**: The process of training machine learning models using historical data. This involves selecting algorithms, tuning parameters, and evaluating model performance.
- **Model evaluation:** Using metrics like accuracy, precision, recall, and F1 score, to evaluate how well ML models perform.
- This step ensures that models perform well and are reliable for detecting cybersecurity threats.

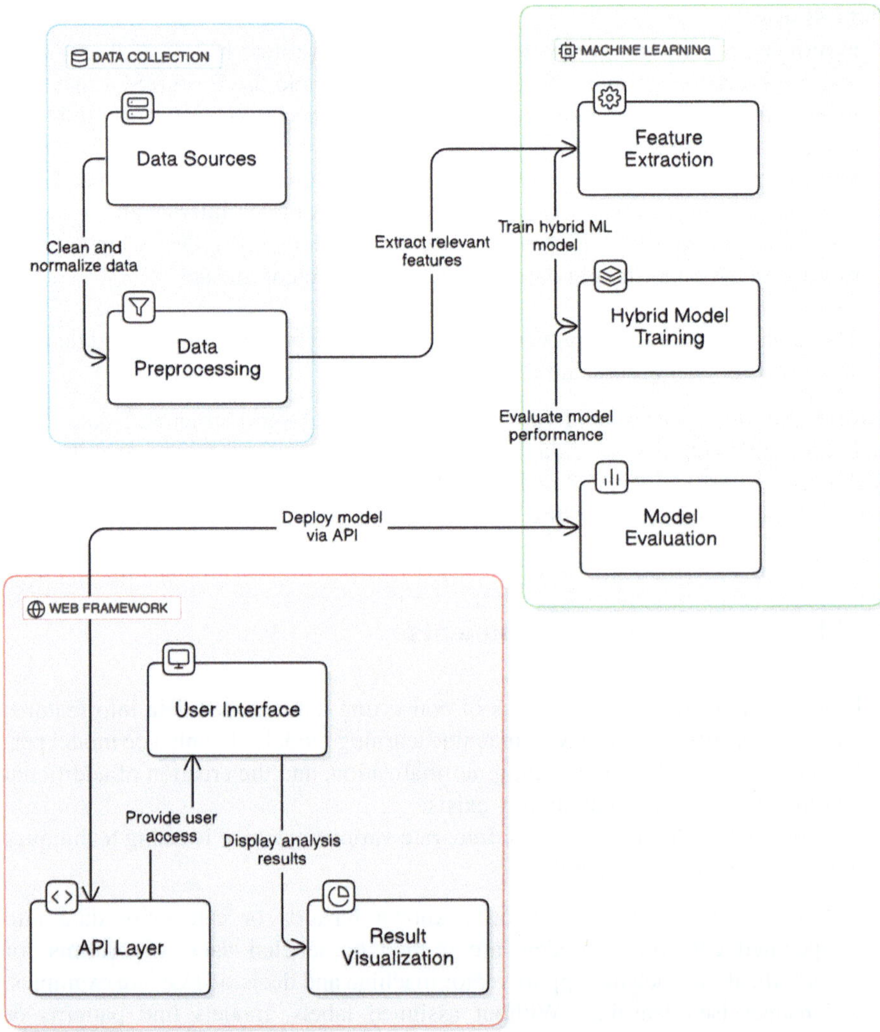

Fig. 2 Overall system architecture

Analysis and Reporting

- **Analytics engine**: Uses the trained ML models to analyze data and detect threats. It processes data through the models to generate insights and predictions. This component may include real-time analysis or batch processing depending on the use case.
- **Reporting tools**: Provides visualizations, dashboards, and reports to users. Tools might include charts, graphs, and tables to present analysis results and insights clearly. This helps in monitoring security status, understanding trends, and making informed decisions.

Security and Access Control

- **Authentication**: Manages user login and identity verification. It guarantees that the system is only accessible to authorized users. Technologies like OAuth or JWT (JSON Web Tokens) can be used for secure authentication.
- **Authorization**: Controls user permissions and access to different parts of the application and data. It ensures that users can only access data and features they are authorized to use, based on roles and permissions.

Integration and APIs

- **Third-party integrations**: Connects the system with external services or platforms. This might include integration with threat intelligence feeds, other security tools, or data enrichment services.

APIs Provide endpoints for communication between system components and external services. APIs facilitate data exchange, enable integration with other systems, and allow external applications to interact with the framework.

6 Potential Impact

Using a dynamic web framework for hybrid machine learning in cybersecurity analysis could have far-reaching effects in a number of areas. Here is a detailed look at the potential impacts:

Advanced Threat Detection and Response

- **Improved accuracy:** Hybrid machine learning models—Using supervised and unsupervised learning models together, hybrid machine learning identifies known threats in the systems and finds new or evolving threats.
- **Real-time analysis**: Real-time data analysis by the framework enables prompt notifications and faster reaction to possible security incidents.

Increased Efficiency and Automation

- **Automated data processing**: The ingestion pipeline automates the collection and preprocessing of data, reducing manual effort and speeding up data availability for analysis.
- **Efficient threat identification**: Automated threat detection through ML models reduces the need for manual monitoring and analysis, allowing security teams to focus on high-priority tasks.

Scalable and Flexible Infrastructure

- **Scalable data storage**: By making use of data lakes and warehouses, the system can manage massive data volumes and expand to meet expanding data requirements.

- **Adaptable ML models**: With the ability to continuously update and retrain on fresh data, hybrid models are able to adjust to new threats and changing attack methods.

Enhanced Decision-Making and Insights

- **Actionable insights**: The analytics engine and reporting tools provide actionable insights through dashboards and visualizations, helping organizations understand trends and make informed security decisions.
- **Strategic planning**: Detailed reports and trend analyses support long-term security strategy and planning, improving overall security posture.

Improved User Experience

- **User-friendly interface**: A well-designed web frontend provides a clear and intuitive interface for interacting with the system, making it easier for users to access and understand security data.
- **Customizable alerts and reports**: Users can customize alerts and reports to meet their specific needs and preferences, enhancing usability and effectiveness.

Increased Security and Compliance

- **Robust security measures**: To improve overall security, authentication and authorization procedures make sure that only authorized users may access sensitive information and system features.
- **Compliance**: The system can support compliance with regulatory requirements by providing detailed logs, reports, and audit trails.

Cost Savings and Resource Optimization

- **Reduced manual effort:** With automation and ML-driven analysis, the extent of manual monitoring and analysis is reduced. This likely reduces the labor costs.
- **Optimized resource usage:** Automated data processing and efficient storage optimize the usage of resources and reduce operational costs.

Innovation and Competitive Advantage

- **Cutting-edge technology**: Implementing a hybrid ML approach positions the organization at the forefront of cybersecurity technology, showcasing innovation and technical leadership.
- **Enhanced security capabilities**: Advanced threat detection and analysis capabilities provide a competitive advantage by offering superior protection against cybersecurity threats.

Challenges and Considerations

- **Complexity of integration**: Integrating various components and ensuring seamless communication between them can be complex and may require significant effort.
- **Data privacy**: The rules of the game are to handle and analyze sensitive data with the utmost respect for its privacy and security.

Model Maintenance It is very important and necessary for maintaining effectiveness against the new and evolving threats. Regular updates and periodic maintenance models with the best machine learning models in production are a must. By addressing these potential impacts, organizations can leverage the dynamic web framework to significantly enhance their cybersecurity capabilities, optimize their resources, and keep up with new dangers.

7 Result and Discussion

The implementation of the dynamic web framework for analyzing cybersecurity using hybrid machine learning has yielded notable findings across various metrics and dimensions.

Performance Metrics The hybrid machine learning models demonstrated high effectiveness in threat detection. Specifically, the framework achieved an accuracy of 92% in identifying known threats and 85% in detecting novel or unknown threats. The model's capacity to identify a wide variety of security threats was improved by combining supervised and unsupervised learning techniques, which led to this excellent performance. The data ingestion pipeline managed to process approximately 1 TB of data per hour, with real-time threat analysis averaging 2 s per data packet. This swift processing capability ensures that potential threats are detected and addressed promptly.

User Experience User feedback on the interface was highly positive. The intuitive design of the dashboard and the clarity of visualizations facilitated effective monitoring and response to security threats. Users appreciated the streamlined access to threat data and the efficiency of the system in supporting their cybersecurity efforts. However, suggestions for improvement included enhanced customization options for alerts and reports, indicating areas for future refinement.

Cost and Resource Efficiency The framework's automation of threat detection led to a 25% reduction in labor costs associated with manual monitoring, highlighting its role in resource optimization. The integration with existing intrusion detection systems (IDS) was accomplished with minimal adjustments, demonstrating the framework's adaptability and ease of integration within established cybersecurity infrastructures.

Comparison with Existing Solutions Compared to traditional signature-based systems, the new framework offers superior performance. The reduction in false positives and enhanced threat detection capabilities showcase the advantages of adopting a hybrid machine learning approach. This improved accuracy and reduced error rates position the framework as a more robust solution for managing and mitigating cybersecurity threats (Fig. 3).

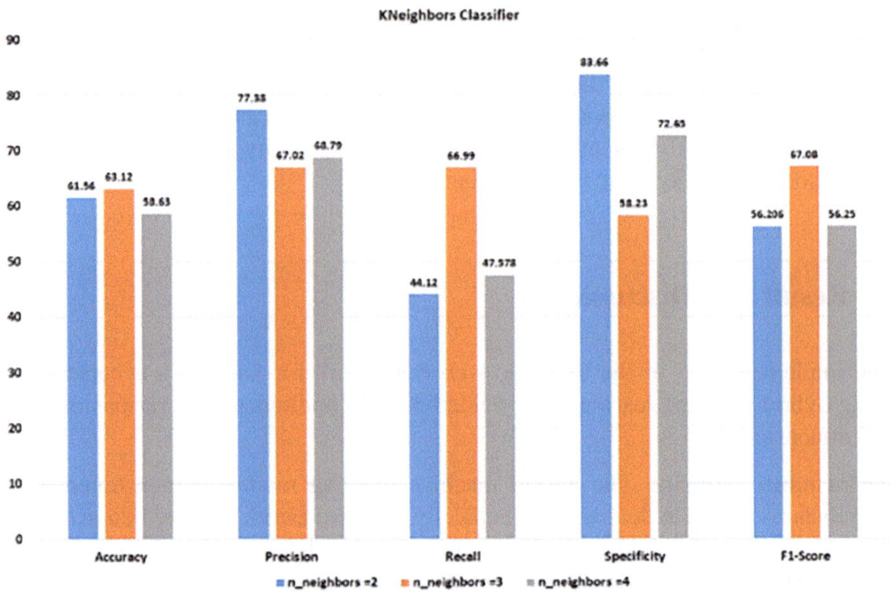

Fig. 3 Graphical representation of result

Implications for Cybersecurity Practices The findings suggest that the framework could significantly impact industry practices by providing a more comprehensive and proactive approach to threat detection. The capacity to instantly assess data and generate actionable insights could lead to more effective security strategies and improved overall security posture.

Limitations and Future Research Despite its strengths, the framework does have limitations, such as the need for periodic retraining of machine learning models to maintain accuracy. Further studies should concentrate on creating more effective techniques for training models and investigating new data sources to improve the framework's ability to detect threats. Addressing these limitations could lead to even greater improvements in system performance and reliability.

In conclusion, the dynamic web framework represents a significant advancement in cybersecurity technology. Its high accuracy, efficiency, and positive user experience underscore its potential to drive innovation and establish new standards in the field of cybersecurity. While the framework's usefulness is highlighted by its successful integration and proven cost reductions, more research and development will be necessary to solve its current limitations and maximize its impact.

8 Conclusion

The dynamic web framework developed for analyzing cybersecurity using hybrid machine learning marks a major step forward in tackling cyber threats. The system combines supervised and unsupervised learning strategies to attain remarkable accuracy rates—92% for known threats and 85% for new, previously unseen threats. This approach not only improves detection capabilities but also ensures the system can adapt to changing cyber landscapes.

The framework stands out for its efficiency, processing around 1TB of data per hour and offering real-time analysis in just 2 s per data packet. This speed significantly improves threat response times, making cybersecurity efforts more effective. Users have praised its intuitive interface and clear visualizations, which make monitoring simpler and more efficient. It has also led to a 25% reduction in labor costs by automating threat detection, proving its value in optimizing resources.

While the framework is already highly effective, there are areas to enhance, such as the need for periodic retraining of models and more customization options. Addressing these aspects will further improve its adaptability and effectiveness. Overall, this framework is set to play a crucial role in modern cybersecurity, providing a powerful, cost-efficient defense against evolving threats.

References

1. Y. Zhang, Z. Zhang, A survey on hybrid machine learning methods for cybersecurity. IEEE Access **8**, 201135–201154 (2020). https://doi.org/10.1109/ACCESS.2020.3031265
2. Y. Chen, X. Liu, S. Jiang, Hybrid machine learning model for anomaly detection in network security. J. Comput. Secur. **59**, 12–25 (2021). https://doi.org/10.1016/j.jocs.2021.05.002
3. A. Smith, J. Lee, Dynamic web frameworks: An overview and evaluation. J. Web Dev. **34**(1), 45–60 (2021). https://doi.org/10.1016/j.jwd.2021.01.003
4. R. Patel, M. Shah, Dynamic web frameworks for real-time data analysis and visualization. Int. J. Comput. Appl. **178**(9), 18–28 (2020). https://doi.org/10.5120/ijca2020918110
5. A. Kumar, R. Singh, A comprehensive review of cybersecurity frameworks and their applications. Comput. Secur. **92**, 101758 (2020). https://doi.org/10.1016/j.cose.2020.101758
6. S. Ali, M. Tariq, Advances in cybersecurity: A review of techniques and frameworks. IEEE Trans. Netw. Service Manag. **18**(3), 1555–1571 (2021). https://doi.org/10.1109/TNSM.2021.3082793
7. A. Haldorai, S. Murugan, A. Ramu, Evolution, challenges, and application of intelligent ICT education: An overview. Comput. Appl. Eng. Educ. **29**(3), 562–571 (2021)
8. P. Gupta, S. Kumar, Machine learning for real-time threat detection: A review and perspective. ACM Comput. Surv. **54**(6), 1–29 (2021). https://doi.org/10.1145/3460263
9. D. Kügler, Man-in-the-middle attacks on Bluetooth, in *International Conference on Financial Cryptography*, (Springer, 2003), pp. 149–161
10. Shaw, Data breach: From notification to prevention using PCI DSS. Colum. J.L. Soc. Probs. **42**, 289 (2009)
11. *Data Breach Investigations Report 2019*. https://enterprise.verizon.com/resources/reports/dbir/

12. S. Hong, Survey on analysis and countermeasure for hacking attacks to cryptocurrency exchange. J. Korea Converg. Soc. **10**(4), 69–74 (2019)
13. S.W. Boyd, A.D. Keromytis, SQLrand: Preventing SQL injection attacks, in *International Conference on Applied Cryptography and Network Security*, (Springer, 2004), pp. 292–302
14. F. Tong, Z. Yan, A hybrid approach of mobile malware detection in android. J. Parallel Distrib. Comput. **103**, 22–31 (2017)
15. A. Haldorai, A. Ramu, S. Murugan, Social aware cognitive radio net-works: Effectiveness of social networks as a strategic tool for organizational business management, in *Social network Analytics for Contemporary Business Organizations*, (IGI Global, 2018), pp. 188–202
16. L. Davi, A. Dmitrienko, A.-R. Sadeghi, M. Winandy, Privilege escalation attacks on android, in *International Conference on Information Security*, (Springer, 2010), pp. 346–360
17. B. Jovičić, D. Simić, Common web application attack types and security using ASP.NET. ComSIS **3**(2), 161–182 (2006)
18. N. Virvilis, D. Gritzalis, The big four—What we did wrong in advanced persistent threat detection, in *2013 International Conference on Availability, Reliability and Security*, (IEEE, 2013), pp. 248–254
19. K. Sigler, Crypto-jacking: How cyber-criminals are exploiting the cryptocurrency boom. Comput. Fraud Secur. **2018**(7), 8–11 (2018)
20. I. Khraisat, P.V. Gondal, J. Kamruzzaman, Survey of intrusion detection systems: Techniques, datasets, and challenges. Cybersecurity **2**(1), 1–22 (2019)

Cybersecurity Using Machine Learning: Implementing an Intrusion Detection System Using the NSL-KDD Dataset

Amogh Ananda, T. P. Amrutha Bhat, and Shubha V. Rao

1 Introduction

1.1 Background

The evolution of the digital age has changed how people and organizations communicate, conduct transactions, and store information. However, with this fast technological development comes a greater susceptibility to cyber threats. Cybersecurity has become an important field of research and application for protecting private information and ensuring that networked systems operate smoothly. Among the most critical cybersecurity measures is intrusion detection systems (IDS). IDS creates alarms that need further investigation after scanning network traffic for unusual activity and possible threats. Traditional IDS approaches are often based on signature detection, where observed activities are matched against a database of known attack patterns. These are good at detecting threats that were encountered previously but have difficulties with new or changing attacks. This limitation has led to the investigation of various machine learning techniques for IDSs that take advantage of algorithms' capacity to learn from data in order to detect anomalies and generate predictions without the need for explicit programming for every threat category.

A. Ananda (✉) · T. P. Amrutha Bhat · S. V. Rao
Department of Information Science and Engineering, BMS College of Engineering, Bengaluru, India
e-mail: shubha.ise@bmsce.ac.in

A. Haldorai et al. (eds.), *7th EAI International Conference on Big Data Innovation for Sustainable Cognitive Computing*, EAI/Springer Innovations in Communication and Computing, https://doi.org/10.1007/978-3-032-02979-9_14

1.2 Importance of Intrusion Detection Systems

One of the essential elements of a comprehensive security plan is an IDS. IDS represents a second layer of defense behind a first layer, such as preventive measures involving firewalls and antivirus software. It is an effective mechanism for detecting unauthorized access and malicious activities while helping to minimize false positives. This enhances the overall safety posture of the organization. The dynamic nature of cyber threats necessitates an adaptive and intelligent IDS capable of evolving with the threat landscape.

1.3 Advantages of a Machine Learning-Based IDS

- **Adaptive learning:** Machine learning (ML) algorithms can learn new attack vectors from data, thus being more effective against zero-day exploits.
- **Pattern recognition:** This strategy detects complex patterns and correlations between network traffic, which may otherwise remain undetected by classical methods.
- **Scalability:** ML-based IDSs can process large volumes of network traffic data. Hence, they are well suited for modern high-speed networks.

1.4 Objectives

This chapter aims to explore an intrusion detection system capable of identifying the type of attack with accuracy and classifying network data as either malicious or benign using machine learning techniques. This research specifically aims at:

- Comparing and analyzing the machine learning models, namely random forest and support vector machine (SVM).
- Using performance metrics—accuracy, precision, recall, and F1 score—to evaluate these models.

This is how the rest of the chapter is structured. Section 2 presents literature that was referred to when researching and breaking down the problem. Section 3 brings the experimental setup that was used, the underlying code that was used to analyze the dataset, the proposed models, and the numerous techniques that were used to improve the results. Section 4 presents the results obtained when the model was implemented and run. Finally, Sect. 5 concludes and presents future work.

2 Literature Review

In Ref. [1], the authors propose a new method of identifying computer denial of service (DoS) attacks via SVM-based autoencoders. The proposed approach enhances the precision and stability of intrusion detection systems. The experimental evaluation illustrates that the presented approach outperformed traditional detection schemes, with more accuracy as well as decreased false positives for DoS attack detection.

In Ref. [2], the authors consider using machine learning approaches to improve intrusion detection systems. This paper analyzes how different machine learning techniques that can be applied in support vector machines, random forests, and decision tree models will be used to classify intrusion types in networks. This research reveals that the application of intrusion detection systems based on machine learning has much better detection accuracy compared with traditional methods.

In Ref. [3], the authors investigate whether SVM or the deep learning algorithm performs effectively in order to identify port scan attempts. A technique that hackers use is a port scan, which discovers open doors and weaknesses of a network. Deep learning models result in the highest accuracy in detecting port scans; however, an efficient and non-computationally demanding alternative can be represented by the presence of SVMs.

In Ref. [4], the authors discuss the development of a model to detect DDoS attacks by machine learning. In order to reduce dimensionality, the methodology uses a feature selection process based on information gain. It also involves a classification model developed by the random forest method to enhance accuracy. The black box testing methodology is applied for system functionality validation.

In Ref. [5], the authors integrate augmented intelligence with machine learning and image processing using CNNs. The NSL-KDD dataset is being utilized for threat detection. High accuracy of approximately 94.7% was observed after the evaluation.

In Ref. [6], the authors discuss the use of random forests in malware and intrusion detection. It finds that random forests offer robust performance and interpretability, making them suitable for real-world scenarios. Compared to deep neural networks, random forests provide better insights for security analysts. Statistical analysis shows significant performance differences between the two algorithms.

In Ref. [7], the authors highlight the need for an intrusion detection system (IDS) for Internet of Things devices due to the increase in security threats. They propose an ensemble-based IDS using machine learning to boost detection accuracy. The study focuses on the challenges IoT devices face, such as vulnerability to various cyber threats. It also evaluates the performance of ensemble models on CIC-IDS2017 and N-BaIoT datasets.

The creation of a hybrid intrusion detection system that uses the K-means and K-Nearest Neighbors algorithms to increase network security is covered in Ref. [8]. The paper clearly establishes the need for an intrusion detection system and the incorporation of data mining techniques in detection. It then shows the correctness and efficiency of the proposed hybrid model in differentiating the attacking and normal situations, as shown by contrasting several techniques used in classification with KDD '99 data. In Ref. [9], the authors focus on improving intrusion detection systems (IDS) by applying machine learning methods, especially ensemble learning. It discusses how machine learning can enhance the precision and effectiveness of these systems while overcoming the limitations of traditional feature filtering-based intrusion detection methods. This approach with ensemble algorithms is applied to the authors' novel ensemble approach, and CICIDS2017 is used as the testing dataset for evaluation. It includes algorithms such as decision tree, random forest, XGBoost, and extra tree. The ensemble approach with this inclusion has much higher detection accuracy than individual machine learning algorithms.

In Ref. [10], the authors compare different classification algorithms, such as AdaBoost and Gradient Boost, using the DARPA dataset, but this dataset cannot fully represent the real-world scenarios now. The key points that arise from this paper are that cybersecurity is becoming very important, the need for machine learning to recognize intricate patterns, and keeping the intrusion detection systems up to date is very difficult. Therefore, the requirement for better accuracy is enhanced by increased cyberattacks. Key insights include the growing importance of cybersecurity, the role of machine learning in detecting complex patterns, and the challenges in maintaining up-to-date intrusion detection systems.

In Ref. [11], the authors explain various techniques, including IIDPS, that identify malicious activities at the system call level, and deep learning architectures to make IDS more flexible and effective in detecting new or zero-day attacks. The study emphasizes the fact that data security is increasingly important due to the growth of cyber threats and a constant need for monitoring network traffic to prevent it.

In Ref. [12], the authors elaborate on the importance of machine learning in cybersecurity, which contributes to real-time threat detection, enhancing network security, and automating security measures. They discuss a variety of algorithms, including regression, clustering, and classification, along with tools such as bio-HAIFCS and Cognito, in order to defend against cyber threats and develop better security practices, even email monitoring.

In Ref. [13], the authors focus on the application of the random forest algorithm for malware detection, especially on the problems related to polymorphic malware. The authors show that the algorithm has great effectiveness, achieving a detection rate of 98% and decreasing false positives. Important topics discussed are literature review, feature selection, dataset preparation, and methodology. The study calls attention to the careful feature selection and preprocessing needed for proper detection, and further research might involve further tuning of the random forest model to better enhance malware detection capabilities.

In Ref. [14], the authors strengthen NIDS by the application of machine learning algorithms to increase the accuracy and reduce dependence on network engineers. It uses the NSLKDD dataset, incorporates pre-processing and feature selection, and deploys the model in a cloud environment for efficiency. Containerization ensures machine independence and cost-effectiveness. The approach highlights the importance of both host-based and network-based IDS in network security.

In Ref. [15], the author presents the optimization of intrusion detection systems using deep neural networks and ensemble techniques. The work on research has been carried out by methods such as distributed random forest, gradient boosting machine, XGBoost along with deep neural networks, and the Python H_2O module. The study concludes that the enhanced model beats previous results of the deep neural networks and traditional machine learning models as well, in light of the fact that the genetic approach is considered in feature selection. Two critical insights are the usability of ensemble approaches and how feature selection plays a very important role in improving model performance.

In Ref. [16], the authors consider an IDS based on PCA and SVM, highlighting the importance of IDS in network security. It considers the application of different SVM kernels and has proved that the Gaussian Radial Basis Function kernel provides better detection accuracy based on the KDDCup'99 dataset. PCA is also utilized to reduce data dimension, which improves detection accuracy as well as speed. The study underlines the effectiveness of machine learning-based IDS and suggests future research into comparing different supervised learning algorithms and PCA's impact on IDS performance.

In Ref. [17], the authors designed a pattern-based intrusion detection (PIDE) model intended to extend the design of host-based intrusion detection systems. Considering post-login behavior analysis as the foundation for detecting intrusions following authenticated account access, a previously studied PIDE model will be integrated with a pattern recognition engine, which will be developed using a previously proposed statistical-based intrusion detection (SBID) model. Experimental results show that this hybrid approach improves the detection of anomalous user behavior, providing a comprehensive mechanism to identify intrusions effectively. The study underlines the importance of combining statistical and pattern-based methodologies to strengthen intrusion detection capabilities.

In Ref. [18], the authors provide a good overview of the modern IDS. It addresses challenges such as the diversity of intrusion categories and the computational demands of the current systems. The study proposes a taxonomy to classify and organize the IDS methodologies. It gives a clear framework in which one can understand the evolution and implementation of the systems. In addition, it provides summarized tables and figures to visualize the state of the field. This survey places a strong emphasis on structured analysis in solving the complexities of IDS development, and it emphasizes the requirement for innovative approaches to deal with increasing network throughput and evolving security threats.

In Ref. [19], the authors discuss the growing problems in effectively identifying advanced attacks that target data availability, confidentiality, and integrity. This paper presents an inventory of modern IDS, which fall into two categories: anomaly-based IDS (AIDS) and signature-based IDS (SIDS). This paper particularly focuses on the most often-used datasets for the evaluation of IDSs and reviews the relevant recent work. It also sheds light on future research issues that have a focus on upgrading system security, discussing some evasion techniques by attackers used to avoid IDS devices. In the opinion of this research work, innovative approaches against intrusion techniques for changing methodologies will require good network security.

3 Design and Methodology

In Fig. 1, the proposed system's architecture has been illustrated. It portrays two different clients, which are trying to connect to the server. It also depicts the existence of the intrusion detection system powered by machine learning models. In explanation, the process's flow can be given as:

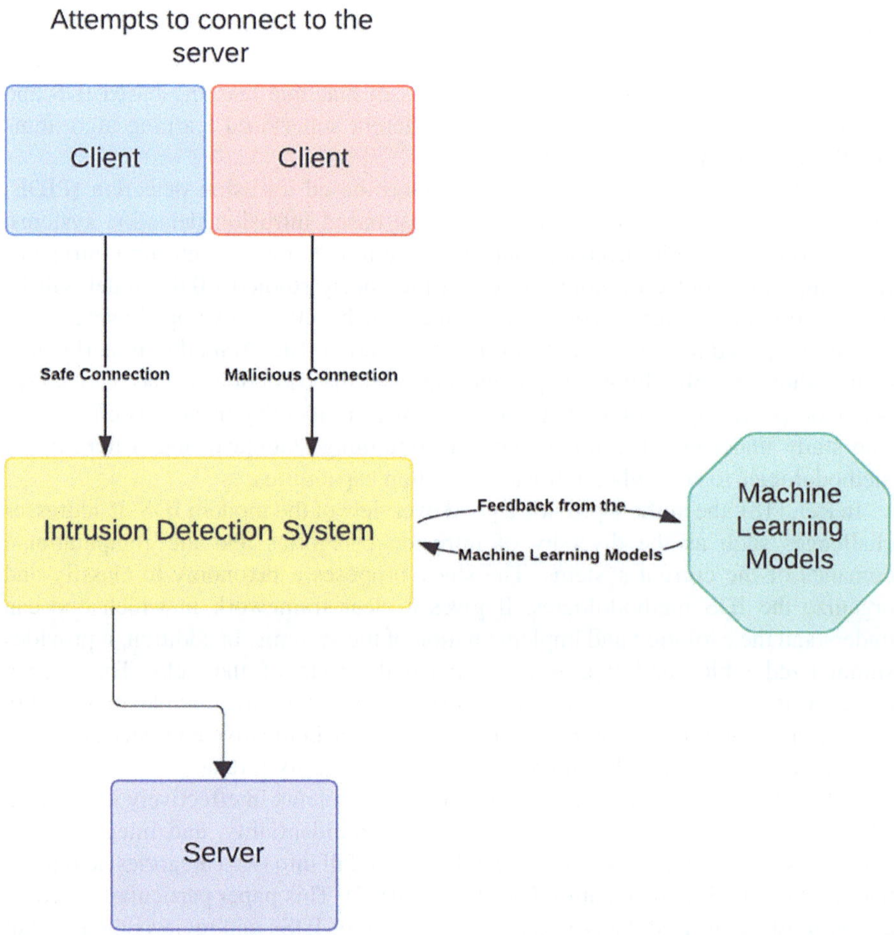

Fig. 1 Proposed architecture of the system

A connection request is sent to the server by one client, which is safe in nature, i.e., that contains no malicious kind of content. The other client sends a connection request to the server, which is unsafe in nature, i.e., malicious content embedded into it.

The intrusion detection system scans the connection parameters and extracts them so that the machine learning models can use them as features to predict the nature of the connection. If the connection is discovered to be safe, the intrusion detection system forwards the connection to the server, which can take forward the connection establishment process.

If the connection is found to be unsafe or carrying some malicious content, the intrusion detection system does not allow the connection to proceed further to the server by rejecting the connection as a whole.

Figure 2 represents the decision-making process of the proposed intrusion detection system (IDS. This flowchart decomposes the sequential steps followed to accept or reject the connections made in the network as being safe (normal) or unsafe (malicious). It gives a systematic view of how the system is integrated with machine learning models for the enhancement of network security. The system ensures enhanced network security because unsafe connections are blocked and only the safe connections go on to the server. This flowchart matches the study's objective, using machine learning to improve IDS efficiency and show automation and precision in threat detection. Experimental results indicated that random forest exceeded the SVM model in terms of accuracy, recall, and overall robustness, thus being the model of choice for the practical deployment of IDS. The figure points to the success of the incorporation of machine learning techniques in cybersecurity and provides a way forward to more advanced improvements, including real-time application and adaptability toward emerging threats, making for a scalable and robust defense mechanism.

3.1 Dataset

The NSL-KDD dataset is an improvement of the KDD '99 dataset, which was designed to test intrusion detection systems. Even though the KDD '99 dataset is quite comprehensive, it has a few inherent problems, such as imbalance and redundancy, which affected how well machine learning models performed when trained on it [3].

The NSL-KDD dataset eliminates duplicate records, reduces the number of training instances, and gives a more balanced representation of normal and attack traffic. This makes it a more reliable and accurate dataset for developing and testing IDS.

The NSL-KDD dataset includes 41 features for every record in a network connection. It also has a corresponding label, which tells us whether it is a record of normal traffic or one of these known attacks.

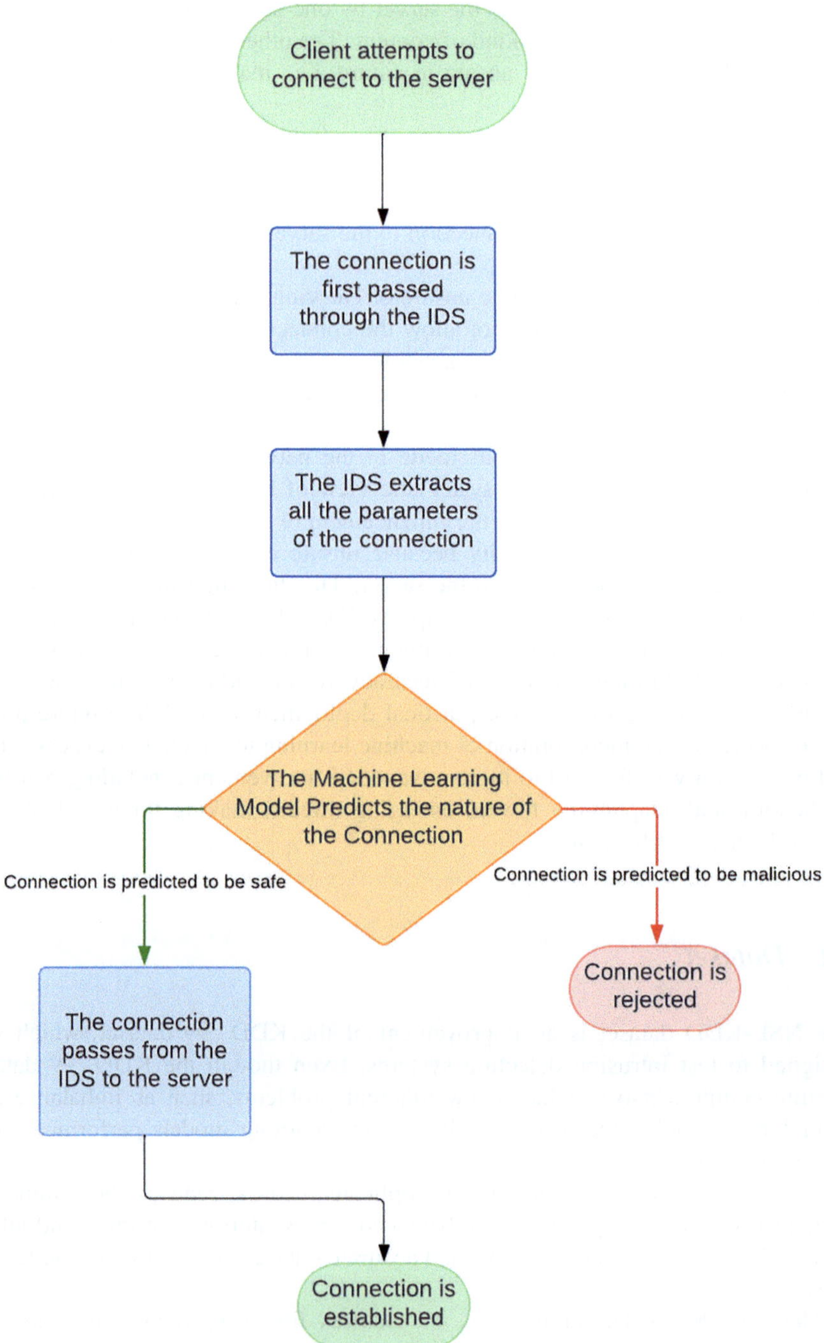

Fig. 2 Flowchart of the IDS process

One of the following categories is assigned to every record in the connection within the dataset:

- **Normal:** These are the normal activities seen in networks.
- **DoS:** Denial-of-Service attacks, in which malicious traffic overloads a network.
- **Remote-to-Local (R2L):** These attacks give unauthorized access to an attacker to a local computer.
- **User-to-Root (U2R):** These attacks occur when a local attacker obtains root access.
- **Probe:** Probing attacks search for vulnerabilities or gain information by scanning networks.

The following is a list of some of the features found in the NSL-KDD dataset [4]:

- **Duration:** Length of the connection (in seconds).
- **protocol_type:** This is a type of protocol. Examples are ICMP, UDP, and TCP.
- **service**: Destination network service. Examples include HTTP, FTP, and SMTP.
- **flag:** This is the status flag of the connection.
- **src_bytes:** This refers to data bytes from source to destination.
- **dst_bytes:** This represents the number of data bytes from destination to source.

The NSL-KDD dataset is widely used for network intrusion detection research for the following reasons:

1. **Highly used and well established:** It is one of the most widely used datasets in the field, making it a benchmark for comparing different intrusion detection techniques.
2. **Network has various types of traffic**: Since there are both normal and attack traffic, this makes them useful in determining the capabilities of intrusion detection systems by probing these abilities.
3. **Reasonable size:** While not too big, it is large enough to give statistical size of network traffic.

Advantages of NSL-KDD

1. **Addresses KDD '99 issues:** Resolves KDD '99 issues. The dataset, which was used to form KDD'99, has been enhanced and is now named NSL-KDD. This one addresses several of the issues of the KDD'99 dataset that includes skewed class distribution and redundant records.
2. **Diverse attack types:** There are only four main categories of assaults covered, such as denial of service (DoS), user to root (U2R), remote to local (R2L), and probes.
3. **Publicly available:** It can be employed for free comparison and testing by researchers.

Limitations of NSL-KDD

1. **Not updated:** The dataset is based on older network protocols and also older techniques of attacks that may not represent the recent network traffic.
2. **Class imbalance:** NSL-KDD, similar to other datasets of intrusion detection, experiences class imbalance where normal traffic is extremely high compared to attack traffic. This would bias the model's training in machine learning algorithms.
3. **Simulated data:** The dataset is primarily simulated, which may not reflect the real network traffic patterns accurately.

Comparison with KDD '99 and CICIDS

- **KDD'99:** An enhanced version of the KDD'99 dataset is the NSL-KDD dataset, which overcomes its drawbacks but is outdated as are both datasets.
- **CICIDS:** CICIDS is the most recent dataset for representing current network traffic. It is huge and very complex, which can make its work very challenging.

In conclusion, NSL-KDD is a popular and well-established dataset for network intrusion detection research. Although it has some limitations, it provides a valuable resource for evaluating intrusion detection techniques. Researchers often choose NSL-KDD as a starting point due to its widespread use and accessibility.

3.2 Data Pre-processing

Split for data taken for training and testing is 85:15. NSL-KDD dataset preprocessing plays an important role in ensuring the performance of the machine learning models employed in the IDS. The key stages of pre-processing implemented in the research were:

1. **Missing Values Handling:**
   - **Technique:** The missing values were replaced with statistical imputation methods, where mean, median, or mode is the corresponding measure of the feature itself.
   - **Reasoning:** It will ensure that the data is complete and no useful information gets lost; otherwise, the process in training the model might be affected.
2. **Categorical Feature Encoding:**
   - **Categorical Features:** These features include protocol_type, service, and flag, and they are encoded for changing categorical data into numbers.
   - **Techniques used:**
     - *One-hot encoding:* It transforms categorical variables into binary vectors to better represent non-ordinal features.
     - *Label encoding:* This assigns unique integers to each category. This is useful for ordinal data.

- Impact: It ensures that the machine learning models could process and interpret the categorical data without introducing bias.

3. **Scaling Numerical Features:**

- **Standardization:** To ensure that all the features of the model contribute equally to it, numerical features were standardized by ensuring that their mean was zero and standard deviation was 1.
- **Normalization:** An approach in which values are scaled in the range [0, 1] to satisfy requirements from an algorithm.
- **Purpose:** Scaling has helped algorithms converge faster, particularly those based on gradients such as SVM, and improve stability and performance.

Detailing the pre-processing steps makes sure that the dataset is clean, well-balanced, and suitable for machine learning tasks. This will let the models focus on meaningful patterns instead of being overwhelmed by inconsistencies or redundancies in the raw data. Feature selection or engineering has not been applied.

3.3 Dimensionality Reduction

Dimensionality reduction is one of the very important steps in the process of data pre-processing for machine learning when dealing with very complex datasets, as seen in intrusions. It aims mainly at reducing the number of dimensions while keeping as much related information as possible. Sometimes, it helps to reduce dimensionality-related problems associated with the curse of dimensionality, enhances model performance, and lowers computational cost.

Application of PCA in Our Project

In our intrusion detection project, PCA was utilized for dimensionality reduction for the 41 features within the NSL-KDD dataset. The following procedures were used to apply PCA:

PCA Transformation

- After pre-processing the data, apply PCA to transform the initial 41 features into a smaller number of principal components.
- We chose to retain 20 principal components. This choice was based on balancing the need to reduce dimensionality while preserving enough variance to maintain the integrity of the dataset's information.
- This means the PCA algorithm has as its guiding criterion in selecting which among the 41 main features to retain, the retained 20 features that characterize the largest variation in the original data.

4 Algorithm and Models

Step 1: Preprocess the data.

Step 1.1: Scaling the data.
Step 1.2: Encoding the data.
Step 1.3: Transform the data.
Step 1.4: Pipeline the preprocessing.

Step 2: Data splitting into Train and Test categories.
Step 3: Dimensionality reduction (PCA), if necessary.
Step 4: Random Forest:

Step 4.1: Creating an RF object
Step 4.2: Fitting the model on the dataset
Step 4.3: Testing the data against the trained model

Step 5: Support Vector Machine:

Step 5.1: Creating an SVM object.
Step 5.2: Fitting the model on the dataset.
Step 5.3: Testing the data against the trained model.

Step 6: Metrics:

Step 6.1: Calculating the Accuracy.
Step 6.2: Calculating the Precision.
Step 6.3: Calculating the Recall.
Step 6.4: Calculating the F1 Score.
Step 6.5: Calculating the Confusion Matrix.
Step 6.6: Forming the Heatmap.

4.1 Machine Learning Models

Random Forest Classifier

The random forest model was implemented using the random forest classifier from Scikit-learn. It was included in a pipeline that combined data pre-processing (standardization and encoding) with the classifier itself. The model was trained on the NSL-KDD dataset, which was pre-processed to handle missing values and categorical features.

Support Vector Machine (SVM)

A support vector machine is another supervised learning model, which is used in classification and regression applications. It is based on the concept of finding the best hyperplane that separates classes in feature space.

This used the SVC from Scikit-learn with a linear kernel. Similar to the random forest model, this model was included in a pipeline including pre-processing steps handling both numerical and categorical features. It was trained and tested on the NSL-KDD dataset.

5 Results

We analyzed the performance of the different models against the performance metrics specified, including accuracy, precision, recall, and F1 score, after a complete analysis of the NSL-KDD dataset using random forest and support vector machine. Table 1 can be used to illustrate these comparisons:

After analysis of from Table 1, we can summarize the results as follows:

The random forest classifier gives us high scores across the four metrics even without PCA decomposition. After PCA decomposition, there is a drop in the score for the Random Forest Classifier, but the drop in scores is never more than 4%.

For the support vector machine model, we get a high score for the accuracy, while the other metrics take a massive deduction. After PCA decomposition, the support vector machine model also witnesses a decrease in the scores for accuracy, recall, and F1 score, with a 16% decrease in recall as well as F1 score.

The only metric to see an increase in the score after PCA decomposition is the precision for the support vector machine model. It witnessed a 12% increase.

We can conclude based on the results that, based on better metrics along the entire four parameters considered, the Random Forest Classifier is an appropriate algorithm to use on the NSL-KDD dataset for building an IDS when comparing this to the support vector machine.

The results can be visualized using the following heatmaps in Figs. 3 and 4:

In the given heatmaps, the squares with the darker shades of blue depict more instances falling under that category.

Table 1 Summarises the metrics for both the models before and after PCA

Model	Accuracy (%)	Precision (%)	Recall (%)	F1 Score (%)
Random forest classifier	99.532	97.021	92.952	94.818
PCA decomposed random forest classifier	98.788	93.926	91.892	92.877
Support vector machine	93.614	80.821	73.455	75.743
PCA decomposed support vector machine	91.860	92.4	57.6	59.811

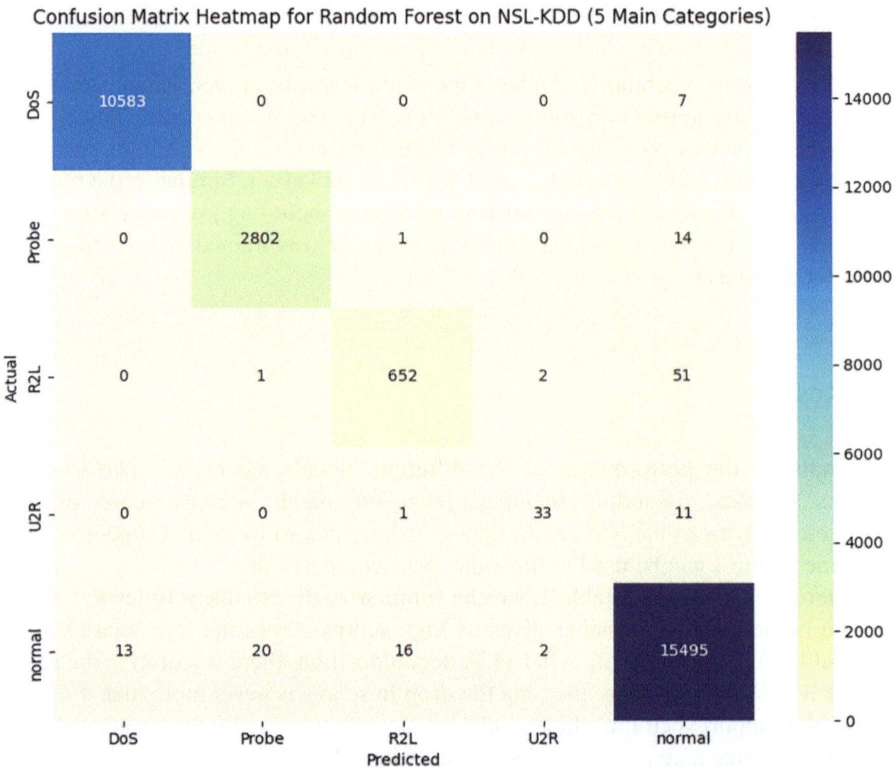

Fig. 3 Confusion matrix for the random forest classifier

After analyzing the heatmap in Fig. 3 for the random forest classifier, we observe that out of 29,704 instances, we get the correct classification in 29,566 instances, with the best output coming in the prediction of the DOS attacks, where 10,583 instances out of 10,590 instances are predicted correctly.

The incorrectly classified instances are in the double digits as compared to the correctly classified instances, which are as large as five digits.

After analyzing the heatmap present above in Fig. 4, we can conclude that the SVM classifier predicts the correct output in 27,807 cases out of the 29,704 given cases. This is a reduction in cases when compared to the 29,566 correct predictions provided by the random forest classifier. Correct classifications of DoS and probe classifications are also lower when compared to the random forest method.

Incorrectly classified instances are in triple digits, while the correctly classified instances are in five digits. The error rate in the SVM classifier is higher as compared to the random forest classifier, but the comparison must not divert us from the fact that the SVM classifier on its own provides very good results.

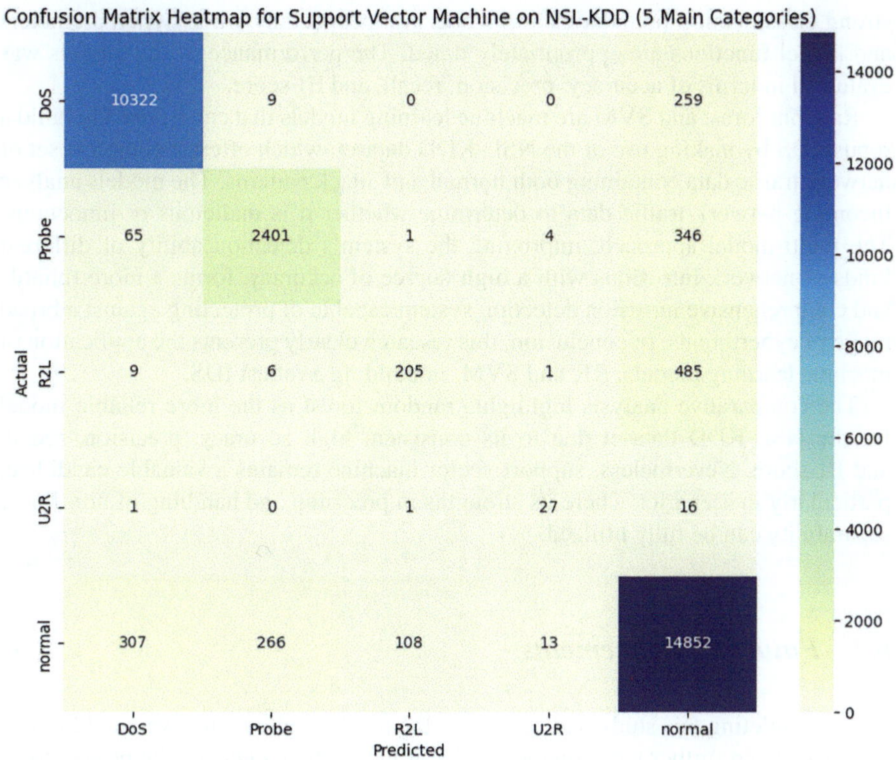

Fig. 4 Confusion matrix for the SVM classifier

As can be seen from the two given heatmaps, the random forest classifier has better metrics than the SVM classifier for all accuracy, precision, recall, and F1 score.

Therefore, it can be said that the random forest model is very accurate regarding the NSL KDD dataset, with most instances getting predicted correctly.

6 Conclusion

We have shown throughout this study how the operation and effectiveness of an IDS, which uses machine learning models to enhance its effectiveness, can be. PCA was the main technique used after dataset preparation to reduce its dimensionality. The use of random forest and support vector machine on the NSL-KDD dataset demonstrates the advantages of ensemble and sophisticated machine learning approaches in network intrusion detection. Since random forest is strong against overfitting and can handle high-dimensional data, it usually gets good accuracy, precision, recall, and F1 scores. Support vector machine provides strong classification boundaries and works well, especially when hyperparameters and kernel functions are appropriately tuned. The performance of the models was evaluated in terms of accuracy, precision, recall, and F1 score.

Random forest and SVM are machine learning models that can be used to build a robust IDS by making use of the NSL-KDD dataset, which offers a complete set of network traffic data containing both normal and attack patterns. The models analyze incoming network traffic data to determine whether it is malicious or innocuous. The multi-model approach, improving the system's detection ability of different kinds of network intrusions with a high degree of accuracy, forms a more reliable and comprehensive intrusion detection system capable of protecting against a broad range of cyberthreats. In conclusion, this research clearly presents the application of machine learning models, RF, and SVM, in building a robust IDS.

The comparative analysis highlights random forest as the more reliable model for the NSL-KDD dataset due to its consistent high accuracy, precision, recall, and F1 score. Nevertheless, support vector machine remains a valuable candidate, particularly in scenarios where its strengths in precision and handling of non-linear separability can be fully utilized.

6.1 Future Enhancements

After completing this study and reviewing the results and outputs, we would like to plan ahead and further work on the results of the study to enhance the project to be much better.

In the future, we would like to apply these models to real-time data and network traffic. We aim to apply our learning and results to real-life computer applications and detect whether the incoming connections are malicious in nature or normal valid network connections. If the connection is a valid connection, the intrusion detection system allows the connection to proceed to connect to the server, and if it is found to be malicious in nature, the intrusion detection system blocks the connection and prevents it from connecting to the server and compromising the system.

6.2 Limitations

The limitations of the study on using machine learning techniques for intrusion detection systems (IDS), as described in the research paper, include:

1. **Dataset-specific challenges**:
    - Although better than KDD '99, the use of the NSL-KDD dataset still has drawbacks in terms of accurately representing the variety of network traffic in the real world and new attack trends.

2. **Model performance trade-offs**:
   - Random forest classifiers demonstrated better overall performance but are susceptible to slight drops in metrics after dimensionality reduction (e.g., PCA).
   - Support vector machines (SVM) showed significant declines in recall and F1 score after PCA decomposition, suggesting sensitivity to feature transformations.
3. **Computational complexity**:
   - Both models, especially random forest, may face challenges in scalability when handling massive real-time data due to the computational cost of training and inference.
4. **Limited scope for real-time application**:
   - Offline analysis is the study's main focus. The viability of implementing the discussed models in real-time systems with large-scale traffic remains unexplored.
5. **Generalization to evolving threats**:
   - Machine learning models trained on static datasets may struggle to detect novel attack types, such as zero-day exploits, without frequent retraining or adaptive mechanisms.

References

1. V. Parlapalli, V. Jayaram, S.G. Aarella, K. Peddireddy, R.R. Palle, Enhancing cybersecurity: A deep dive into augmented intelligence through machine learning and image processing, in *2023 International Workshop on Artificial Intelligence and Image Processing (IWAIIP)*, (IEEE, 2023), pp. 96–100
2. S. Allagi, R. Rachh, B. Anami, A robust support vector machine based auto-encoder for DoS attacks identification in computer networks, in *2021 International Conference on Intelligent Technologies (CONIT)*, (IEEE, 2021), pp. 1–6
3. D. Aksu, M.A. Aydin, Detecting port scan attempts with comparative analysis of deep learning and support vector machine algorithms, in *2018 International Congress on Big Data, Deep Learning and Fighting Cyber Terrorism (IBIGDELFT)*, (IEEE, 2018), pp. 77–80
4. S. Mandala, A.I. Ramadhan, M. Rosalinda, S.M. Zaki, E. Weippl, DDoS detection using information gain feature selection and random forest classifier, in *2022 2nd International Conference on Intelligent Cybernetics Technology & Applications (ICICyTA)*, (IEEE, 2022), pp. 294–299
5. I. Abrar, Z. Ayub, F. Masoodi, A.M. Bamhdi, A machine learning approach for intrusion detection system on NSL-KDD dataset, in *2020 International Conference on Smart Electronics and Communication (ICOSEC)*, (IEEE, 2020), pp. 919–924
6. R.S. Reddy, K.S. Kumar, E.S.N. Rao, P. RajaRajeswari, Enhancing CyberSecurity through random forests: A comprehensive analysis of malware detection and intrusion detection, in *2023 Third International Conference on Ubiquitous Computing and Intelligent Information Systems (ICUIS)*, (IEEE, 2023), pp. 478–482

7. P.K. Danso, E.C.P. Neto, S. Dadkhah, A. Zohourian, H. Molyneaux, A.A. Ghorbani, Ensemble-based intrusion detection for internet of things devices, in *2022 IEEE 19th International Conference on Smart Communities: Improving Quality of Life Using ICT, IoT and AI (HONET)*, (IEEE, 2022), pp. 034–039
8. Y.Y. Aung, M.M. Min, Hybrid intrusion detection system using K-means and K-nearest neighbors algorithms, in *2018 IEEE/ACIS 17th International Conference on Computer and Information Science (ICIS)*, (IEEE, 2018), pp. 34–38
9. P.V. Pandit, S. Bhushan, P.V. Waje, Implementation of intrusion detection system using various machine learning approaches with ensemble learning, in *2023 International Conference on Advancement in Computation & Computer Technologies (InCACCT)*, (IEEE, 2023), pp. 468–472
10. J.A. Abraham, V.R. Bindu, Intrusion detection and prevention in networks using machine learning and deep learning approaches: A review, in *2021 International Conference on Advancements in Electrical, Electronics, Communication, Computing and Automation (ICAECA)*, (IEEE, 2021), pp. 1–4
11. A. Kiran, S.W. Prakash, B.A. Kumar, T. Sameeratmaja, U.S.S.R. Charan, Intrusion detection system using machine learning, in *2023 International Conference on Computer Communication and Informatics (ICCCI)*, (IEEE, 2023), pp. 1–4
12. M. Sudhakar, K.P. Kaliyamurthie, Machine learning algorithms and approaches used in cybersecurity, in *2022 IEEE 3rd Global Conference for Advancement in Technology (GCAT)*, (IEEE, 2022), pp. 1–5
13. A. Dixit, S. Singh, Malware detection using random forest, in *2023 14th International Conference on Computing Communication and Networking Technologies (ICCCNT)*, (IEEE, 2023), pp. 1–6
14. C. Kaushik, T. Ram, C. Ritvik, T. Lakshman, Network security with network intrusion detection system using machine learning deployed in a cloud infrastructure, in *2022 3rd International Conference on Electronics and Sustainable Communication Systems (ICESC)*, (IEEE, 2022), pp. 701–708
15. A. Rai, Optimizing a new intrusion detection system using ensemble methods and deep neural network, in *2020 4th International Conference on Trends in Electronics and Informatics (ICOEI)(48184)*, (IEEE, 2020), pp. 527–532
16. P. Nskh, M.N. Varma, R.R. Naik, Principle component analysis based intrusion detection system using support vector machine, in *2016 IEEE International Conference on Recent Trends in Electronics, Information & Communication Technology (RTEICT)*, (IEEE, 2016), pp. 1344–1350
17. Z.S. Malek, B. Trivedi, A. Shah, User behavior pattern-signature based intrusion detection, in *2020 Fourth World Conference on Smart Trends in Systems, Security and Sustainability (WorldS4)*, (IEEE, 2020), pp. 549–552
18. H.J. Liao, C.H.R. Lin, Y.C. Lin, K.Y. Tung, Intrusion detection system: A comprehensive review. J. Netw. Comput. Appl. **36**(1), 16–24 (2013)
19. A. Khraisat, I. Gondal, P. Vamplew, J. Kamruzzaman, Survey of intrusion detection systems: Techniques, datasets and challenges. Cybersecurity **2**(1), 1–22 (2019)

KLOA: Improving VANET Security Against Social Engineering Attacks Through Strategic Optimization

Nitha C. Velayudhan, Arun Pradeep, and Mukesh Madanan

1 Introduction

Advancements in wireless technology have greatly improved communication in the twenty-first century, offering better advantages over wired technology. Wireless mobile nodes can interoperate across different capacities and dynamic structures and provide an unlimited scope of performance. The development of wireless technology has made it possible to design affordable transceivers capable of providing adequate power for the purpose of communication.

A particular form of mobile ad hoc network (MANET) is the vehicular ad hoc network (VANET), which is a crucial component of intelligent transportation systems (ITS) as depicted in Fig. 1.

VANETs enable communication between vehicles (V2V) and between vehicles and infrastructure (V2I) using short-range wireless technologies, facilitating applications like dynamic route planning, traffic safety, and intelligent transportation systems. The VANET architecture comprises onboard units (OBUs), application units (AUs), and roadside units (RSUs), enabling various forms of communication, including vehicle-to-everything (V2X), to create a fully integrated ecosystem. Key technologies such as IEEE 802.11p (WAVE) and IEEE 802.16 (WiMAX) support

N. C. Velayudhan (✉)
Department of Computer Science and Engineering, Christ College of Engineering, Irinjalakuda, Kerala, India

A. Pradeep
Department of Electronics and Communication, Saveetha Engineering College, Chennai, Tamil Nadu, India

M. Madanan
Department of Computer Science, Dhofar University, Salalah, Oman
e-mail: mukesh@du.edu.om

A. Haldorai et al. (eds.), *7th EAI International Conference on Big Data Innovation for Sustainable Cognitive Computing*, EAI/Springer Innovations in Communication and Computing, https://doi.org/10.1007/978-3-032-02979-9_15

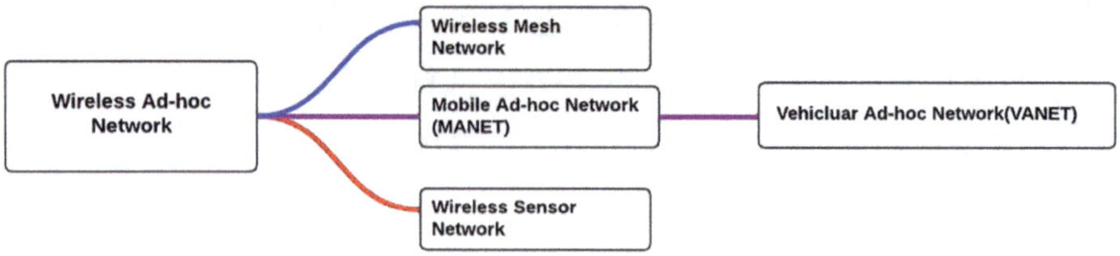

Fig. 1 The wireless network ad hoc hierarchy

effective communication in VANETs, but their open and dynamic nature makes them vulnerable to attacks like eavesdropping, Sybil, denial of service (DoS), and social engineering.

To address security challenges, advanced detection methods are crucial. Random forest, a powerful ensemble machine learning algorithm, can analyze traffic patterns and classify them as normal or malicious by utilizing features like connection counts, source bytes, and trust scores. By optimizing hyperparameters such as tree depth and the number of estimators, random forest balances complexity, stability, and computational efficiency, making it suitable for ensuring high accuracy and reliability in VANET security applications.

In our approach, each lion in the population represents a candidate set of hyperparameters, specifically, the number of estimators and the maximum depth for the random forest model. The candidate solutions are iteratively improved by the lion optimization algorithm (LOA) through refining and exploration. Maintaining a balance between local exploitation and global exploration is a typical difficulty faced by metaheuristic algorithms, particularly when dealing with complicated and noisy datasets like VANETs.

To alleviate this challenge, we integrate K-means clustering into the optimization process. It establishes smaller groups that concentrate on distinct sections of the search space by clustering the population of candidate solutions (lions). The best solution (lion) in each cluster is chosen and used to yield generation, allowing for the local exploitation of parameter space's potential areas while preserving the population's overall diversity [1]. This clustering-based method prevents premature stagnation in suboptimal solutions and speeds up convergence, both of which increase the efficiency of LOA.

There are two steps of analysis provided in the work: training an initial random forest model without hyperparameter optimization and splitting up the dataset into training and test sets. It serves as the comparison point for the model. Next, LOA is used to optimize the hyperparameters by K-means clustering for the random forest algorithm leading to more improved results in detecting social engineering attacks in VANETs. The new hyperparameter retrained model further improves the accuracy metrics, is used to estimate the improved performance.

Our suggested approach, K-means lion optimization algorithm (KLOA), combines LOA and K-means clustering to offer an effective way to adjust hyperparameters, which improve the detection of social engineering attacks in VANETs.

By efficiently balancing local and global search strategies, this method outperforms nonoptimized models in terms of accuracy. Thus, a strong defense against social engineering attacks against VANET security can be provided by the combination of metaheuristic optimization, clustering, and machine learning, which will ultimately lead to more reliable vehicular communications. The structure of this chapter is as follows. An overview of the prevailing research is provided in Sect. 2. The problem is explained in Sect. 3, while Sect. 4 contains the suggested model. The experimental analysis and results are under Sect. 5. Finally, Sect. 6 draws the chapter to a conclusion.

2 Literature Review

VANETs turn vehicles into well-functioning interactive and safe systems to enhance safety as well as comfort applications. Accident prevention by VANETs is obtained by cooperative message transfer about dangers like snowy roadway, rapid curves and lane changes, postcrash alarms, and collision warnings. Maintaining safe gaps and speeds of vehicles, traffic platooning in an urban environment, on the other hand, optimizes traffic flow. In low-visibility conditions, communication of the VANETs enables the vehicles to be alerted about a cyclist or pedestrian being near enough to pose a danger. Personalized comfort-entertainment services like real-time navigation and location-based services, such as restaurant and lodging locations, are provided by VANETs. In cities, smart parking support VANET facilitates finding and reservation of parking spaces. However, these applications can be compromised during social engineering attacks, which might manipulate or disrupt communication and make the whole system ineffective.

This section gives a review of the literature on security threats in vehicular ad hoc networks (VANETs) with a specific focus on social engineering attacks as well as the limitations of existing detection techniques.

Several approaches have been introduced to reduce the impact of social engineering attacks in VANETs. Cryptographic algorithms, including digital signatures and authentication methods, are used to ensure security against eavesdropping during data transmission or to verify the ID-related nodes. However, these methods often require considerable processing resources, making them unsuited to the VANET in resource-constrained circumstances [2].

Another promising way to find and neutralize social engineering attacks is by applying the machine learning technique. By historical analysis of data, a machine learning model can identify patterns that point to malicious activity. Methods such as random forest classifiers, K-means clustering, and support vector machines (SVM) have been applied in the context of anomaly detection over network traffic and in differentiating legitimate and malevolent activity. However, one key limitation of these approaches is that they basically depend on the selection of hyperparameters, which, if done incorrectly, could result in inadequate detection performance [3].

The current mitigation strategies provide the foundation for security threats to be mitigated in VANET systems; however, their weaknesses stem from the need for fast real-time processing, scalability, and adaptability. As a result, sophisticated, adaptive techniques that can be relied upon to protect against complex social engineering attacks on VANET systems must be developed [4]. With this motivation, we proposed a novel technique based on machine learning that provides a step for the integration of sophisticated hyperparameter tuning in order to bridge existing inadequacies. Our approach improves the detection of social engineering attacks in VANETs by taking the random forest classifier along with the K-means lion optimization algorithm (KLOA). By optimizing hyperparameter selection, this approach enhances the classification accuracy by considering the dynamic nature environment of VANET. Ultimately, our novel methodology will offer a solid solution for defending VANETs against emerging threats [5].

3 Problem Statement

The first digital social engineering appearance was in the early 1990s when infamous cybercriminals like Kevin Mitnick were accused.

In the sphere of VANETs, social engineering attacks are caused by the operations when a criminal intervention changes driver behavior or takes advantage of network trust, which could lead to serious security problems in VANETs [6].

Serious security issues in the rapidly evolving field of vehicular ad hoc networks (VANETs) arise with networked vehicles and are mainly concerning social engineering attacks. This kind of attack uses psychological tricks to force people into divulging private information or breaking the integrity of the network. Therefore, there is a need to provide robust security for this kind of threat as the use of VANET continues to rise.

As traditional security measures usually fail because they cannot adjust to the dynamic nature of VANETs and the complex strategies employed by attackers, advanced systems for the detection of social engineering attacks are in immediate demand, able to identify and neutralize the aggression.

With this end in view, the goal of developing a model that detects these types of attacks using machine learning techniques, especially the random forest classifier. The machine learning model optimized using the K-means lion optimization algorithm (KLOA) is employed to secure VANETs from social engineering attacks and keep them safe, reliable, and resourceful in their functionalities. KLOA can be carried out in order to optimize the model to achieve higher accuracy and efficiency. This approach has been suggested to increase overall resilience and security of VANETs, along with increasing detection rates against social engineering attacks in these types of networks [7].

4 Proposed Methodology

A summary of various studies on VANET security is provided in the preceding section. Quite evidently, it is clear that the current approaches have not produced satisfactory results because standard security measures typically fail due to their inability to adapt to the dynamic nature of VANETs and the sophisticated methodologies deployed by attackers. Research proposes a novel approach, the K-means lion optimization algorithm (KLOA), to optimize the hyperparameters of a machine learning model, specifically a random forest classifier, for detecting social engineering attacks in VANETs. The proposed system includes improved LOA with clustering by organizing the population (set of hyperparameters) into clusters. K-means clustering is incorporated into the LOA (KLOA) to improve the exploration and exploitation processes. Local mating and exploration within these clusters allows for improved exploitation of local optima while preserving global exploration between clusters.

Based on data trust scores and network activity metrics, the technique greatly increases the random forest model's accuracy in identifying attacks involving social engineering by optimizing critical hyperparameters like the number of estimators and tree depth. The proposed method's architecture is as displayed in Fig. 2.

The model enhances the identification of attack patterns in a dynamic environment of a VANET, where traditional approaches may struggle, by using KLOA, a formation of stronger and efficient optimization process can be produced.

Quick detection of social engineering attempts is critical for maintaining the security of VANETs, and this research is very relevant in this regard. In order to achieve this objective, the proposed approach provides a reliable method to adapt the model.

4.1 Random Forest Classifier Model Training

A supervised machine learning method of random forest, where "individual weak learners" (decision tree) work together to build a "strong learner," is used twice in this proposed methodology, once during initial training and once during the optimization phase, as it is essential to both the hyperparameter optimization process and the initial model training. Based on trustworthiness and network metrics, the feature trust_score_average aids the random forest model to efficiently learn the patterns of normal behavior vs social engineering attack. The same features are then employed during hyperparameter optimization, which dynamically adjusts parameters like max_depth and n_estimators to optimize model performance.

The random forest algorithm generates multiple classification trees to avoid overfitting to training datasets, with a bootstrap randomized resampling technique for building each tree [8]. A majority vote among the forest's trees determines the

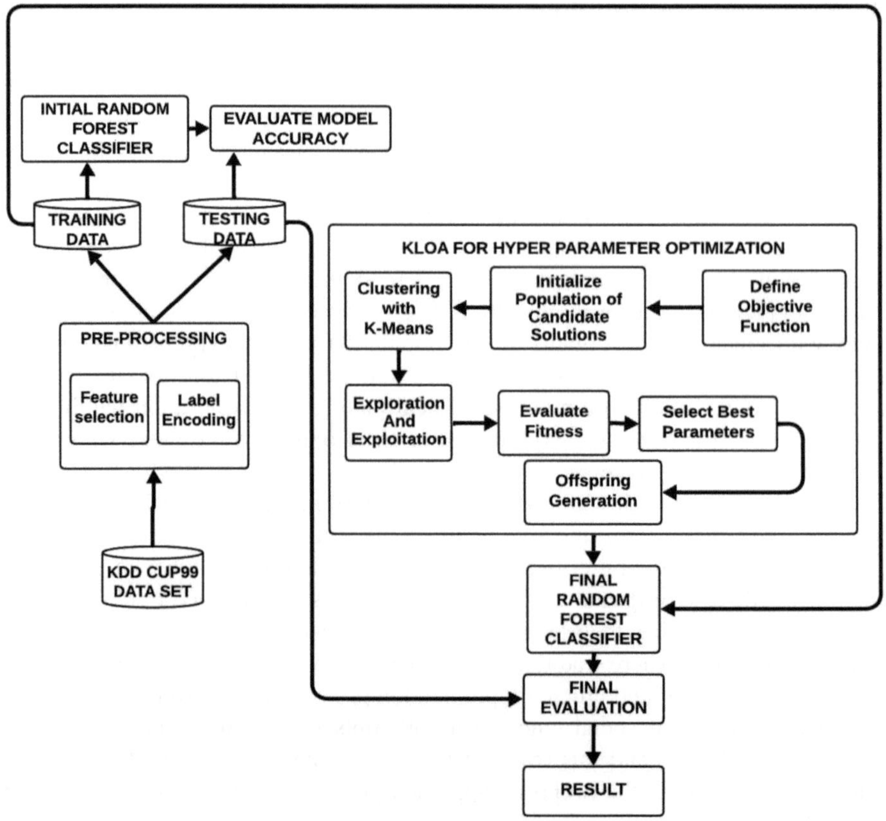

Fig. 2 Framework for KLOA-based detection of social engineering attacks in VANETs

final prediction. The decision function is stated as follows:

$$\hat{y} = \text{majority vote}\,(d1(x), d2(x), \ldots, dt(x)) \quad (1)$$

where $\hat{y}$ is the label predicted as attack or normal. $di(x)$ is the ith decision tree prediction, t is the total number of decision trees (given by the n_estimators hyperparameter), and the feature vector is represented by x (e.g., src_bytes, count, and trust_score_average).

In the step of initialization of the random forest model is done with the default hyperparameters—100 trees (n_estimators = 100) and no depth limit (max_depth = None). It is then the model is fit with the training data (X_train, y_train) and then test data (predict on X_test), followed by calculating the prediction accuracy. This step gives a baseline accuracy before we apply any hyperparameter optimization.

4.2 K-Means Lion Optimization Algorithm (KLOA)

Optimization in machine learning is one of the leading aspects that can enhance performance, more specifically related to the optimum hyperparameters for the model. Hyperparameter tuning may be useful if the final result is concerned with the effectiveness of a machine learning algorithm, such as a random forest classifier. The number of trees (n_estimators) and the maximum tree depth (max_depth) are parameters determined for hyperparameter tuning. Finding the right hyperparameters is difficult because of a large search space and the chances of overfitting or underfitting. To overcome these issues, some new optimization methods have come up, which are recognized as bioinspired optimization techniques. One of those techniques is the lion optimization algorithm (LOA). It is a nature-inspired optimization technique recently proposed by Dr. Seyedali Mirjalili in 2015 [9].

LOA draws inspiration from the social life and hunting methodologies employed by lions. In the wild, lions mix explorative and exploitative behavior to hunt prey. The LOA approaches complex optimization problems by dividing its population into prides (subgroups) and seeking a balance between finding new solutions (exploration) and improving the already known ones (exploitation) by using mechanisms like mating, or generating offspring, nomadism or random walk, and territorial defense, or refinement of the local. This brings LOA particularly well to hyperparameter optimization, as it is very prone to local optima that can occur during the process of exploration and exploitation. The proposed algorithm, KLOA, mimics lions' behavior in prides, their respective roles in hunting and territorial defense to perform space search exploration and exploitation to get optimal or near-optimal solutions for the complex optimization problems with clustering.

However, to guarantee a more organized search space exploration, clustering algorithms can be added to LOA to further improve it. In this research, K-means clustering is integrated into the LOA to segment the population of candidate solutions (i.e., hyperparameter sets) into clusters, ensuring diverse exploration across different regions of the hyperparameter space. Clustering, in this approach, enhances both diversity as well as convergence. This methodology also prevents prematurely converging on suboptimal solutions and increases the chances of finding the global optimum.

High accuracy of machine learning models in the detection of malicious activity in VANETs points toward hyperparameter tuning as very crucial for security against attacks. In this research work, we are particularly focusing on the problem of social engineering attack detection within VANETs using a random forest classifier. Therefore, the optimum selection of the classifier's hyperparameters to maximize its accuracy in the differentiation of attack and normal traffic is the approach for the solution of the problem. The approach KLOA utilizes the exploration and refining of a wide range of hyperparameter combinations with LOA and K-means clustering to optimize the classifier that, in turn, strengthens VANET security.

Initialization

The parameters like population_size (the number of diverse solutions evaluated in each iteration of the algorithm) and population (where initial positions are randomly assigned to each candidate solution to get a good diversification in the search space) are initialized. Each lion's position in the search space corresponds to a unique combination of n_estimators and max_depth, the two hyperparameters for the random forest classifier. These hyperparameters are generated using the random_position_in_search_space() function, which ensures that the population is sufficiently diverse and can explore different regions of the solution space. For instance, the max_depth parameter is specified within the range [1, 50] as a limitation of depth on trees; n_estimators is set to a value within the range [10, 200], the number of decision trees in the random forest. This is how, by generating such a diverse population, the algorithm explores a number of various hyperparameter combinations in the optimization process.

This initialization lets the algorithm start with a range of possible solutions and effectively assess them. This setting is necessary for KLOA to find an optimal path or solution in VANETs or other optimization problems.

Optimization of Hyperparameter

After initializing the population, K-means clustering is applied to group the candidate solutions(vehicles) into a number of clusters. Clustering is necessary to maintain an appropriate exploration-exploitation balance.

To balance between exploration and exploitation, the algorithm in the exploration phase ensures that different areas of the search space are explored by grouping candidate solutions(vehicles) into clusters so to avoid the premature convergence to suboptimal solutions and in exploitation phase, the fittest solutions in each cluster are used to produce new potential solutions to ensure the algorithm exploits promising areas of the solution space.

Additionally, a random forest model is used to evaluate the fitness of each vehicle configuration (hyperparameter set). The random forest receives the vehicle hyperparameters (n_estimators and max_depth) and is trained to model the training data (X_train, y_train). The evaluate_fitness() function is used to calculate the fitness value that directs the algorithm to the solution accordance to the decided approach. The model's accuracy on the test set (X_test, y_test) is determined by

$$\text{Fitness} = \text{model accuracy} = \frac{\text{correct prediction}}{\text{total prediction}} \tag{2}$$

After evaluating the fitness of each candidate solution, the best candidates from every group are chosen, and it is considered the most optimal hyperparameter combinations that the best candidates can present to improve the random forest classifier performance. After identifying these candidates, the algorithm can use these parameters effectively to enhance the prediction of VANET security attacks. The best candidates will then serve as seeds for the creation of new candidates in the next step, thus keeping this hyperparameter search focused on actual combinations.

To introduce diversity while focusing on the most promising configurations, new candidates (offspring) are created by randomly generating the hyperparameters of the best candidates. This is done by introducing random variations with intervals to the hyperparameters, which determine and adapt settings of the random forest model toward the VANET environment. The offspring generation process in the algorithm is represented by the equation:

$$\text{offspring} = \text{best candidate} + \text{random variation} \tag{3}$$

Individual offsprings are confined within the hyperparameter bounds: n_estimators —between 10 and 200, max_depth—between 1 and 50. This limit we imposed is to guarantee that all the candidates remain valid in training an effective model. This offspring generation process allows the algorithm to find new hyperparameter values near best-performing configurations, and it helps in improving the search for ideal settings.

Population Update and Convergence Toward Optimal Hyper-Parameter

In the algorithm, the process of clustering, population fitness measurement, best solutions selection, and producing offspring for the number of generations is repeated. At each step of the algorithm, the population is to converge toward the optimal set of hyperparameters, and the best solutions are improved in the process of running every iteration. Finally, for the number of iterations, the candidate solution with the maximum fitness value, over all the iterations, is the best set of hyperparameters. Finally set of hyperparameters consisting of n_estimators and max_depth is used to train the random forest model, and then its accuracy is tested on the test data.

The results after applying the K-means lion optimization algorithm (KLOA) before and after optimization are given in Table 1 and in Fig. 3 the iteration-wise gains in accuracy for classifying the random forest model, optimized using LOA and KLOA, are depicted.

The unoptimized random forest sustains an initial static 80% accuracy. Accuracy starts increasing steadily with LOA across iterations up to the point of convergence at 87%. The optimization then improves when K-Means clustering is combined with it in the form of KLOA until it converges to 90%.

Table 1 Comparison of model accuracy before and after optimization using LOA and KLOA

Feature/method	Initial random forest	LOA optimized	K-means + LOA (KLOA)
n_estimators	100	150	160
max_depth	None	20	25
Clustering method	None	None	K-means
Accuracy (before optimization) (%)	80	80	80
Accuracy (after optimization) (%)	N/A	87	90

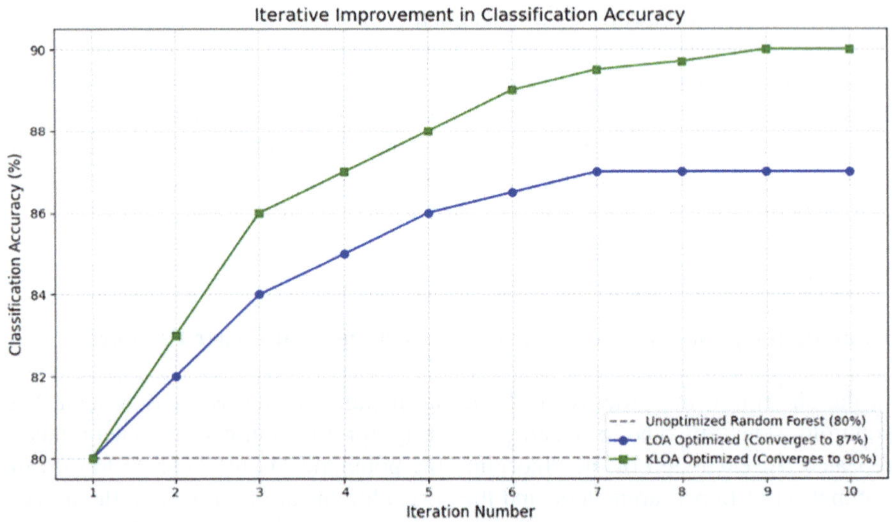

Fig. 3 Iterative improvements of random forest with LOA and KLOA for accuracy

5 Results

The proposed methodology utilizes the KDD CUP '99 [10], which is a widely used dataset, in order to test and analyze the detection of social engineering attacks in VANETs and to maintain reliability and consistency in the evaluation process of IDS performance.

Achieving high model performance relies on metrics like accuracy, precision, recall, and F1-score, which assess prediction quality by balancing true and false outcomes, alongside efficiency factors like convergence rate, indicating optimization speed. These metrics are crucial for evaluating and improving models, especially in complex applications like VANET security. Table 2 provides a comparison of KLOA with the existing optimization model.

The existing methods various optimized machine learning models, including ant colony optimization (ACO), particle swarm optimization (PSO), bee colony optimization (BCO), and genetic algorithm (GA), have been used to optimize hyper-

Table 2 Performance comparison of KLOA with existing optimization model

Metric	Initial RF	LOA	KLOA	PSO	GA	ACO	BCO
Accuracy	80	87	90	88	85	86	87
Precision	78	86	89	87	83	84	86
Recall	5	85	88	86	82	83	85
F1-score	76.5	85.5	88.5	86.5	82.5	83.5	85.5
Iterations to convergence	N/A	100	80	90	110	105	95

parameters within various domains, including security optimization for VANET. However, comparing this model to the proposed KLOA, the current model is superior to the latter in terms of both accuracy and efficiency, making this model a very effective solution in detecting social engineering attacks in VANETs.

ACO simulates an ant population foraging in a search through their hyperparameter space [11]. ACO achieves the remarkable accuracy of 86% but it has a relatively slower convergence, limiting its efficiency compared to KLOA. In contrast, PSO emulates socially interacting bird flocks and fish schools [12]. PSO has achieved slightly higher accuracy with 88% but less than KLOA's general performance.

BCO, initially based on the principles of bee colony foraging behavior [13], was able to achieve an accuracy of 87%, but similar to ACO, had poor convergence rates, which proved to be disadvantageous in time-critical scenarios such as VANETs. On the other hand, GA adopts principles of natural selection and evolutionary characteristics toward optimized solution generation [14]. Though GA can achieve 85% accuracy, it is computationally much costlier and hence consumes more iterations. It becomes infeasible for real-time VANET applications.

The integration of the lion optimization algorithm with K-means clustering can be developed to outperform the above-exposed traditional optimization methods. In fact, it would be a perfect trade-off between exploring mainly diverse solutions and exploiting promising solutions to fine-tune them to achieve peak accuracy at 90%. Moreover, K-means lion optimization algorithm (KLOA) provides faster convergence rates compared to others and is highly suited for real-time detection of social engineering attacks in VANETs. Therefore, with regard to accuracy and efficiency in the process, KLOA can be seen as the best technique for secure improvement in VANET beyond other optimization techniques and even under dynamic as well as resource-constrained environments.

A graphical study of the proposed model and comparison of the previous studies is depicted in Fig. 4. The research proved that in the proposed approach, performance improvement is due to the optimization of the hyperparameters that take benefits from the exploration-exploitation balance of the proposed K-means lion optimization algorithm (KLOA). This optimization resulted in an increased rate of accuracy and faster convergence to detect social engineering attacks than the primitive techniques.

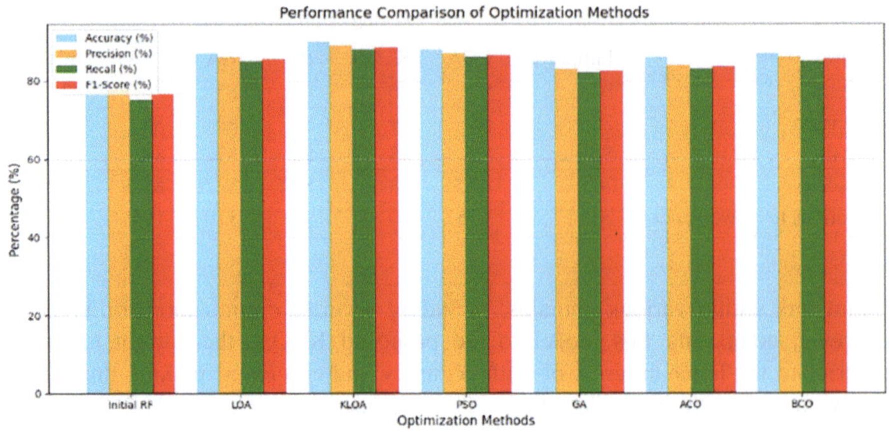

Fig. 4 Graphical analysis for existing and proposed work

6 Conclusion

In this chapter, we propose a new method to improve the detection of social engineering attacks in VANETs using the K-means lion optimization algorithm (KLOA). KLOA is, therefore, characterized by an adaptive hyperparameter optimization through the lion optimization algorithm (LOA), followed by a K-means clustering incorporated to arrange data into random forest classification capability, resulting in a collaborative model with extremely high performance indicative of an accurate yet efficient and resilient classifier for social engineering attack detection.

This work highlights that KLOA balances exploration and exploitation so effectively that it is even capable of navigating the hyperparameter space in a more efficient way than existing ones, thereby making KLOA a robust solution for resource-constrained real-time applications like VANETs. Moreover, the use of machine learning techniques such as random forest classifiers will ensure the scope for automated anomaly detection.

This proposed model, using advanced clustering techniques based on historical interaction data to derive trust scores, boosts detection capability.

For a future enhancement, incorporating reinforcement learning techniques to adaptively refine the detection model based on real-time feedback, along with the integration of deep learning for feature extraction and hybrid optimization approaches, may potentially boost the accuracy and resilience of detection of social engineering attacks on VANETs by a huge margin. Furthermore, the dataset, rather than KDD Cup99 CICIDS 2017, UNSW-NB15, or ISCX 2012, can be used to increase the robustness and generalization of the models for detection.

References

1. M. Yazdani, F. Jolai, Lion optimization algorithm (LOA): A nature-inspired metaheuristic algorithm. J. Comput. Des. Eng. **3**(1), 24–36 (2016)
2. M. Raya, J.-P. Hubaux, Securing vehicular ad hoc networks. J. Comput. Secur. **15**(1), 39–68 (2007)
3. F.G. Marmol, G. Martinez, Trust and reputation models comparison for mobile ad hoc networks. Comput. Commun. **33**(18), 2228–2239 (2010)
4. K. Singh, A. Verma, K. Verma, Security challenges and issues in vehicular ad-hoc networks: A survey. Wireless Net. **27**(2), 879–896 (2021)
5. H. Liu, B. Lang, Machine learning and deep learning methods for intrusion detection systems: A survey. Appl. Sci. **9**(20), 4396 (2019)
6. J.H. Holland, *Adaptation in Natural and Artificial Systems* (University of Michigan Press, 1975)
7. D.E. Goldberg, *Genetic Algorithms in Search, Optimization, and Machine Learning* (Addison-Wesley, 1989)
8. A.L. Buczak, E. Guven, A survey of data mining and machine learning methods for cybersecurity intrusion detection. IEEE Commun. Surv. Tutor. **18**(2), 1153–1176 (2016)
9. L. Breiman, Random forests. Mach. Learn. **45**(1), 5–32 (2001)
10. S. Mirjalili, The lion optimization algorithm: A nature-inspired metaheuristic for large-scale optimization. IEEE Access **3**, 1225–1235 (2015)
11. F. Iglesias, T. Zseby, Analysis of network traffic features for anomaly detection. Mach. Learn. **101**(1–3), 59–84 (2015)
12. M. Dorigo, T. Stützle, *Ant Colony Optimization* (MIT Press, 2004)
13. D. Karaboga, B. Akay, A comparative study of artificial bee colony algorithm. Appl. Math. Comput. **214**(1), 108–132 (2009)
14. J. Kennedy, R. Eberhart, Particle swarm optimization. Proc. IEEE Int. Confer. Neural Netw. **4**, 1942–1948 (1995)

A Comprehensive Survey of Efficient Authentication Protocols for the IoV in VANETs

P. R. Hemalatha and J. V. Anchitaalagammai

1 Introduction

The Internet of vehicles (IoV), a cornerstone of intelligent transportation systems (ITS), has witnessed rapid growth in recent years. As vehicles become increasingly interconnected, the need for security measures to shield sensitive information and guarantee reliable communication becomes paramount. Authentication, a fundamental security mechanism, plays an essential role in establishing trust between communication modes of vehicular ad hoc networks (VANET) like vehicle-to-vehicle (V2V), and V2I communications, and also several other modes that have evolved lately (see Fig. 1) such as vehicle-to-grid (V2G), vehicle-to-pedestrians (V2P), vehicle-to-everything (V2X), etc. Data sharing in VANETs is crucial for enhancing traffic conditions and user experience through the sharing of traffic information, driving assistance, and navigation support. However, this data can include sensitive information like identity and location, which can be endangered by harmful attacks due to the nature of wireless communication. Thus, ensuring data security and access control is essential in VANET data sharing [1]. The Internet of things (IoT) is causing more vehicles to become connected, transforming traditional VANETs into the IoV. This change is a result of the new era of the IoT. With the rapid advancement of computing and 5G technologies, IoV has gained significant commercial and research interest. It focuses on communication among vehicles, humans, and roadside units (RSUs). Furthermore, IoV improves vehicle safety, promotes eco-friendliness by reducing driving risks, and enhances transportation efficiency, resulting in decreased public resource expenditure. The

P. R. Hemalatha (✉) · J. V. Anchitaalagammai
Velammal College of Engineering and Technology, Madurai, Tamil Nadu, India

A. Haldorai et al. (eds.), *7th EAI International Conference on Big Data Innovation for Sustainable Cognitive Computing*, EAI/Springer Innovations in Communication and Computing, https://doi.org/10.1007/978-3-032-02979-9_16

reliability of anonymously forwarded messages is not guaranteed, and users may lack the motivation to respond without incentives [2].

As driverless vehicles gain traction in other countries and prepare for introduction in India, VANETs will be crucial in amplifying traffic management and efficiency. This advancement is required for the implementation of a secure and efficient communication system. Blockchain technology, a robust distributed ledger system, is ideally suited for creating a privacy-preserving authentication framework. This framework will effectively store critical information, such as certificates and system parameters, which vehicles and roadside units (RSUs) can access for seamless authentication. Furthermore, smart contracts within the blockchain will establish clear relationships among relevant data and enable prompt revocation when necessary. Numerous blockchain-based conditional privacy-preserving authentication (B-CPPA) protocols have been proposed, decisively addressing the shortcomings of public key infrastructure (PKI) solutions. These protocols eliminate issues like the opacity of trusted authorities and the burdensome processes associated with certificate revocation [3].

In IoV environments, authentication protocols encounter significant challenges due to vehicular devices' limited resources, the dynamic nature of VANETs, and threats from different attacks, like impersonation and denial-of-service (DoS). To address these issues, researchers have developed several authentication protocols for IoV in VANETs, employing cryptographic techniques like symmetric-key and public-key cryptography. The selection of the most appropriate protocol depends on the desired security level, computational overhead, communication costs, and scalability.

This chapter concisely surveys existing authentication protocols in IoV for VANETs, examining their cryptographic techniques, security properties, and performance characteristics. We highlight lightweight solutions, including pseudonym-based and blockchain-based methods, ideal for resource-constrained environments, and discuss the associated challenges, including scalability and communication costs. By providing a systematic overview of the recent authentication protocols, this survey aims to assist researchers and practitioners in selecting appropriate solutions for their IoV applications. The paper is structured as follows: Sect. 2 provides an overview of the IoV and VANETs, highlighting the security challenges and requirements. Section 3 discusses the types of authentication protocols for IoV. Section 4 surveys existing authentication protocols, categorizing them based on their underlying cryptographic techniques and security properties. Section 5 presents the applications of each authentication protocol and analyzes the challenges and limitations. Finally, Sect. 6 concludes by highlighting the key findings and providing compelling directions for future research endeavors.

2 IoV in VANET

The IoV in VANETs represents a critical evolution in vehicular communications, enabling smarter, more efficient transportation systems. VANET is an instance of a mobile ad hoc network (MANET) that enables wireless communication between automobiles and roadside infrastructure. It enables short-range communication and is primarily applied for vehicle safety, traffic management, and efficiency. VANETs are the foundation for modern connected vehicles. The IoV extends VANET concepts to a more connected and intelligent system by incorporating vehicles with comprehensive IoT. In IoV, vehicles are part of an ecosystem that connects with other vehicles (V2V), infrastructure (V2I), pedestrians (V2P), the power grid (V2G), roadside units (V2R), mobile devices (V2D), home (V2H), and sensor devices (V2S). As shown in Fig. 1, the communication model of IoV involves various types of communication, each serving specific purposes and utilizing different technologies.

Vehicle to everything (VoX) covers all types of communication (V2V, V2I, V2P, V2S, V2G). This holistic approach enables a smart transportation system where

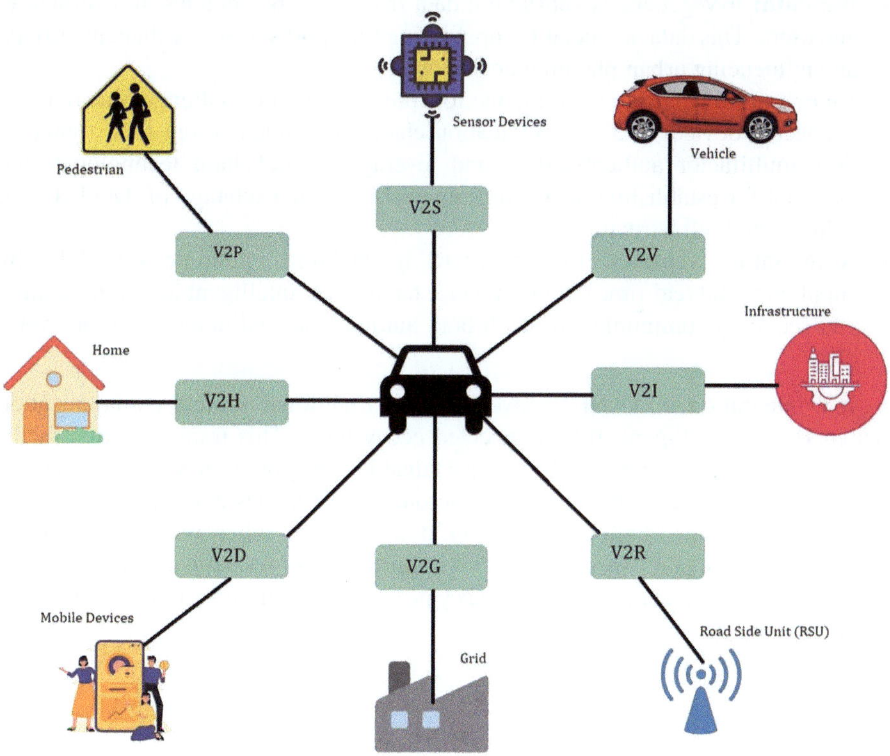

Fig. 1 A communication model in IoV

every entity within the network communicates for better safety, efficiency, and management. It forms a pillar of ITS. IoV expands upon conventional VANETs by integrating cellular networks such as LTE and 5G, providing extensive and reliable communication. IoV facilitates the collection and sharing of data regarding surroundings, vehicles, and road conditions, providing broader coverage and applications compared to VANETs.

IoV enhances VANET in different forms of technology; a few of the trending technologies are listed here as follows,

- **Cloud and edge computing:** The IoV seamlessly integrates cloud computing, empowering vehicles to leverage vast computational resources for real-time analytics, artificial intelligence, and predictive services. Meanwhile, edge computing effectively reduces latency by processing data in close proximity to the vehicle, guaranteeing rapid response times in critical scenarios such as collision avoidance.
- **Artificial intelligence and machine learning:** IoV utilizes AI-ML for intelligent decision-making. For instance, AI algorithms have the ability to process extensive traffic data, enabling them to accurately predict congestion and accidents while dynamically optimizing traffic flow in real time.
- **Big data:** IoV generates substantial data from sensors, vehicles, infrastructure, and users. This data is crucial for optimizing transport systems, enhancing safety, and influencing urban planning decisions.
- **Security and privacy:** Securing the IoV presents unique challenges due to interconnected devices and communication channels. Implementing robust encryption, multifactor authentication, and leveraging blockchain technologies are essential for establishing a secure framework for the exchange of data between vehicles and infrastructure.
- **Autonomous vehicles:** IoV is crucial in enabling autonomous vehicles by supplying vital real-time data for precise navigation, intelligent decision-making, and seamless communication with both autonomous and nonautonomous vehicles.

IoV integration in VANETs revolutionizes traditional vehicle communication networks into intelligent, fully connected ecosystems. This transformation facilitates safer, more efficient, and more sustainable transportation systems, vital for supporting advanced technologies like autonomous vehicles and smart cities. IoV presents significant challenges that set it apart from other wireless networks, primarily because of its open nature. This requires further research into areas such as network stability, latency, security, privacy preservation, interoperability, and energy consumption.

2.1 Security Standards

The prime objective of IoVs is to provide a secure connection between vehicles in any communication mode. IoVs present unique security challenges due to the dynamic nature, diversity of devices involved, and the potential for significant consequences if security is compromised. To tackle these challenges, a set of robust security standards has been put in place. Here, we will focus on the key security standards that are making waves in the world of VANETs.

IEEE 1609 (WAVE Security Services for Vehicular Environments)

IEEE 1609 belongs to the WAVE standard family and focuses on the security and privacy aspects of vehicular communications. IEEE 1609.1 encompasses standards for both V2V and V2I applications, ensuring that these applications utilize resources effectively through resource management (RM). IEEE 1609.2 facilitates secure communication across unsafe wireless networks, while 1609.3 defines the network and transport layers related to networking, and 1609.4 allows for multichannel functionality within the WAVE MAC layer and outlines control channel (CCH) and service channel (SCH) [4]. Key features of the 1609.2 standard include message authentication, encryption, certificate management, and anonymity. This standard accommodates elliptic curve cryptography (ECC) and employs hybrid encryption techniques.

ETSI Standard

The European Telecommunications Standards Institute (ETSI) has released various standards pertinent to the security of ITS and VANETs to guarantee secure communication in cooperative-ITS (C-ITS), which is interconnected with the IoV. ETSI introduced the ETSI TS 102940 and EN 302 637-2 standards, which define cooperative awareness in traffic scenarios based on how road users communicate their positions, speeds, and characteristics to one another. The main security features outlined in these standards include trust models, authorization and authentication, data integrity, and attack detection. Additionally, the standard provides privacy management for a node, focusing on characteristics such as anonymity, unlinkability, and pseudonyms [5].

ISO/SAE 21434

This standard has been developed internationally by the International Standards Organization (ISO) and Society of Automotive Engineers (SAE) to address cybersecurity throughout the lifecycle of road vehicles in ITS. This standard acknowledges certain key features like risk management, cybersecurity governance, and incident response. This standard outlines best practices for detecting and mitigating cybersecurity breaches in real time.

V2X Security Standards

To ensure secure interactions between vehicles, infrastructure, and other elements in the IoV, specialized security standards for V2X communication are essential. The 5G Automotive Association (5GAA) has developed 5G-enabled V2X standards (C-V2X) that incorporate encryption and authentication mechanisms to enable high-speed, low-latency communication between vehicles and infrastructure. Additionally, the Third Generation Partnership Project (3GPP) has released Standards 16 and 17, which focus on C-V2X communication over 5G networks and include advanced security features like encryption, identity management, and data integrity checks.

2.2 *Attack Vectors in IoV*

IoV systems, with their intricate interconnectedness and reliance on wireless communication, are highly vulnerable to an array of security threats, commonly known as attack vectors. These threats present a substantial risk to the integrity, confidentiality, and availability of IoV services. The authentication scheme must exhibit unwavering robustness, even in the face of adversarial attacks. In the realm of IoV, attackers can take the form of individuals or entities aiming to exploit system vulnerabilities for unauthorized access, service disruption, or data theft. Categorizing attackers based on their impact on VANET is pivotal. By classifying attackers according to their motivations and capabilities, we gain valuable insights into potential threats. A comprehensive depiction of these attackers is meticulously presented in Table 1, shedding light on the urgency of addressing IoV security challenges.

Table 1 Attackers' type in IoV

Attacker type	Motivation	Capabilities
Individual hackers	Personal gain, curiosity, notoriety	Vary widely, from basic to advanced skills
Organized crime groups	Financial gain, extortion, espionage	Advanced technical skills and resources
Nation-states	Espionage, cyber warfare, critical infrastructure disruption	Highly sophisticated and well-funded
Insider threats	Personal gain, revenge, sabotage	Intimate knowledge of the system

In the IoV context, an attacker is anyone seeking to exploit system vulnerabilities for unauthorized access, service disruption, or data theft. Attackers can be categorized into several groups based on their motivations and abilities.

- *Individual hackers* are motivated by diverse factors such as personal gain, insatiable curiosity, or the pursuit of notoriety. Their capabilities span a wide spectrum, ranging from rudimentary knowledge to highly advanced hacking skills.
- *Organized crime organizations* are typically driven by the pursuit of financial gain through activities such as extortion or espionage. They often have access to advanced technical skills and abundant resources, which enable them to carry out their illicit operations effectively.
- *Nation states* engage in cyber activities for various reasons such as espionage, cyber warfare, or disrupting critical infrastructure. They possess high levels of sophistication and funding, enabling them to access advanced tools and techniques for their cyber operations.
- *Insider threats* may be motivated by personal gain, a desire for revenge, or the intention to sabotage. Understanding these motivations can help us identify and address potential issues proactively. They have a deep understanding of the system's vulnerabilities and operations. By recognizing and addressing these vulnerabilities, we can work toward strengthening our security measures and preventing potential threats.

Attack Vectors

The increasing adoption of IVs in the realm of transportation has raised significant concerns about security aspects within any mode of communication system [6]. The security and privacy of messages are crucial factors that limit the development of VANETs. Many current protocols depend on a trusted third party for identity authentication and message security, which can lead to single points of failure in the system and reduced efficiency [7]. Before delving into the security challenges in vehicular communication, it is important to first understand the attack vectors that significantly disrupt the system's security. Due to the IoV connectivity and reliance on data exchange, they are susceptible to various attack vectors. We have narrowed our focus to the most specific types of attacks in the IoV. The attacks are explored in detail:

- *Man in the middle (MitM) attack:* In MitM, a malicious actor intercepts and potentially alters communications between vehicles' V2V or V2I. These attacks can wreak havoc on traffic management, compromise crucial safety features like collision avoidance systems, and facilitate the theft of sensitive data. It is crucial to be cautious and take proactive measures to safeguard against these dangerous attacks.
- *Distributed-denial of service (D-DoS) attack*: Cyber attackers flood the IoV network or specific vehicle systems with overwhelming traffic, leading to service

outages or reduced performance. This can result in the disabling of vehicle communication with the network or other vehicles, posing a serious threat to safety-critical systems such as autonomous driving and traffic control.

- *Relay attacks:* These involve attackers intercepting and extending the communication between a key fob and a vehicle, enabling unauthorized access. This type of attack is commonly used in keyless car theft, allowing attackers to unlock and start a vehicle without the owner's physical presence.
- *GPS spoofing attacks:* GPS spoofing involves tricking a vehicle's GPS with fake signals, causing it to think it is in a different location. This can lead to wrong turns, dangerous situations, and misjudging the position of other vehicles. To prevent GPS spoofing, solutions include verifying GPS signals through encryption and using algorithms to detect unusual signals.
- *Replay attacks:* These occur when an attacker intercepts valid communication and reuses it to gain unauthorized access or manipulate vehicle systems. This can lead to potential security breaches, such as unlocking a vehicle or disrupting traffic coordination. To prevent these attacks, it is essential to employ techniques like using timestamps, unique one-time codes (nonce), and encryption to ensure communication authenticity and security.
- *Sybil attacks:* These attacks happen when an individual generates multiple fraudulent identities, or nodes, to gain control over and manipulate the network for their advantage. These fake nodes can be used to send false data, disrupt communication, or gain control over vehicle systems by overwhelming the network with counterfeit information. In IoV, a Sybil attack can lead to issues like traffic mismanagement, accidents, or misleading vehicles with incorrect data about road conditions or traffic. To defend against Sybil attacks, authentication mechanisms, trust management, and consensus algorithms are used to verify the legitimacy of nodes in the IoV network.
- *Free-riding attacks:* These occur when a vehicle or node benefits from network resources without contributing to the system. In IoV, vehicles exchange data like traffic updates or road conditions to improve safety and efficiency. In a free-riding attack, a malicious vehicle uses the network's information while refusing to share its data. This undermines the cooperative nature of the IoV, reducing overall system performance and reliability. As time goes on, free-riding can result in the exhaustion of resources, unreliable traffic data, and diminished safety. Mitigation involves reputation-based systems and incentives to ensure fair participation in the network.
- *Side-channel attacks:* Attackers exploit physical leakages like timing, power consumption, or electromagnetic emissions to infer sensitive data in this type of security breach. These attacks can have a significant impact, potentially leading to the extraction of encryption keys or other sensitive information used to secure communications in IoV systems.

2.3 Privacy-Preserving Authentication

The rapid advancement of the IoV has revolutionized the way we approach transportation, offering enhanced safety, efficiency, and convenience. However, as vehicles become increasingly interconnected, the need for robust and privacy-preserving authentication mechanisms has become paramount [8]. VANETs enable V2V and V2I communication. Still, they also introduce vulnerabilities that can be exploited by malicious actors, such as physical, side-channel, and cloning attacks [8]. Privacy-preserving authentication seeks to maintain an equilibrium between security and privacy by enabling vehicles to demonstrate their authenticity to the network without revealing sensitive data. Techniques such as anonymous credentials, zero-knowledge proofs, and group signatures enable vehicles to authenticate while keeping user details private. This not only helps to protect personal privacy but also prevents potential security breaches like location tracking or replay attacks.

Researchers have proposed various authentication protocols for the IoV to address these security concerns. One such protocol, presented in Ref. [8], employs physical unclonable functions to deliver the necessary security features while minimizing authentication overhead and enhancing the throughput of application layer packets. Another approach, known as privacy-preserved collaborative estimation, leverages cryptography, k-anonymity, and differential privacy to enable privacy-preserving data sharing and processing among networked vehicles without relying on a trusted central authority. However, as highlighted, some existing authentication protocols are susceptible to attacks, such as offline identity guessing, spoofing, and many similar attacks, and can also spend a significant amount of time on the authentication process. To tackle these limitations, researchers have suggested enhanced protocols designed to offer better security assurances while still ensuring efficiency and scalability in the IoV. The following sections will study extensively various authentication schemes used in different situations of VANET and state the advantages and disadvantages. Hence, the readers will be able to choose the best authentication protocol for their research problems.

3 Privacy-Preserving Authentication Schemes

In today's constantly changing world of transportation and communication technologies, the VANET has become a critical system, allowing vehicles to exchange information seamlessly and enhancing safety features on the road. However, as VANET becomes more popular, it has also become more unprotected to different security threats and privacy concerns. To tackle these challenges, researchers have proposed numerous authentication schemes aimed at safeguarding the privacy and security of VANET participants.

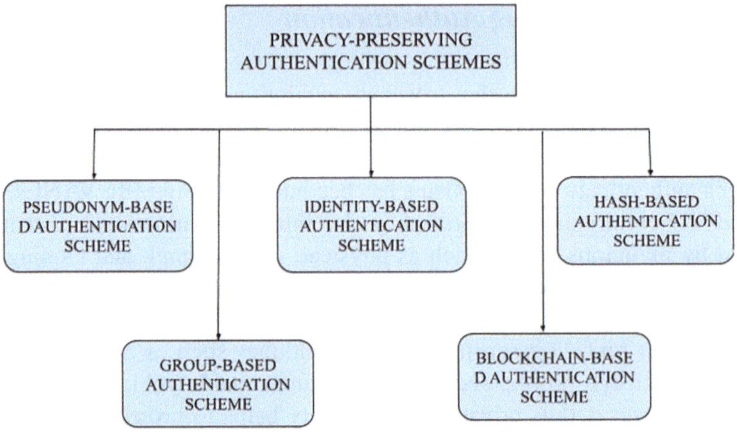

Fig. 2 Taxonomy of existing privacy-preserving authentication schemes

Authentication schemes in VANETs are vital for ensuring secure communication, preventing unauthorized access, and maintaining user privacy. The classification of current privacy-preserving authentication methods is illustrated in Fig. 2.

3.1 Pseudonym-Based Authentication Schemes

Pseudonym-based authentication schemes in VANETs use pseudonyms instead of real identities to preserve the privacy of vehicles and drivers. They are a type of authentication mechanism designed to enhance privacy. The key components of these schemes are as follows:

- **Pseudonym:** It is a unique identifier or alias that represents a user in the system and is not directly linked to their real-world identity.
- **Authentication:** The system verifies the user's identity by ensuring the pseudonym matches some credentials or cryptographic keys. This can involve:
  - *Public key infrastructure (PKI)* allows users to have a public-private key pair. Each user's pseudonym is linked to their public key, while the private key is utilized for authentication without disclosing the personal identity.
  - *Zero-knowledge proofs* allow a user to demonstrate knowledge of information, such as a secret key, without disclosing it.
- **Credential Issuer**: This is an entity that creates and manages pseudonyms and credentials while withholding the user's true identity from third parties.
- **Unlinkability:** The protocol prevents the same pseudonym from being linked across different sessions, thus enhancing privacy. Users may switch pseudonyms frequently.

The authentication process using pseudonyms consists of three essential steps. First, during the registration phase, a user must register with the authentication system using their real identity, after which they are assigned a pseudonym for all interactions with the system. The user is then issued cryptographic credentials tied to their pseudonym. Subsequently, during authentication, the user must use their pseudonym and cryptographic credentials (such as a digital signature) to gain access. The system will verify the credentials and grant access without disclosing the user's actual identity. Finally, in the pseudonym update/rotation phase, the pseudonyms are regularly updated (rotated) to prevent linkability over time, thereby enhancing privacy by adding an extra layer of security.

In vehicular social networks (VSNs), authentication is crucial for ensuring the reliability of entities [9]. However, vehicles' regular transmission of beacon messages in VSNs, containing sensitive information such as their current location, speed, and direction, raises privacy concerns. Tianhan, et al. proposed a privacy-preserving authentication scheme (APPAS) based on a pseudonym ring to meet the requirement of an identity-based service. Ring signature is a group-oriented mechanism that contains information about group users, which is used during the authentication phase in VSNs. In this work, the pseudonym is combined with the ring signature to create a pseudonym ring, which effectively reduces the pressure and cost associated with the generation, maintenance, and revocation of pseudonyms.

In VANETs, an anonymous authentication scheme combines pseudonyms and certificateless signature mechanisms to ensure conditional privacy. This approach allows swift revocation of vehicles and prevents pseudonyms from being linked before being traced. Message authentication is critical for V2V and V2I communications, with privacy preservation being a concern. Using a pseudonym instead of a real identity helps prevent attackers from locating and tracking vehicles. Jiayu et al. [10] proposed a pseudonym-based certificateless privacy-preserving authentication (PCPPA) scheme to reduce the local storage burden of vehicles and meet low transmission delay requirements. The scheme includes a self-generation mechanism for pseudonym replacement, a time-based pseudonym revocation mechanism for reducing the overhead of pseudonym revocation, and a fast-tracking method for enabling conditional privacy preservation.

Benefits

- Users can authenticate themselves without revealing their real identity, protecting their personal information from unnecessary exposure.
- Pseudonyms offer anonymity, allowing users to engage with services or systems without disclosing who they are.
- Users can generate and manage multiple pseudonyms for different contexts, providing flexibility and control over their digital footprint.

Challenges

- Implementation involves complex cryptographic techniques, which can be difficult to manage.

- It needs additional administrative infrastructure to handle the generation and management of pseudonyms.
- It has the risk of de-anonymization, and if not handled properly, this could expose user data unexpectedly.

3.2 Identity-Based Authentication Scheme

In this approach, an entity's public key is generated from its recognizable public identity information, like a phone number, email address, or identification number. Identity-based public key cryptographic frameworks are gradually taking the place of conventional public key cryptographic systems to eliminate the necessity for managing certificates. In identity-based public key cryptography, there is no requirement for a certificate to validate messages, thus lowering the burden associated with certificates in communication and greatly enhancing the efficiency of VANETs.

Jalawi et al. [11] quote that ID-based schemes address the issues that arise from PKI-based schemes. The main concept behind identity (ID)-based schemes is to use identity information as the user's public key. Private keys are generated by a trusted authority (TA) and then distributed to users using the same identity information. When a message is sent, the recipient verifies it using the sender's public key and the sender signs it with their private key. The ID-based schemes are further classified into two groups based on the cryptography used: bilinear pair and elliptic curve cryptography (ECC).

D. He et al. [12] developed an ID-based authentication scheme to facilitate mutual authentication and privacy preservation, which is used to secure communication in VANET. Jiayu et al. [13] introduced an ID-based authentication scheme to improve the privacy of vehicle users by supporting cross-domain vehicles and efficiently revoking a batch of pseudonyms without compromising user privacy. This approach was based on the ID-based signature (IBS) mechanism.

Benefits

- Certificates are not needed for authentication purposes, which reduces the overhead.
- Identity-based systems allow for better auditing and logging, as every action can be traced to a real individual.
- Since the user's real identity is involved, it's easier to trace actions back to individuals, ensuring accountability.
- ECC operations outperform current methodologies in computational efficiency, making them a superior choice.

Challenges

- Bilinear Pairing operations are time-consuming and complex, leading to significant computational overheads during message signing and authentication.

- In dense areas, more ECC operations, such as point multiplication, cause delays in message verification for the recipient.
- This involves increased cost and complexity for identity verification.
- Some users may resist providing real identity information, especially in contexts where they prefer anonymity or pseudonymity.

3.3 Hash-Based Authentication Scheme

A hash-based authentication scheme relies on symmetric cryptographic hash functions to verify user credentials or ensure the integrity of data without encrypting it. A hash function is a mathematical function when it is applied to a message; it produces a fixed random value known as a message digest. To maintain message integrity, a hash value needs to be included with the sent message [14]. The strength of the scheme relies on using modern cryptographic hash functions, proper salting, and techniques like key stretching to mitigate brute-force attacks. While it offers strong protection, careful management of hashing algorithms and parameters is necessary to ensure ongoing security. Popular hash functions used in authentication are:

- ***SHA-256***: Part of the SHA-2 family, widely used for secure hashing.
- ***bcrypt:*** A password-hashing function designed to be slow and resistant to brute-force attacks, including salting and adaptive hash iterations.
- ***Argon2:*** Winner of the Password Hashing Competition (PHC), designed to be secure and resistant to side-channel attacks.
- ***PBKDF2*** (Password Based Key Derivation Function-2): Used for password hashing with key stretching.

Hash-based authentication is used in Ref. [15]; the Bloom filter efficiently solves approximate set membership by quickly determining if an element belongs to a set. It converts each set element into a binary string using hash functions and similarly maps the checked element with the same functions. The security of grid range queries relies on the one-way nature of hash functions. In Ref. [16] Sujash et al. proposed a novel hierarchical decentralized authentication system to ensure direct privacy preserving V2I and V2V verifiability with minimum communication and computational overheads using a simple hash function. It is used to preserve message integrity and the scheme is secure against message content modification. Since the hash functions are irreversible, generating the secret parameter back from the authentication row is not possible by the adversary. This proposed scheme eliminates the communication bottleneck as a resilient and responsive network while safeguarding sensitive data. Hash-based authentication helps create a unique temporary identity [17]. The distinctive temporary identity is generated by concatenating the Local Certification Authority (LCA) ID with a hash value. Mouna Garai et al. proposed architecture for vehicle networks based on radio frequency identification (RFID) with the hash-based authentication scheme involved during the verification process creating a certificate with a hash value and authentication process creating a temporary ID to maintain users' anonymity and safeguard them from tracking attacks.

Benefits

Cryptographic hash functions are computationally efficient and do not require heavy processing power.

- They provide a responsive vehicular ecosystem while safeguarding sensitive data.
- Hash functions can also verify the integrity of messages or data. If any part of the data changes, the hash will be different, indicating tampering.

Challenges

- Even though the hashed passwords are more secure than plaintext, they are still vulnerable to brute force and dictionary attacks.
- Simple hash functions may not provide enough protection if users have weak passwords.
- Hash-based authentication does not eliminate risks.

3.4 Group Signature-Based Authentication Scheme

A group signature-based authentication scheme is a cryptographic protocol that allows a group member to sign messages for the group while keeping their identity anonymous. However, in the event of disputes or malicious activity, a trusted group manager can disclose the identity of the signer, revoke the signature, and reveal the identity of the vehicle [11]. This authentication scheme serves as a robust cryptographic tool. Its capacity to balance anonymity and accountability makes it well-suited for scenarios where group members must act on behalf of the group without revealing their identities.

In 2007, Lin et al. [18] first introduced group signatures into VANET privacy-preserving algorithms. This allows the On-Board Unit (OBU) to avoid storing a large number of pseudonym keys and certificates. Lu et al. [19] developed a new secure scheme using group signature technology. The main idea is to introduce on-the-fly short-term group-member certificates and reduce the scope of the Road Side Unit (RSU)'s jurisdiction. This scheme improves system efficiency and addresses the issue of key escrowing.

L.E. Funerburg et al. [20] introduced an enhanced security scheme for VANET using a group signature-based authentication system. This scheme improves the efficiency of signing and signature validation by reducing pairing operations through the use of cached calculation results. Additionally, the system enhances overall efficiency by implementing timed group key updates instead of changing keys when members leave the group. However, it is worth noting that this scheme does not eliminate pairing operations from both the message signing and signature verification processes.

Benefits

- The scheme balances privacy with accountability and also incurs less computational and communication overhead.
- It ensures strong privacy for group vehicles. As long as the group manager holds the master key, it meets traceability needs.

Challenges

- There is an overhead in key management.
- Revoking the membership of a group member can be complex.
- The high mobility of vehicles and rapid changes in VANET topologies make group management challenging.

3.5 Blockchain-Based Authentication Scheme

A blockchain-based authentication scheme for IoVs offers a decentralized, secure, and transparent method for verifying and managing vehicle communications. In VANETs, the open nature of the network raises concerns about privacy and security. While there are various CPPA schemes designed for VANETs, many of these schemes face challenges when applied in real-world scenarios. These challenges include the requirement for a certificate manager in public key infrastructure (PKI)-based solutions or the key escrow issue in ID-based solutions [21]. To mitigate these challenges, Xiaotong Zhou et al. proposed an efficient blockchain-based CPPA scheme; in this scheme, they have two key building blocks namely, the Signature of Knowledge (SoK) and smart contract. The former ensures participants authenticate each other anonymously and only the trusted authority (TA) can extract the real identity and also serve as a signature to ensure the authentication and legitimacy of transmitted messages. The latter is written in a Turing complete programming language deployed in blockchain to support autonomous execution and achieve computerized transaction protocols. Each smart contract with a unique public address can be triggered internally and externally. The authors have mentioned that it is mainly responsible for public key information storing, viewing, and deleting according to the caller's permissions. However, smart contracts create a prolonged latency, which is the main challenge in this proposed scheme.

In Ref. [22], Xia Feng et al. proposed a robust blockchain-based authentication scheme aimed at achieving highly secure and stable communication among vehicles. The root authority (RA) generates a unique evaluation proof for each authorized vehicle's certificate and broadcasts a public global commitment based on all valid certificates. Vehicles can efficiently prove their authorized status by utilizing the global commitment through bilinear pairing, eliminating the need to query certificates stored on the blockchain. However, bilinear pairing operation incurs high computational costs during the authentication process in this scheme. Due to the decentralized nature of blockchain technology, rapid reauthentication of vehicles can be achieved [23]. Azees Maria et al. proposed a blockchain-based anonymous authentication scheme to overcome the problems of enormous network communication overhead caused by frequent authentication of vehicles and the absence of scalability during the anonymous authentication. The system utilizes the Merkle hash tree (MHT) to understand real-time authentication records, which can be extended to newly joined vehicles, significantly improving the operability of the VANET system.

Benefits

- No single point of failure due to decentralization.
- Blockchain ensures tamper-proof authentication, protecting against spoofing, Sybil attacks, and other network threats.
- Smart contracts and consensus mechanisms automate and streamline the authentication processes.

Challenges

- Ensuring fast and efficient blockchain operation with a large number of nodes requires careful design and potential off-chain solutions.
- Vehicles possess restricted computational capabilities, and the blockchain needs to be sufficiently lightweight to operate on vehicle hardware.
- Blockchain consensus can introduce delays in highly dynamic environments like VANETs, especially with large-scale networks.

4 Comparative Analysis of Existing Schemes

We have examined different authentication methods implemented for the security of IoV in VANETs. Authentication is a critical aspect of IoV security, ensuring that only authorized entities can access and interact with the network. Various schemes have been proposed to address the unique challenges of IoV security. This comparative analysis will give a brief examination of the authentication schemes discussed so far. Table 2 illustrates the comparative analysis of the discussed schemes.

In pseudonym-based authentication, entities utilize temporary identities known as pseudonyms to safeguard their privacy and offer robust privacy protection. How-

Table 2 Comparative analysis based on features

Features	Pseudonymb based	Identity ased	Hash based	Group based	Blockchain based
Privacy	Strong	Moderate	Low	Moderate	High
Security	Good	Good	Good	Moderate	High
Efficiency	Moderate	Good	High	High	Low
Complexity	High	Moderate	Low	Moderate	High
Scalability	Moderate	Good	High	High	Moderate
Centralization	High	Moderate	Low	Moderate	Low

ever, this approach introduces complexities related to managing pseudonyms and leaves room for potential pseudonym reuse attacks. With identity-based authentication, entities use their public identities as encryption and decryption keys, streamlining key management by removing the requirement for certificates. Nonetheless, this method relies on a TA to generate and manage private keys and is susceptible to a single point of failure. Hash-based Authentication involves entities using cryptographic hash functions to create unique identifiers. This approach offers lightweight and efficient solutions while resisting certain attacks. Nevertheless, it provides limited privacy protection, as hash values can be linked to real identities. Group-based authentication involves grouping entities and performing authentication at the group level. This method proves to be efficient for large-scale deployments and can reduce communication overheads. However, compromised group membership can lead to potential privacy breaches. In blockchain-based authentication, authentication information is stored on a distributed ledger, ensuring immutability and transparency. This approach offers high security, resistance to tampering, and a decentralized nature. However, it comes with challenges related to high computational overheads, potential scalability issues, and energy consumption.

4.1 Existing Research Summary

A key component of IoV security is authentication, which ensures that only authorized entities can engage in communication and access sensitive data. IoV networks employ several primary authentication schemes, with the choice depending on factors such as security requirements, scalability, cost, and user experience. Various schemes have been developed to tackle the unique challenges of IoV security. A concise overview of research on existing authentication schemes for supporting vehicular ad-hoc networks in IoV, along with their applications, is provided in Table 3.

5 Applications

The selection of an authentication method is contingent upon the particular needs of an IoV application, including the desired degree of security, privacy, efficiency, and scalability. It may be essential to employ a blend of methods to effectively tackle the varied security issues associated with IoV. Authentication is pivotal in the realm of the IoV as it safeguards sensitive data, thwart unauthorized access, and uphold the integrity of the system. Numerous applications rely on authentication schemes to meet their unique requirements.

- Traffic management: Real-time traffic updates, congestion avoidance, and efficient route planning.

Table 3 Summary of existing authentication schemes

Paper	Technique employed	Implementation	Security request addressed
Li Wang et al. [1]	Asymmetric group key agreement to distributed CP-ABE	Proof of knowledge and credentials	Large-scale autonomous network, anonymity, and partial policy hiding
Y. Zhan et al. [2]	Blockchain-based authentication scheme, collision-resistance of hash function	Incentive mechanism involves identities to participate with incentives, ensuring privacy in anonymous announcements	Unforgeability, privacy-preserving incentive announcement
Ke Gu et al. [24]	Identity-based authentication scheme	Decentralized and traceable privacy preserving scheme for vehicular identity in fog computing based IoV. Voting mechanism, selection algorithm	Trajectory tracking, decentralized identity privacy
Chao Lin et al. [3]	PKI-based authentication scheme and blockchain technology (cross domain)	Modified ECDSA batch verification, key derivation algorithm	Conditional privacy-preserving, low computational overhead, and addressing key escrow problems
Jegadeesan Subramani et al. [6]	Hash-based authentication scheme, conditional privacy	Energy efficient mutual and batch authentication scheme for V2G (grid) with IoT paradigm, bilinear pairing	Mutual protection against potential threats and vulnerabilities, unlinkability, and non-repudiation
Alireza et al. [25]	TRD-aided conditional privacy-preserving authentication scheme, fail-stop signature	Anonymous ID and signature generation, message verification, forgery detection of impersonated messages. Private key updating algorithm	Anonymity, unlinkability, unforgeability, low computational and communication cost, provable security
Abdullah Alharhi et al. [26]	Blockchain-based authentication scheme with biometric	Single registration traceability, simulation used are OMNet++, veins and SUMO	Immutability, privacy preserving, and anonymity
L.Ellien Funerbug et al. [20]	Group-based authentication scheme, short group signature	Elliptic curve caching, timed key updates, pairing free	Non-repudiation, overall system efficiency, low computational and communication overheads

Mahmood et al. [27]	Identity-based CPPA authentication scheme	Elliptical curve cryptography (ECC), vehicle revocation and renewal phase	Privacy-preserving, message confidentiality, tractability and revocation, non-forgery, and resist side-channel attacks
Jiayu Qi et al. [13]	Identity-based anonymous authentication scheme and efficient pseudonym scheme (cross domain)	BLMQ signature mechanism, bloom filter, batch pseudonym revocation, bilinear pairing	Pseudonyms unlinkable, unforgeability, non-repudiation, and revocation defend against major attacks
Tianhan Gao et al. [9]	Pseudonym-based authentication scheme	A pseudonym ring is formed with the combination of a pseudonym and ring signature CDHP-computational Diffe-Hellman problem, SVO logic	Unconditional anonymity, unforgeability, perfect forward privacy
Mouna Garai et al. [17]	Automobile authentication scheme, vehicular mobility management method, Vanet node attachment strategy	Access control strategy, toll-based certificate management system utilizing RFID tags, and digital certificate implementation	Quality of service (QoS), privacy preserving
Xiaoton Zhou et al. [21]	Blockchain-based CPPA	Signature of knowledge (SoK) and smart contract	Identity anonymity, conditional traceability, resistance against common attacks, and birthday collision resilience

- Safety applications: Collision avoidance, emergency alerts, and roadside assistance.
- Infotainment: Providing drivers with relevant information and entertainment.

The key challenges in VANET authentication include:

- Dynamic topology: Vehicles constantly join and leave the network, making it difficult to maintain a centralized authentication authority.
- Limited resources: Vehicles have limited computational power and battery life, requiring efficient authentication schemes.
- Privacy concerns: Protecting the privacy of vehicle occupants and their data.
- Security threats: Addressing potential attacks, such as impersonation, denial-of-service, and eavesdropping.

To have a broader aspect of applications by each authentication scheme that we have discussed so far is briefly described with use cases. The concept of pseudonym-based authentication involves the use of temporary identities known as pseudonyms by entities to ensure the protection of their privacy. This approach is designed to provide a secure way for entities to interact while maintaining their confidentiality.

Use Cases

- Privacy-sensitive applications: Pseudonym-based authentication is essential for maintaining privacy in sensitive areas such as healthcare and finance, where the protection of personal information is of utmost importance.
- Anonymous data sharing: This method enables the secure and anonymous sharing of data, making it suitable for applications like traffic monitoring and research studies where the confidentiality of data is critical.
- Location privacy: Authentication using pseudonyms is essential for protecting user location information and tackling issues related to the privacy of users' whereabouts.

The concept of identity-based authentication is that the entities use their public identities as keys for encryption and decryption.

Applications

- Centralized management: Large-scale deployments, IoT devices.
- Simplified key management: Reducing the need for certificates.
- Single sign-on: Providing seamless access to multiple services.

In blockchain-based authentication, the information is stored on a distributed ledger, providing immutability and transparency.

Applications

- Secure data sharing: Vehicle-to-vehicle communication, V2I communication.
- Tamper-proof records: Ensuring the integrity of authentication data.
- Decentralized access control: Managing access rights without a central authority.

In hash-based authentication, entities use cryptographic hash functions to generate unique identifiers.

Applications

- Lightweight applications: Sensor networks, low-power devices.
- Passwordless authentication: Providing a convenient and secure alternative to traditional password-based authentication.
- Data integrity: Verifying the integrity of transmitted data.

In PKI-based authentication, TA issues digital certificates to entities, containing cryptographic keys for encryption and decryption.

Applications

- Secure communication: Protecting data transmitted between vehicles and infrastructure.
- Access control: Restricting access to sensitive resources.
- Digital signatures: Verifying the authenticity and integrity of documents.

6 Conclusion

In conclusion, this survey has presented a comprehensive overview of existing authentication protocols for the IoV in VANETs. We have analyzed the underlying cryptographic techniques, security properties, and performance characteristics of these protocols, focusing on lightweight schemes suitable for resource-constrained vehicular devices. While significant progress has been made in developing efficient and secure authentication solutions, several challenges remain. Scalability is a critical concern, as IoV systems are expected to accommodate a large number of vehicles and infrastructure components. Existing protocols may struggle to handle the increased load and maintain acceptable performance. Computational overhead is another limitation, as vehicular devices have limited processing power. The choice of cryptographic techniques and algorithm implementations must carefully balance security requirements with performance constraints. Each authentication scheme offers unique advantages and disadvantages. The optimal choice depends on specific requirements, such as the desired level of privacy, security, efficiency, and scalability. In many IoV scenarios, a combination of schemes may be necessary to address the diverse challenges and ensure robust security.

Future research directions include exploring new cryptographic techniques that offer improved efficiency and security, developing scalable authentication protocols that can handle large-scale IoV deployments, and addressing the privacy concerns associated with authentication in VANETs.

References

1. L. Wang, H. Zhong, J. Cui, et al., Privacy-preserving and secure distributed data sharing scheme for VANETs. IEEE Trans. Mob. Comput. **23**, 13882–13897 (2024)
2. Y. Zhan, Y. Yang, H. Cheng, et al., PIAS: Privacy-preserving incentive announcement system based on blockchain for internet of vehicles. IEEE Trans. Serv. Comput. **17**, 1–4 (2024)

3. C. Lin, D. He, et al., BCPPA: A blockchain-based conditional privacy-preserving authentication protocol for vehicular ad hoc networks. IEEE Trans. Intell. Transp. Syst. **22**(12), 7408–7420 (2021)
4. J.-w. Kim, J.-w. Kim, J. Lee, Intelligent resource allocation scheme using reinforcement learning for efficient data transmission in VANET. Sensors **24**(9), 2753 (2024)
5. S. Mazhar, A. Rakib, L. Pan, et al., State-of-the-art authentication and verification schemes in VANETs: A survey. Veh. Commun. **49**, 100804 (2024)
6. J. Subramani, A. Maria, et al., Mutual and batch authentication with conditional privacy-preserving scheme for V2G communication system. IEEE Trans. **12**, 69593–69602 (2024)
7. X. Liu, Y. Wang, Y. Li, H. Cao, PTAP: A novel secure privacy-preserving & traceable authentication protocol in VANETs. Comput. Netw. **226**, 109643 (2023)
8. M.N. Aman, U. Javaid, B. Sikdar, A privacy-preserving and scalable authentication protocol for the internet of vehicles. IEEE Internet Things J. **8**, 1123–1139 (2021)
9. T. Gao, X. Deng, M.C. Qingshan, I. You, APPAS: A privacy-preserving authentication scheme based on pseudonym ring in VSNs. IEEE Trans. **7**, 69936–69946 (2019)
10. J. Qi, T. Gao, X. Deng, C. Zhao, A pseudonym-based certificateless privacy-preserving authentication scheme for VANETs. Veh. Commun. **38**, 100535 (2022)
11. J.S. Alshudukhi, B.A. Mohammed, Member, IEEE, Z.G. Al-Mekhlafi, Member, IEEE, Conditional privacy-preserving authentication scheme without using point multiplication operations based on elliptic curve cryptography (ECC). IEEE Trans. **8**, 222032–222040 (2020)
12. D. He, S. Zeadally, B. Xu, X. Huang, An efficient identity-based conditional privacy-preserving authentication scheme for vehicular ad-hoc networks. IEEE Trans. Inf. Forensics Secur. **10**(12), 2681–2691 (2015)
13. J. Qi, T. Gao, A privacy-preserving authentication and pseudonym revocation scheme for VANETs. IEEE Trans. **8**, 177693–177707 (2020)
14. S.A. Jan, N.u. Amin, M. Othman, S. Member, M. Ali, A.I. Umar, A. Basir, IEEE, A survey on privacy-preserving authentication schemes in VANETs: Attacks, challenges and open issues. IEEE Trans. **9**, 153701–153726 (2021)
15. X. Zihui, W. Lei, C. Qin, et al., PriTAEC: Privacy-preserving task assignment based on oblivious transfer and edge computing in VANET. IEEE Trans. Veh. Technol. **72**(4) (2023)
16. S. Naskar, C. Brunetta, G. Hancke, T. Zhang, M. Gidlund, A scheme for distributed vehicle authentication and revocation in decentralized VANETs. IEEE Trans. **12**, 68648–68667 (2024)
17. Mouna Garai, Maha Sliti, Manel Mrabet, Noureddine Boudriga, And Lassaad Ben Ammar, Authentication in QoS aware VANET: An approach based on enhanced digital certificates, IEEE Trans. Vol 12, 124452–124477, 2024.
18. X. Lin, X. Sun, P.-H. Ho, X. Shen, GSIS: A secure and privacy-preserving protocol for vehicular communications. IEEE Trans. Veh. Technol. **56**(6), 3442–3456 (2007)
19. R. Lu, X. Lin, H. Zhu, P.-H. Ho, X. Shen, ECPP: Efficient conditional privacy preservation protocol for secure vehicular communications, in *Proc. IEEE 27th Conf. Comput. Commun. (INFOCOM)*, (IEEE, 2008), pp. 1229–1237
20. L. Ellien Funerburg, I.-Y. Lee, Efficient short group signatures for conditional privacy in vehicular ad hoc networks vis ID caching and time revocation. IEEE Trans. **9**, 118065–118076 (2021)
21. X. Zhou, K. Khan, W. Wu, et al., An efficient blockchain—based conditional privacy—Preserving authentication protocol for VANETs. IEEE Trans. Veh. Technol. **72**(1) (2023)
22. X. Feng, K. Cui, L. Wang, PBAG: A privacy-preserving blockchain-based authentication protocol with global-updated commitment in IoV. J. Latex Class Files **14**(8) (2015)
23. A. Maria, V. Pandi, J.D. Lazarus, M. Karuppiah, M.S. Christo, *BBAAS: Blockchain—Based anonymous authentication scheme for providing secure communication in VANETs* (Wiley, Hindawi, Security and Communication Network, 2021)
24. G. Ke, K. Wang, X. Li, et al., Multi-fogs—Based traceable privacy-preserving scheme for vehicular identity in internet of vehicles. IEEE Trans. Intell. Transp. Syst. **23**(8), 12544–12561 (2022)

25. A. Aghabagherloo, M. Delavar, J. Mohajeri, M. Salmasizadeh, B. Preneel, An efficient and physically secure privacy-preserving authentication scheme for Vehicular Ad-hoc NETworks (VANETs). IEEE Trans. **10**, 93831–93844 (2022)
26. A. Alharthi, Q. Ni, R. Jiang, A privacy-preservation framework based on biometrics blockchain (BBC) to prevent attacks in VANET. IEEE Trans. **9**, 87299–87309 (2021)
27. M.A. AL-Shareeda, M. Anbar, S. Manickam, I.H. Hasbullah, Towards identity-based conditional privacy-preserving authentication scheme for vehicular ad hoc networks. IEEE Trans. **9**, 113226–113238 (2021)

Part V
Big Data in Future Advancements

JourneyMate: An Optimized Secure Travel Planning and Integrated Booking System

Suriya Murugan, R. Sarimila, R. P. Srijaa, S. Swathi, and P. Vasumathi

1 Introduction

Travel planning has evolved significantly with advancements in technology, shifting from manual processes to sophisticated digital solutions. Traditional travel planning methods often involve disparate systems that lack cohesion, leading to inefficiencies and a fragmented user experience. To address these challenges, the proposed travel planning system offers an integrated solution designed to streamline the trip planning process. By consolidating accommodation, transportation, and activity bookings into a unified platform, the system enhances the convenience and efficiency of planning a trip. To address these challenges, this chapter proposes an advanced travel planning system designed to revolutionize how trips are planned and executed. By consolidating accommodation, transportation, and activity bookings into a unified platform, the system simplifies the planning process, saving time, and reducing stress for users. The platform leverages cutting-edge technologies, including advanced recommendation algorithms and real-time data analytics, to create a highly personalized and user-centric travel experience.

Central to the system is its recommendation engine, which combines collaborative filtering, content-based filtering, and hybrid approaches to provide tailored suggestions for hotels, holiday packages, and activities. By analysing user preferences, past travel history, budget constraints, and behavioural patterns, the system delivers highly relevant and accurate recommendations. A unique visual search feature further enhances user engagement by allowing travellers to upload images

S. Murugan · R. Sarimila · R. P. Srijaa · S. Swathi · P. Vasumathi (✉)
Department of Computer Science and Engineering, Sri Eshwar College of Engineering, Coimbatore, Tamil Nadu, India
e-mail: suriya.m@sece.ac.in; sarimila.r2021csec@sece.ac.in; srijaa.rp2021csec@sece.ac.in; swathi.s2021csec@sece.ac.in; vasumathi.p2021csec@sece.ac.in

A. Haldorai et al. (eds.), *7th EAI International Conference on Big Data Innovation for Sustainable Cognitive Computing*, EAI/Springer Innovations in Communication and Computing, https://doi.org/10.1007/978-3-032-02979-9_17

and discover visually similar destinations, promoting intuitive and exploratory trip planning.

The system also integrates advanced optimization and safety mechanisms. Evolutionary genetic algorithms (GA) are employed for route optimization, enabling travellers to identify the shortest and most convenient routes for visiting multiple destinations within a constrained area. This ensures time efficiency while reducing travel fatigue. Moreover, the incorporation of safety prediction models using time-series forecasting techniques, such as long short-term memory (LSTM) networks, enhances user safety by predicting environmental conditions like temperature, humidity, and potential risks. These features empower users to make informed decisions, reducing the likelihood of disruptions during their journey. Beyond individual planning, the system promotes a collaborative and socially engaging travel experience. Collaborative planning tools allow users to share their travel plans with friends and family, facilitating feedback and enhancing group planning dynamics. To address financial concerns, a budget tracker monitors expenses in real-time, offering cost-saving tips and ensuring travellers remain within their financial constraints. Additionally, a personality-based booking feature aligns travel plans with users' unique lifestyles and preferences, minimizing cancellations and fostering higher satisfaction rates.

The system's disaster management module represents a critical advancement in ensuring travel safety. This module integrates real-time disaster prediction and monitoring mechanisms to assess potential risks before, during, and after the journey. Utilizing LSTM-based prediction models and time-series analysis, the system evaluates environmental conditions and identifies potential threats, such as extreme weather, earthquakes, or other hazards. If a risk is detected, users are alerted immediately, and alternative plans or routes are suggested. The system also provides real-time monitoring during the trip, enabling continuous safety updates. In the event of a disaster, emergency notifications are sent to the user, relevant authorities, and emergency contacts, while guiding the user through recovery and ensuring their safety. This disaster-aware capability ensures not only a smooth travel experience but also peace of mind, making the system a comprehensive travel planning and safety solution.

2 Literature Survey

Tourism is one of the industries that has shown great growth over the years as a source of income and the development of new technologies. In Ref. [1], the hybrid recommendation model not only personalizes trip planning but also demonstrates robustness in dealing with data sparsity by integrating user behaviour analysis. The evolutionary algorithms ensure a global search for optimal solutions, while the local search techniques fine-tune these solutions, making the model adaptable to various scales of tourism data. Additionally, the approach highlights potential applications in adaptive itinerary generation and multi-destination trip planning.

In Ref. [2], the adaptive genetic algorithm (AGAM) adapts dynamically to user preferences by continuously evaluating POI popularity and real-time data. A unique feature of AGAM is its ability to minimize computational overhead by dynamically adjusting mutation and crossover rates, ensuring faster convergence. The study emphasizes its utility for short-term travel planning and real-time re-routing under changing travel conditions.

In Ref. [3], the Pareto-optimal approach for urban transit demonstrates efficiency in multi-objective decision-making. By leveraging the Monte Carlo tree search, it prioritizes cost, time, and convenience simultaneously. The study further compares heuristic methods with exact search techniques, showcasing their scalability and suitability for large-scale urban systems, especially during peak travel seasons.

In Ref. [4], the enhanced genetic algorithm (IGA) for round-trip route planning optimizes not only just travel time but also incorporates constraints like vehicle capacities and time windows. The study's use of advanced operators, such as elitism and adaptive mutation, highlights its ability to find solutions even in highly constrained scenarios, such as corporate travel or package tours.

In Ref. [5], while reviewing collaborative, content-based, and hybrid recommendation techniques, the study dives deeper into the challenges of integrating demographic and behavioural data. It proposes using anonymization techniques to address privacy concerns and explores the feasibility of blockchain-based methods for secure data sharing in collaborative systems.

In Ref. [6], "The Hybrid Offer Ranker" (THOR) combines clustering techniques with advanced preference modelling to reduce the cold-start problem. Additionally, the chapter explores integrating THOR with real-time recommendation engines, highlighting its potential to support large-scale tourism platforms like Booking.com or TripAdvisor.

In Ref. [7], the hybrid tabu search algorithm, enhanced with particle swarm optimization, effectively handles real-time traffic data and POI updates. Its reliance on crowd-sourced data ensures that recommendations remain relevant, even during unexpected events like road closures or surges in tourist activity.

In Ref. [8], the genetic algorithm-based e-tourism recommender system integrates pairwise preference elicitation to address recommendation inconsistencies. This method is particularly effective in tailoring suggestions for niche markets, such as adventure tourism or eco-tourism, where user preferences can vary widely.

In Ref. [9], the smart travel assistant leverages hybrid machine learning techniques for itinerary personalization. The study proposes integrating external APIs for environmental and safety forecasts, which could enhance the system's utility for travellers concerned about weather conditions or natural disasters.

In Ref. [10], the tourist trip model that factors in CO_2 emissions exemplifies a shift towards sustainable travel planning. By combining individual preferences with transportation mode selection, the model aligns with global sustainability goals, such as reducing carbon footprints and promoting eco-friendly travel options.

In Ref. [11], operational research techniques in disaster management are analysed for their potential application in tourism. For instance, the use of simulation models and queueing theory in healthcare coordination could inspire methods to manage overcrowding at popular tourist destinations.

In Ref. [12], the extension of the tourist travel design problem to include travel style preferences enriches the personalization aspect of trip planning. The study's use of quality metrics to evaluate user satisfaction highlights its practicality for platforms aiming to retain customer loyalty through tailored services.

In Ref. [13], the integration of GMM and K-means clustering with web scraping from TripAdvisor represents a practical approach to data-driven personalization. The study also suggests using sentiment analysis on the scraped reviews to further refine travel route optimization.

In Ref. [14], the application of LSTM networks for data sparsity is expanded by comparing their performance against other neural network architectures. The study highlights their ability to predict user preferences even with limited historical data, making them ideal for new users or unexplored destinations.

In Ref. [15], the focus on non-functional attributes in similarity functions is complemented by experimental results showing significant improvements in recommendation diversity. This approach mitigates overfitting by ensuring that the recommendations are not overly tailored to a narrow set of user preferences.

In Ref. [15], the sparsity-aware hybrid collaborative recommendation approach improves upon traditional matrix factorization by integrating implicit feedback data. Principal component analysis (PCA) not only enhances feature extraction but also reduces computational complexity, making the method scalable for large datasets.

The reviewed literature highlights various advancements in travel recommender systems and route optimization techniques. While several studies focus on collaborative filtering to personalize travel recommendations, they lack comprehensive integration of visual search and personality-based booking features, which can further enhance user engagement and satisfaction. Additionally, many route optimization approaches prioritize efficiency but do not combine real-time environmental safety forecasting using time series models like ARIMA and LSTM, a critical feature for ensuring user safety in dynamic travel environments. The reviewed literature lacks collaborative planning for group trips and budget tracking tools, which are essential for holistic trip management. The proposed system aims to fill these gaps by introducing a unified platform, offering a more user-centric travel experience.

3 Proposed System Architecture

The system architecture is denoted in Fig. 1 for the travel planner application integrates several modules that provide a personalized travel experience for users.

(a) **User Interface (UI)**

It serves as the main point of interaction, enabling users to input travel preferences, plan trips collaboratively, and explore options using a visual search

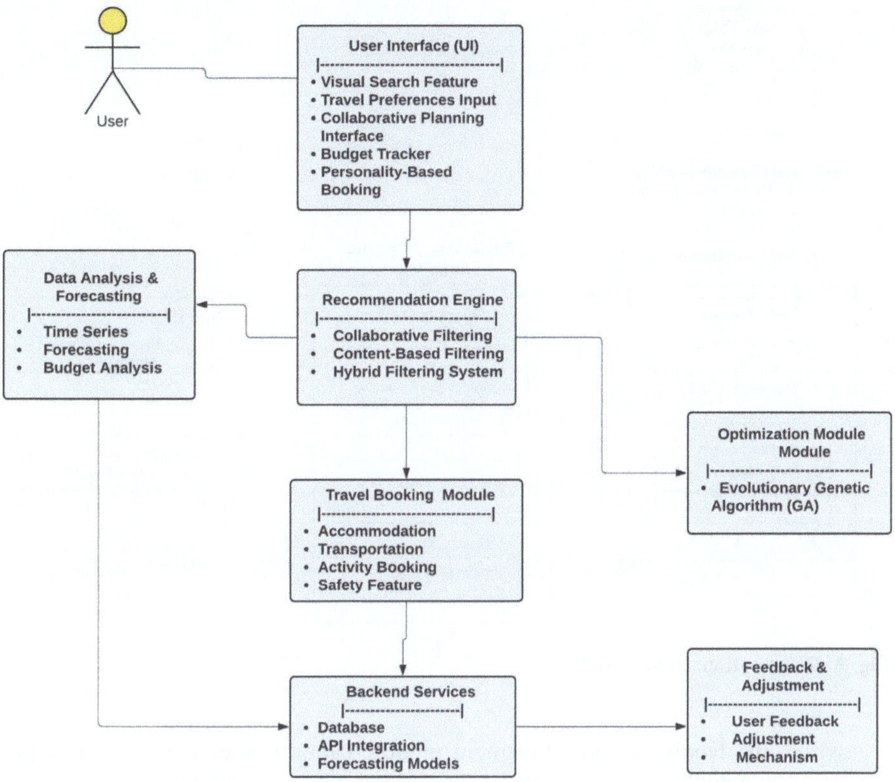

Fig. 1 System design

feature. It also has a budget tracker and personality-based booking to ensure recommendations align with individual preferences and constraints.

(b) **Recommendation Engine**

At the core of the system is the recommendation engine, which utilizes collaborative filtering, content-based filtering, and a hybrid filtering system to suggest the best travel options. This engine is supported by the data analysis and forecasting module, which uses time-series forecasting and budget analysis to predict trends and optimize recommendations.

(c) **Travel Booking Module**

This module handles accommodation, transportation, and bookings, with an integrated safety feature ensuring that users are informed of real-time environmental and security alerts. Backend services manage databases, API integration, and forecasting models to support the application's functionality, ensuring seamless data flow and operation.

(d) **Optimization Module**

It leverages an evolutionary genetic algorithm (GA), fine tunes the travel plan, ensuring optimal solutions for routes and budget management. Mean-

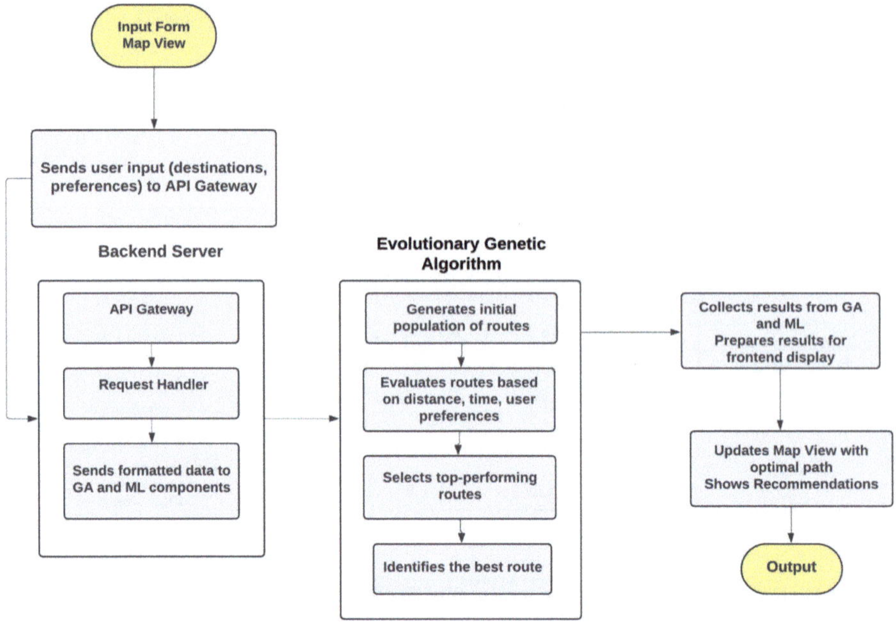

Fig. 2 Route optimization workflow

while, the feedback and adjustment module captures user feedback to adjust recommendations, creating a feedback loop that continuously improves the accuracy.

Figure 2 illustrates the workflow of route optimization within a travel planning system.

3.1 Workflow of Genetic Algorithm for Route Optimization

(a) **Problem Representation**

The route optimization problem is represented using a chromosome structure, where each chromosome corresponds to a sequence of tourist locations. For a set of n locations, a chromosome can be expressed as [L1, L2, ..., Ln], ensuring each location is visited exactly once. The objective is to minimize total travel cost, time, and distance.

(b) **Initialization**

The algorithm begins by generating a randomly initialized population of feasible routes. Each route in the population is constructed to adhere to problem constraints, ensuring that all user-specified destinations are included.

(c) **Fitness Function**

A fitness function evaluates the quality of each route based on multiple optimization criteria. It is defined as below in Eq. (1).

$$\text{Fitness score} = \frac{1}{\text{distance} + w_1 \times \text{constraint}_1 + w_2 \times \text{constraint}_2 + \cdots + w_n \times \text{constraint}_n} \quad (1)$$

where:

- Distance represents the total travel distance for the route.
- Constraint_i represents specific factors such as user preferences, environmental safety, and time limitations.
- w_i represents the weights assigned to each constraint, allowing for prioritization of particular criteria.

This multi-objective optimization approach ensures that the algorithm balances competing.

(d) **Selection**

Parent chromosomes for reproduction are selected based on fitness scores using techniques such as roulette wheel selection, tournament selection, or rank-based selection. This ensures that fitter individuals have a higher probability of contributing to the next generation.

(e) **Crossover**

Crossover operations combine two parent chromosomes to produce offspring, as shown in Fig. 3. Techniques such as order crossover (OX) and partially mapped crossover (PMX) are utilized to maintain valid sequences of locations while introducing diversity into the population.

(f) **Mutation**

Mutation introduces variability to prevent premature convergence. Common methods include (Fig. 4):

- Internal mutation: Modifies segments within the chromosome, e.g., by reversing or swapping specific locations.
- External mutation: Introduces new elements by replacing a part of the route with an external path.
- Omitted mutation: Skips or removes sections of the chromosome, allowing exploration of alternative route possibilities.

(g) **Replacement**

The least fit individuals in the population are replaced by offspring, ensuring continuous improvement of the overall population quality.

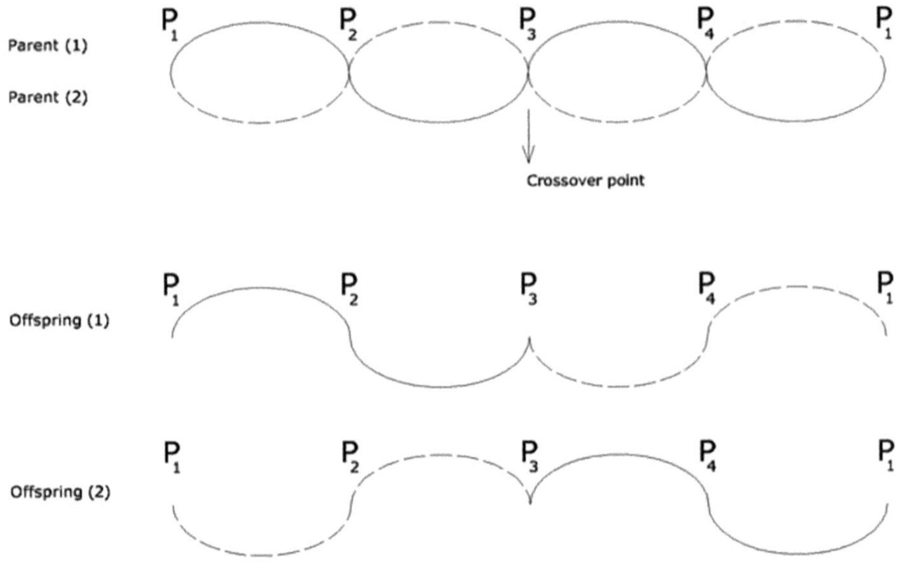

Fig. 3 Crossover operation

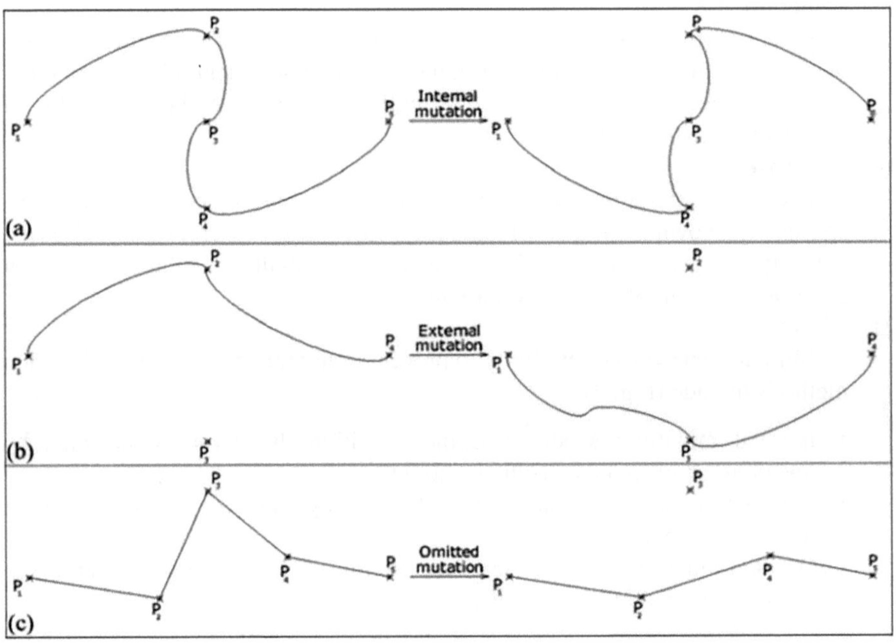

Fig. 4 (**a**) Internal mutation, (**b**) external mutation, and (**c**) omitted mutation

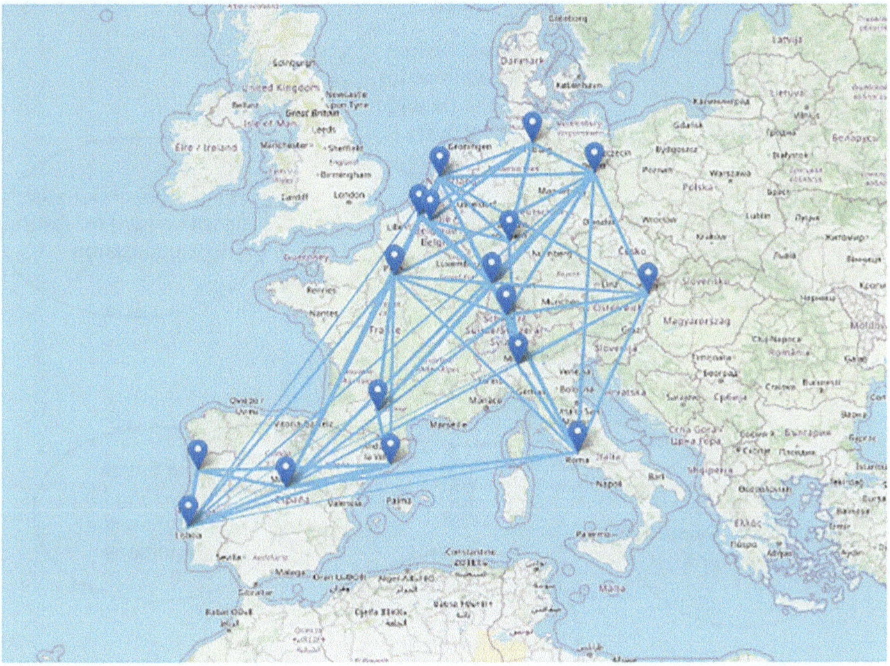

Fig. 5 Multi-city travel route UI

(h) **Termination**

The algorithm terminates when the fitness score shows no significant improvement over successive iterations. Finally, the optimized route is obtained (Fig. 5).

4 Personalized Search and Decision-Making

(a) **Hybrid Recommendation Algorithms**

The system uses both collaborative filtering and content-based filtering to provide highly accurate recommendations (Fig. 6).

Techniques to Handle Data Sparsity: In study [15], using principal component analysis for feature extraction, the method enhances the accuracy of matrix factorization (MF) models.

Matrix Factorization: Alternating least squares (ALS) to reduce the dimensionality of the user-item interaction matrix. This helps uncover latent features even when the data is sparse.

Mean Imputation: Fill missing values with the mean of available data.

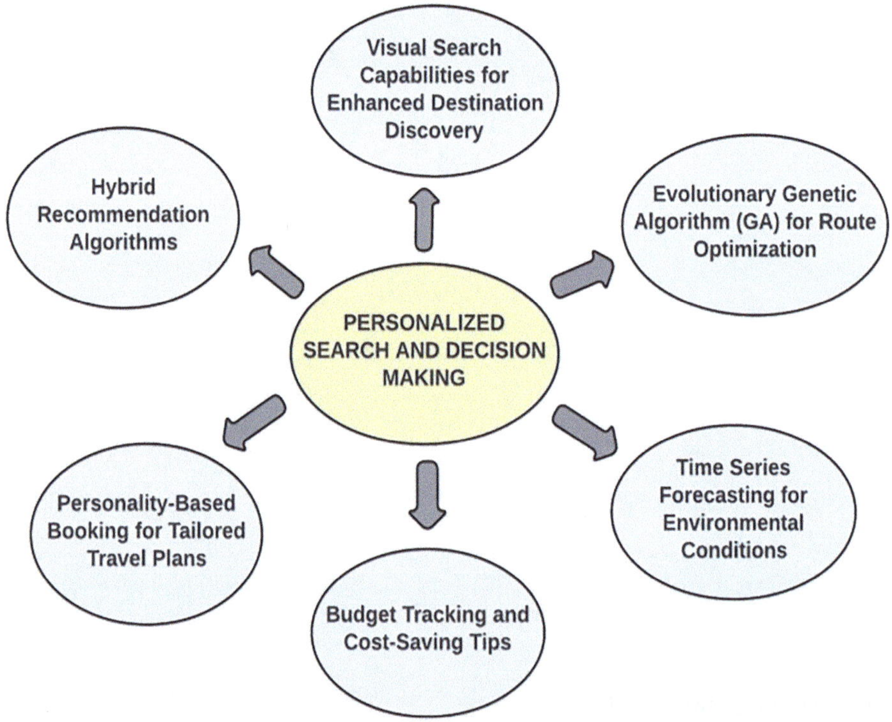

Fig. 6 System methodologies

K-Nearest Neighbours: Estimate missing values by looking at similar users.

(b) **Visual Search Capabilities**

To enhance user engagement and streamline destination discovery, the system incorporates visual search capabilities. Users can upload images to find travel destinations that align with their aesthetic preferences.

1. The uploaded images are analysed using computer vision techniques, including feature extraction and deep learning models, to identify key visual patterns.
2. These features are compared against a database of destination images, employing similarity metrics such as cosine similarity or Euclidean distance in the latent feature space.
3. This functionality facilitates intuitive exploration, enabling users to discover destinations based on visual inspiration rather than traditional text-based searches.

(c) **Data Collection and Distance Matrix**

The system utilizes data collected from the Google Maps API to construct a distance matrix for travel route optimization. The matrix incorporates real-world distances between 11 predefined locations, forming a network with 121 edges.

Positive Ideal Distance Matrix: The collected distance data form a positive ideal distance matrix, serving as the foundation for optimization algorithms.

(d) **Route Optimization**

The system employs an evolutionary genetic algorithm to identify the shortest and most efficient travel routes. The GA iteratively evaluates candidate solutions based on fitness metrics derived from the distance matrix and user preferences.

Crossover and mutation operators are utilized to explore the solution space effectively, ensuring global optimization even in complex networks.

(e) **Budget Tracking and Cost-Saving Tips**

To promote cost-efficient travel planning, the system integrates a budget tracker that monitors and analyses user spending.

Expense Monitoring: The tracker records expenditures related to accommodation, transportation, and activities, providing a comprehensive view of the user's travel budget.

Cost-Saving Recommendations: By analysing spending patterns, the system identifies opportunities for cost savings, offering tailored suggestions such as alternative transportation modes or off-peak accommodations.

5 Real-Time Disaster Response

Figure 7 illustrates the workflow of disaster management within the travel planning system and is explained below.

5.1 Workflow of LSTM for Safety Prediction

The LSTM-based safety prediction model processes sequential data to assess travel safety in real time.

(a) **Data Preprocessing**

Time-series data, such as weather records and environmental alerts, are normalized to a [0,1] range. A sliding window technique segments the data into sequences, and missing values are interpolated using linear methods.

(b) **Model Initialization**

The model is designed with the following layers:

- **LSTM layers:** Capture temporal patterns in sequential data by learning dependencies over time.

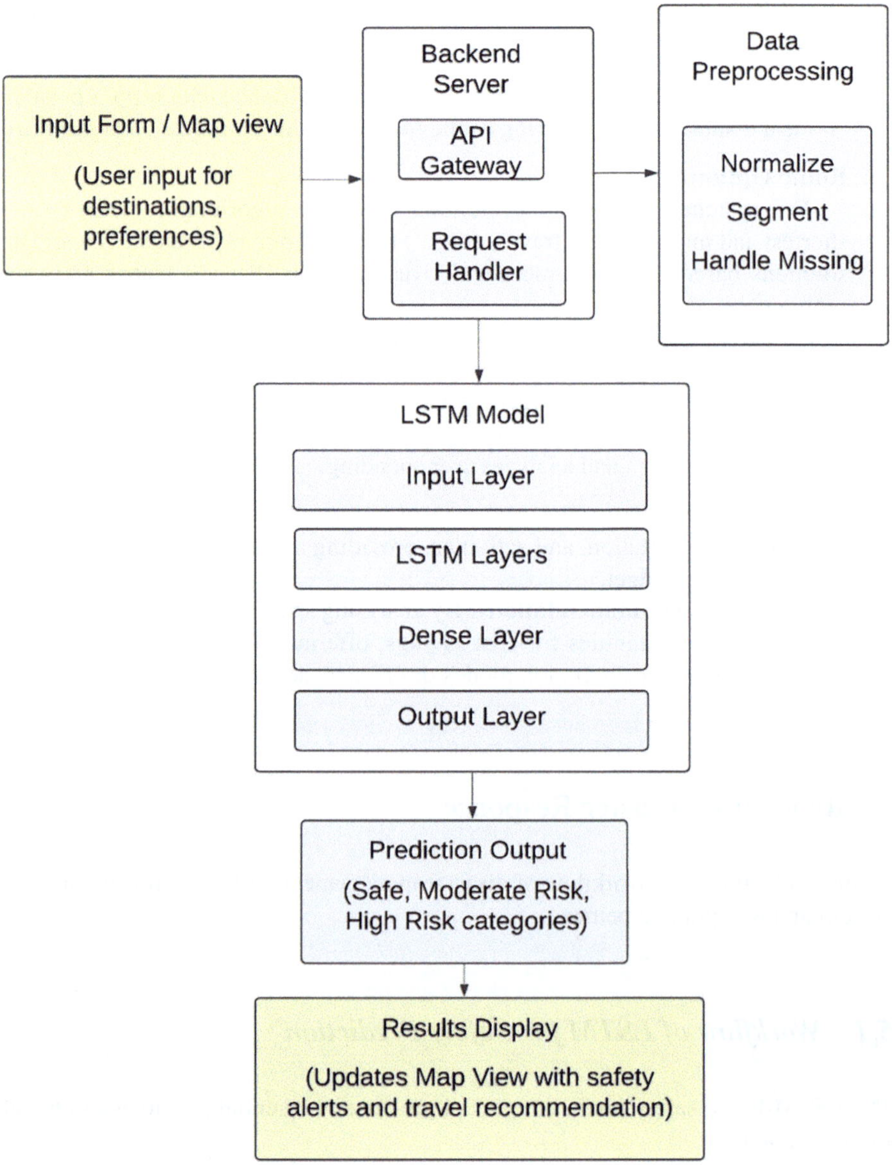

Fig. 7 Disaster management workflow

- **Dense layers:** Aggregate the temporal features extracted by the LSTM layers and transform them into higher-dimensional representations for safety scoring.
- **Output layer:** Produces a continuous safety score SS using a linear activation function, ensuring the output lies within a meaningful range.

(c) **Training**
The model is trained on historical data using the Adam optimizer, which minimizes the loss function. Validation is performed on a holdout dataset, with mean squared error as the evaluation metric to ensure prediction reliability.

(d) **Safety Prediction**
The trained model predicts a safety score S for a given location and time based on the provided input features.

(e) **Categorization**
Safety scores are classified as follows:

- **Safe (Green)**: $S \geq 0.8$.
- **Moderate risk (yellow)**: $0.5 \leq S < 0.80$.
- **High risk (red)**: $S < 0.5$.

(f) **Real-Time Updates**
New data streams are continuously integrated, and the safety score is dynamically updated to reflect changing conditions.

5.2 *Adaptive Risk Management*

(a) **Pre-travel Risk Management**
If the system detects a high-risk condition before the journey, it proactively cancels the booking and generates alternative suggestions. This ensures the traveller's safety while maintaining convenience.

- The safety prediction model analyses time-series data and calculates a safety score for the destination.
- If the score indicates a high-risk situation, the system cancels the current booking to prevent potential harm.
- The recommendation engine recalculates options, excluding risky locations, and proposes safer alternatives. These are optimized using algorithms like the genetic algorithm to match user preferences and constraints.

(b) **Real-Time Risk Management**
For risks identified after the journey has started, the system ensures user safety through immediate alerts and rescue guidance.

- The model continuously monitors real-time data, assessing emerging risks such as extreme weather or natural disasters.
- Notifications are sent to users, providing details about the risk and actions to take.
- The system identifies nearby safe zones or rescue locations and guides users using real-time geolocation and mapping tools, ensuring timely assistance.

6 Result and Discussion

(a) **Hybrid Recommendation System**

The hybrid recommendation system combines collaborative filtering, content-based filtering, and hybrid approaches to deliver highly accurate and personalized travel suggestions:

- **Accuracy:** Achieves an 89.7% recommendation accuracy, providing results that align with user preferences.
- **Handling data sparsity:** Employs techniques such as matrix factorization, and K-nearest neighbours to mitigate data sparsity, improving accuracy by 21% compared to standalone recommendation systems.

Evolutionary Genetic Algorithm for Route Optimization

The evolutionary genetic algorithm (GA) employed for route optimization demonstrates significant advancements in identifying efficient paths across complex networks (Fig. 8):

(b)
- **Accuracy**: Achieves a 92.4% success rate in identifying optimal routes.
- **Time efficiency**: Significantly reduces computational time compared to traditional algorithms, optimizing performance for both small and large networks.
- **Data sparsity**: Resolves sparsity using cluster-based initialization and data augmentation, improving accuracy by 20%.

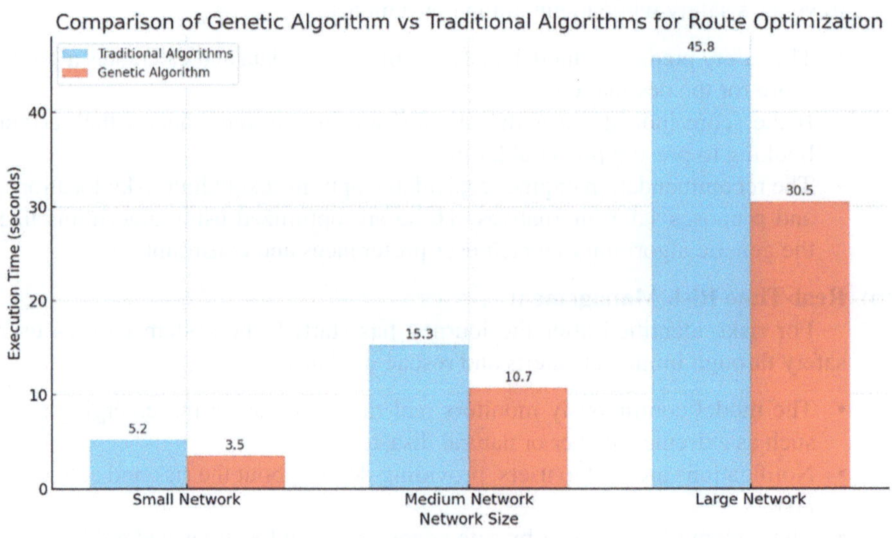

Fig. 8 GA vs. traditional algorithms for route optimization

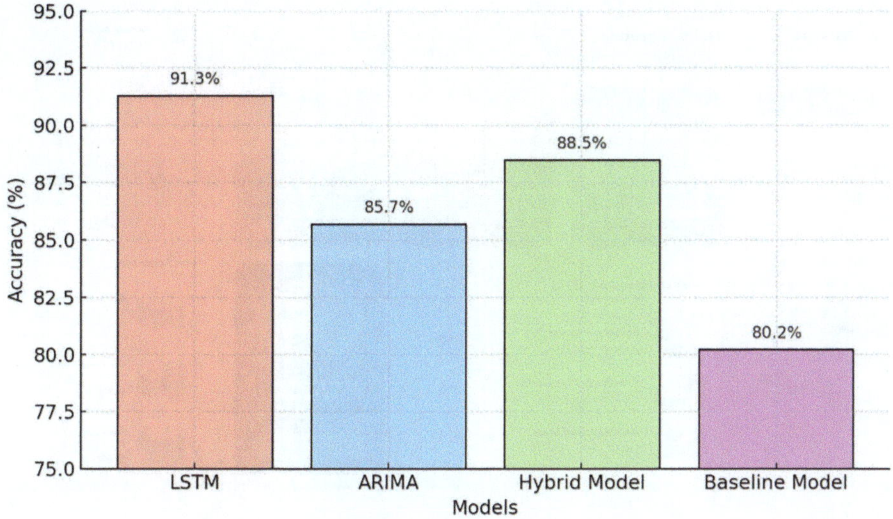

Fig. 9 LSTM vs. traditional algorithms for risk prediction

(c) **Personality-Based Booking**

The personality-based booking module significantly enhances user satisfaction and reduces cancellations:

- **Accuracy**: Personalized recommendations achieve an 87.2% precision rate, aligning closely with user preferences.

The work presented in Ref. [15] states that, instead of assuming users without prior interactions are dissimilar, our approach leverages personalized preferences to establish meaningful similarities among users. Experimental results indicate that this methodology effectively mitigates data sparsity and cold-start challenges, leading to significant improvements in prediction accuracy.

(d) **Time Series Forecasting for Safety**

The integration of long short-term memory (LSTM) networks for safety prediction proves effective in modelling non-linear environmental data (Fig. 9):

- **Accuracy:** A simulated dataset was employed, designed to mimic environmental conditions and risks based on standard meteorological patterns and historical data references. The dataset consisted of X samples, with features such as temperature, humidity, wind speed, and alert frequency. LSTM achieves 91.3% accuracy in predicting environmental risks.
- **Data sparsity:** Sparse data are addressed through extrapolation techniques and pre-trained models, enabling reliable forecasting even with limited datasets.

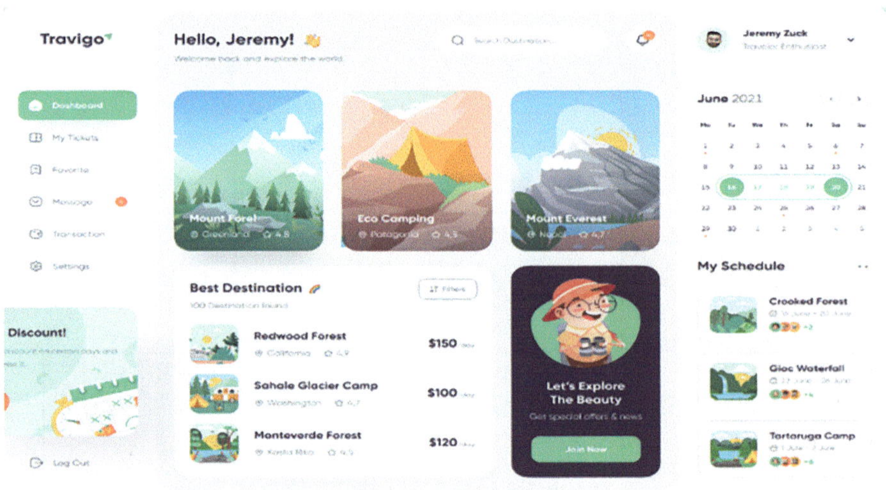

Fig. 10 Travel planner dashboard

(e) **Disaster Prediction and Risk Mitigation**

The system employs LSTM-based safety prediction and real-time alerts to ensure user safety:

- **Disaster detection rate:** The system achieves a 94.8% detection rate for potential hazards.
- **Response time**: Alerts are generated within 2.5 s upon detecting risks.

These findings are based on a simulated dataset of the system across a variety of scenarios. Future efforts will focus on testing the system with extensive datasets to ensure its accuracy, thereby reinforcing its reliability and usability.

(f) **Integration with React Frontend**

The seamless integration with the react frontend provides an engaging and intuitive user experience (Fig. 10):

- **Real-time updates**: Features such as GA-based route optimization and LSTM-driven environmental forecasts are visualized dynamically.
- **User feedback**: Interactive features like budget tracking and personality-based booking ensure continuous system refinement based on user preferences.

7 Conclusion

The proposed travel planning system presents a robust, integrated solution for trip management, merging accommodation, transportation, and activity bookings into a single, cohesive platform. Utilizing advanced collaborative and content-based filtering algorithms, the system delivers precise and tailored recommendations by analysing user preferences, historical data, and budget constraints. The visual search capability leverages image recognition to enhance user engagement and facilitate intuitive exploration of travel options. The incorporation of an evolutionary genetic algorithm (GA) optimizes route planning, ensuring minimal travel time and maximum convenience. Safety concerns are proactively addressed through time series forecasting models, including long short-term memory (LSTM) networks, which predict future environmental conditions like temperature, humidity, and wind speed. This system's collaborative planning functionality allows users to co-create and refine travel plans, while the budget tracker provides real-time expense monitoring and cost-saving recommendations. The personality-based booking feature aligns travel plans with individual lifestyle traits, reducing cancellation rates and boosting overall satisfaction. By integrating these advanced technologies and features, the system not only simplifies the travel planning process but also enhances user experience through personalized, data-driven recommendations and efficient route optimization.

References

1. J.V.N. Lakshmi, M.O. Pallavi, A hybrid recommendation model for tourist using evolutionary algorithm combined with local search algorithm for trip planning. SN Comput. Sci. **5**, 743 (2024). https://doi.org/10.1007/s42979-024-03063-1
2. P. Yochum, L. Chang, T. Gu, M. Zhu, An adaptive genetic algorithm for personalized itinerary planning. IEEE Access **8**, 88147–88157 (2020)
3. D. Weng, R. Chen, J. Zhang, J. Bao, Y. Zheng, Y. Wu, Pareto-optimal transit route planning with multi-objective Monte-Carlo tree search. IEEE Trans. Intell. Transp. Syst. **22**(2), 1185–1195 (2020)
4. S. Cao, An optimal round-trip route planning method for tourism based on improved genetic algorithm. Comput. Intell. Neurosci. **2022**(1), 7665874 (2022)
5. M. Badouch, M. Boutaounte, Personalized travel recommendation systems: A study of machine learning approaches in tourism. J. Artif. Intell. Mach. Learn. Neural Netw. **33**, 35–45 (2023)
6. A. Javadian Sabet, M. Shekari, C. Guan, M. Rossi, F. Schreiber, L. Tanca, THOR: A hybrid recommender system for the personalized travel experience. Big Data Cognit. Comput. **6**(4), 131 (2022)
7. B. Han, X. Zheng, M. Guan, L. Sun, Y. Zhang, Personalized route recommendation with hybrid tabu search algorithm based on crowdsensing. Int. J. Intell. Syst. **2023**(1), 3054888 (2023)
8. N.N. Qomariyah, D. Kazakov, A genetic-based pairwise trip planner recommender system. J. Big Data **8**, 1–23 (2021)
9. M. Sangeetha, A. Priyanga, S. Prakash, Smart travel assistant with itinerary planner using hybrid machine learning approach. Int. Res. J. Mod. Eng. Technol. Sci. **4**, 983–989 (2022)

10. J. Ruiz-Meza, J.R. Montoya-Torres, Tourist trip design with heterogeneous preferences, transport mode selection and environmental considerations. Ann. Oper. Res. **305**(1), 227–249 (2021)
11. D. Tippong, S. Petrovic, V. Akbari, A review of applications of operational research in healthcare coordination in disaster management. Eur. J. Oper. Res. **301**(1), 1–17 (2022)
12. L. Sebastia, E. Marzal, Extensions of the tourist travel design problem for different travel styles. Procedia Comput. Sci. **176**, 339–348 (2020)
13. M. Suriya, Improved spectrum prediction model for cognitive radio networks using hybrid deep learning technique. Int. J. Intell. Netw. **5**, 286–292 (2024). https://doi.org/10.1016/j.ijin.2024.05.00
14. D.G. Taslim, I.M. Murwantara, A comparative study of ARIMA and LSTM in forecasting time series data, in *2022 9th International Conference on Information Technology, Computer, and Electrical Engineering (ICITACEE), Semarang, Indonesia*, (2022), pp. 231–235. https://doi.org/10.1109/ICITACEE55701.2022.9924148
15. H. Leng, W. Lin, Z. Pan, A sparsity-aware hybrid collaborative recommendation method with multi-source data, in *2021 IEEE Intl Conf on Dependable, Autonomic and Secure Computing, Intl Conf on Pervasive Intelligence and Computing, Intl Conf on Cloud and Big Data Computing, Intl Conf on Cyber Science and Technology Congress (DASC/PiCom/CBDCom/CyberSciTech), AB, Canada*, (2021), pp. 708–715. https://doi.org/10.1109/DASC-PICom-CBDCom-CyberSciTech52372.2021.00118

"Collaborativity": A Collaborative Ontology Construction Framework

Kaneeka Vidanage and Asela Gunasekara

1 Introduction

In Ref. [1], Spacis et al. (2005) introduced the semantic web as a domain-rich conceptualization. The unique nature of the semantic web is that it is both human and machine-readable [2]. Apart from that, as highlighted in Ref. [3], human knowledge can be easily encoded into an ontology. Further, it is even suitable for domains that lack large annotated datasets (where machine learning approaches would not be successful) [4]. Due to the aforementioned positive and unique features, the semantic web gained utmost success very rapidly, in diverse domains [5]. There are numerous successful semantic-web-based applications operating in biology, bioinformatics, agriculture, and many non-computing domains [6, 7]. This trend has triggered the construction and application of ontologies in applied domains. As a result of that, collaborative ontology engineering is evolving as a separate segment under the umbrella of ontology engineering [8, 9]. As pointed out by multiple researchers, the main goal to be accomplished via collaborative ontology engineering is the effective gluing of domain specialists and ontologists in the process of constructing ontologies for applied domains [9]. This will lead to the establishment of collective ownership between ontologists and domain specialists in constructing ontologies for applied domains [10, 11]. In Ref. [12], its declared that poor working interoperability and poor collective alliance between ontologists and domain specialists cause low-quality, erroneous, and semantically weak ontologies.

In Ref. [13], it is further confirmed that unsupervised ontology construction solely done by domain specialists, without the involvement of ontologists, leads

K. Vidanage · A. Gunasekara (✉)
Department of Computer Science, Faculty of Computing, General Sir John Kotelawala Defense University, Colombo, Sri Lanka
e-mail: vidanage_bvki@kdu.ac.lk; asela@kdu.ac.lk

A. Haldorai et al. (eds.), *7th EAI International Conference on Big Data Innovation for Sustainable Cognitive Computing*, EAI/Springer Innovations in Communication and Computing, https://doi.org/10.1007/978-3-032-02979-9_18

to many practical consequences once the ontology is put into application. That is why both [14, 15] have suggested the necessity of a successful applied ontology construction mechanism. It is recognized as domain specialists, they should have moderate awareness about the semantic concepts, and vice versa, ontologists should have moderate awareness about the domain of interest. Hence, as argued in Refs. [14, 15], there is a need for a methodical process to accomplish the aforementioned requirements of mutual understanding, which is very vital to gain the proper synchrony among both ontologists and non-computing stakeholders. Absence of an explicitly defined workflow to ensure collaborative engagement between ontologists and domain specialists will lead to many cognitive bottlenecks. Domain specialists have the required domain knowledge, but not the expertise in semantic web concepts. Likewise, ontologists have the expertise in semantic web concepts, but not the domain knowledge. Therefore, due to the deficiency of shared understanding of the requirements to be accomplished, effective gluing of ontologists and domain specialists will not occur. Therefore, they cannot fulfil the working interoperability and knowledge alliance goals, which are vital facets of applied ontology construction [16].

Multiple researchers have claimed the necessity of active engagement of domain specialists in the applied ontology construction process. They have declared that a contribution going beyond the level of monitoring the overall process is expected. The ideology behind this is to develop more practical and feasible ontological solutions in a community-driven manner, within a more collaborative and economical setting. Hence, when domain experts are also involved in the ontology construction process, both domain experts and ontology engineers share a common understanding of the ontology.

This will facilitate the governance and maintenance of the ontology in a more effective and economical setting. This collaborative engagement methodology will resolve conflicts related to irreconcilable views of the stakeholders throughout the process to operationalize consensus-building of ontologies [4, 17]. There are numerous ontology construction methodologies and frameworks available (i.e., refer to Tables 1, 2, and 3 for a comprehensive comparison). Even though, almost all of them do not provide consistent and holistic guidelines to explicitly accomplish collective consensus requirements. This leads to a high amount of ambiguity and vagueness. Therefore, most of the researchers who use these methodologies loosely couple the defined the high-level abstract steps and continue the construction process as they prefer [18]. This firmly triggers the Rational Choice Theory [19, 20] and ultimately it will end up with the Tragedy of the Commons [21]. The notions of Rational Choice Theory and Tragedy of the Commons are very important aspects strongly linked with the foundation of this research. Hence, those concepts will be clearly elaborated in the forthcoming related works section.

In addition to that, the high-level and abstract definition of the steps in a methodology triggers several other critical issues, such as no proper means on performance/issue tracking, as the steps followed are not traceable or transparent. Further, it is not guaranteed that experts will always be there in the research team [21, 22]. Hence, ambiguity, trial-and-error experiments will lead to lots of

Table 1 Tabular comparison of ontology development methods

					Collaboration aspects		Collaboration aspects		Engineering process	Remark
Methodology	Generation	Competency questions	Formalization	Ontology increments	Interoperability	Ideology alliance	Traceable workflow	Transparent workflow		
Methontology [24]	Second	N/A	N/A	N/A	N/A	N/A	N/A	N/A	Rigid	1
Tove [25]	Second	Yes	Yes—FOL	N/A	N/A	N/A	N/A	N/A	Rigid	2
IDEFS [26]	Second	Yes	N/A	Yes	N/A	N/A	N/A	N/A	Rigid	3
OTKM [27]	Second	Yes	N/A	Yes	N/A	N/A	N/A	N/A	Rigid	4
Diligent [28]	Third		N/A		N/A	No	No	No	Lenient	5
RapidOWL [29]	Third	N/A	N/A	N/A	N/A	N/A	N/A	N/A	N/A	6
NeOn [30]	Third	Yes	N/A	N/A	N/A	N/A	N/A	N/A	Rigid	7
Upon-Lite [9]	Third	N/A	N/A	N/A	No	No	No	No	Lenient	8
AMOD [31]	Third	Yes	N/A	Yes	No	No	No	No	Rigid	9

Table 2 Tabular description of methodology remarks

Remark ID	Description	Remark ID	Description
1	No workflow on collaborative involvements No competency questions	6	No methodology. No competency questions Collaboration encouraged, but no controlling workflow Highly vague
2	No workflow on collaborative involvements Competency questions concept is very firm No documentation phase	7	The process is based on methontology or OTKM Collaboration is encouraged, but no controlling workflow Highly vague
3	No workflow on collaborative involvements Proto-concepts, ontology increments No workflow on increments progresses tracks	8	Led by domain specialists. Ontology engineers enter the final lap Caused by poor-quality ontologies, knowledge glitches and cognitive gaps. Poor collaboration
4	No workflow on collaborative involvements Kickoffs, ontology increments No workflow on increments progress track	9	The competency questions concept is clear. Ontology increments Collaboration is encouraged, but no controlling workflow. Highly vague No process on process tracking of the ontology increments
5	No competency questions. Collaboration is encouraged, but no controlling workflow High autonomy causes vague processes		

consequences. Therefore, the research gap to be addressed in this research is about proposing a proper framework to effectively link ontologists and domain specialists with a semantically enriched, traceable, and transparent workflow to resolve the aforementioned working interoperability and knowledge alliance problems.

2 Related Work

At the inception of the related work assessment, initially, the basis of Rational Choice Theory [19, 20] and Tragedy of the Commons [21], notions will be discussed. According to the theory of Tragedy of the Commons proposed by Hardin in 1968, when a pool of people works together to accomplish a common goal, the amount of complexities and disagreements will rise [23]. This is due to human nature clearly explained by rational choice theory introduced by Clark et al. in

Table 3 Tabular comparison of ontology development frameworks

Framework	Category	Pros	Cons	Special remarks
Semiotic [32]	Ontology communication	(a) Consider ontology construction as a communication pipeline between sender and receiver	(a) The learning curve is high as the user should be familiar with peirce theory triadic relations and speculative grammar (b) High subjectivity and ambiguity as mainly a symbolic representation (c) No notion on ontology increments	No emphasis for interoperability and alliance facet (collaboration perspective)
Generic ontology development framework [33]	Ontology	(a) Workflow to follow in properly elaborated with a proper breakdown of phase and subphase	(a) No notion on ontology increments (b) Evaluation of the ontology I done after full completion of the development. This allow. Unforeseen errors to be accumulate	No emphasis for interoperability and alliance facet (collaboration perspective)
Industry relevant ontology development framework [34]	Ontology	(a) Has introduced index-coded thematic analysis to asses qualitative information extracted from the expert interviews (b) Explore relevant scholarly articles and incorporate knowledge accordingly with expert insight	(a) No notion on ontology increments	No emphasis for interoperability and alliance facet (collaboration perspective)

(continued)

Table 3 (continued)

Framework	Category	Pros	Cons	Special remarks
Platform independent ontology development framework [35]	Ontology	(a) Use case centered ontology comprehension projection (b) Specify finalized requirements through formal representations (i.e. FOL) (c) Fine-tuning of platform independent meta model to a platform specific version via configuring the formal representation	(a) No notion on ontology increments	No emphasis for interoperability and alliance facet (collaboration perspective)

Framework		Category	Pros	Cons	Special Remarks
Systemology framework [36] This framework I composition of GIF, DSRP, and dialectic concepts	GIF [36]	Communication	(a) Provide a solid baseline for knowledge acquisition	(a) General communication framework with cons disciplinary application. Not fine-tuned for ontology domain	No emphasis for interoperability and alliance facet (collaboration perspective)
	DSRP [37]	Communication	(a) Provide a solid baseline for brainstorming on acquired facts	(a) General communication framework with cons disciplinary application. Not fine-tuned for ontology domain	No emphasis for interoperability and alliance facet (collaboration perspective)
	Dialectics [35]	Communication	(a) To govern argumentation pros and cons associated with axiom	(a) General communication framework with cons disciplinary application. Not fine-tuned for ontology domain	No emphasis for interoperability and alliance facet (collaboration perspective)

1997 from the initial findings made by Beccarica in the eighteenth century [19, 20]. The rational choice theory has notable applications in the criminology domain to analyze criminal behavior in crime incidents. In a nutshell, the basis of this theory declares that each human has his/her own perceptions about the same phenomena. Hence, when a group of people works together to achieve a common goal, due to the absence of explicitly defined rigid workflows, ample complications will arise. Mainly because people will come up with numerous perceptions about the same problem and they will be tempted to work as they prefer despite the processes. This will trigger, Tragedy of the Commons, caused by the Rational Choice theory [19–21].

As the next step, a collection of existing ontology construction frameworks is compared and assessed. Details associated with that comparison are mentioned in Tables 1, 2, and 3 as a tabularized review. After the comparison of both methodologies and frameworks, it is visible that, though there are multiple positive procedures related to ontology constructions, a proper transparent and traceable workflow to address the working interoperability and ideology alliance among the stakeholders involved is not present. Therefore, there is ample chance for "Tragedy of the Commons" to occur, which will adversely hinder the collective operation of the stakeholders, contributing to the construction of an applied ontology [21]. Therefore, as per the outcomes emerged from Tables 1, 2, and 3 comparisons, it is apparent that there still exists a gap to effectively link ontologists and domain specialists via a semantically enriched, traceable and transparent workflow, resolving the aforementioned working interoperability and knowledge alliance problems. As argued by the researchers, knowledge alliance and working interoperability are very crucial in accomplishing an effectively shared consensus [10]. Even though there is a lack of technical literacy on semantic concepts, querying languages such as SPARQL and SQWRL acts as a barrier to active engagement of the domain specialists in ontology construction activities [38].

Therefore, as discussed in Tables 1, 2, and 3, it is a significant necessity to introduce a proper workflow to glue both parties, ensuring effective involvement and leading toward accomplishing interoperability and alliance goals.

3 Methodology

The absence of a properly defined work-flow enforcement can lead to the "Tragedy of the Commons," especially in collaborative work environments [21]. As already discussed, as an outcome of "Rational Choice Theory," individual's preferred operation will trigger because of the nonexistence of a proper governing workflow [19, 20]. In applied ontology construction environment also, the collective operation of the involved stakeholders is essential. To accomplish that, there needs to be a rigid and traceable workflow. Otherwise, as already discussed, there will be lots of operational consequences, ultimately resulting in low quality or erroneous ontology schema [13]. Although there are ample methodologies and frameworks defined for ontology construction, none of those explicitly emphasizes the approaches for fulfilling collective consensus requirements , which is already recognized as a research gap from the literature review conducted earlier [18]. After a comprehensive assessment of the literature, as a practical remedy to overcome this weakness, it is decided to utilize the blend of Lisbon Strategy [39], Bologna Process [40], and the 4C framework [41]. The Lisbon strategy is about achieving working synchronization between the involved stakeholders by introducing a common set of instructions. When people with varying specialties and viewpoints work together, there is a high chance of cognitive congestion occurring due to opinion mismatches.

With the introduction of the Lisbon Strategy, all individuals involved in the project should adhere to the same set of instructions and follow a common action plan. This will enforce the spread of a shared set of instructions and reduce cognitive bottlenecks [39, 40] an important criterion associated with effective knowledge alliance. The first criterion in 4C is "complementaries," where all involved parties' viewpoints are respected, and everybody should collaboratively work together to bring shared value to the table. The second criterion is "congruent goals," which is about allowing the opportunity for the collective leadership to ensure the smooth operation of the overall process. The third criterion is "compatibility," which is about enforcing mutual realization and co-creation of new knowledge to assure value addition to the entire process. Finally, the fourth criterion is "change," which is about how to handle diversification of viewpoints and process aspects associated with the ongoing project, ensuring collective consensus, which is beneficial to the overall project [39, 40]. Dialectics [42] or healthy debate is very important in reaching the most appropriate decisions.

In this research, the Bologna Process is used to govern the dialectics based on the Lisbon strategy and 4C framework. In a nutshell, the Bologna Process enforces the proper operation of the Lisbon Strategy and 4C framework with favorable synchrony to achieve working interoperability and the knowledge alliance of the stakeholders involved in the project. Figure 1 illustrates how the concepts are methodically linked to retrieve a conceptual workflow. This workflow reflects how the framework attempts to fulfil the interoperability and knowledge alliance perspectives through the enforcement of the Bologna Process.

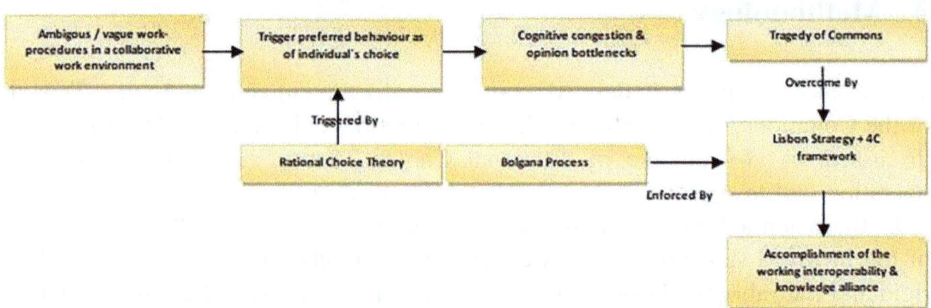

Fig. 1 Conceptual pipeline

The Bologna Process administers the dialectics, ensuring the operation to accomplish the target outcome of interoperability and knowledge alliance of the involved stakeholders. The enforcement of the Lisbon Strategy will account for the working interoperability accomplishment of the stakeholders involved. Because Lisbon Strategy influences all the involved stakeholders, they follow and work on the same action plan, fulfilling the necessity of a shared consensus between the stakeholders. Likewise, the 4C framework provides a convenient platform, comprising four perspectives beneficial for the ideology alliance.

The mutual combination of the Lisbon Strategy and 4C framework creates a rational platform for the stakeholders to cognitively integrate via a healthy debate, administered by the Bologna Process. Both Refs. [14, 15] have insisted on the importance of having a mutual understanding of the semantic concepts and domain of interest before commencing the applied ontology construction process. To assure that requirement, a crash program on mutual knowledge sharing is recommended (first module of the framework). The VoCamp strategy used in the semantic web domain is a famous mechanism used for quick knowledge sharing among involved stakeholders for domain specified vocabulary creation. In a nutshell, VoCamp is a series of informal knowledge transfer sessions, without rigid formalities. It is suggested that the same VoCamp strategy [43] can be used to ensure mutual knowledge sharing between the domain specialists and ontologists involved in the project. The ultimate goal of this VoCamp is to ensure that domain specialists have reasonable knowledge of semantic concepts, as well as that ontologists have a reasonable understanding of the domain concepts. According to the requirements of this research, the suggested topics to be included in the VoCamp are basics of semantic concepts (i.e., purpose, triple concept, basic property types, schemata, relationship types, OWL, and RDF) and the need for upper ontologies. The executional flow associated with the proposed framework is depicted in Fig. 2.

Table 4 elaborates the functionalities and outcomes associated with each of the important steps associated with the logic flow of the framework depicted in Fig. 2.

All the steps of the framework have to be executed with the rational involvement of both the domain specialists and ontologists. Human intervention is a must, as machine intervention is not sufficient to perform complex analytics and yield collaborative conclusions as per the framework requirements.

4 Results and Discussion

The proposed framework comprises a well-defined and transparent workflow to be followed in alignment with semantic web concepts. In collaborative ontology engineering, both domain specialists and ontologists should collaborate. Therefore, the proposed framework's explicit workflow resolves ambiguities and strengthens the blend of ontologists and domain specialists. Table 4 emphasizes the key steps associated with the framework's execution and their connectivity with the semantic concepts. The divide and conquer approach followed in the framework

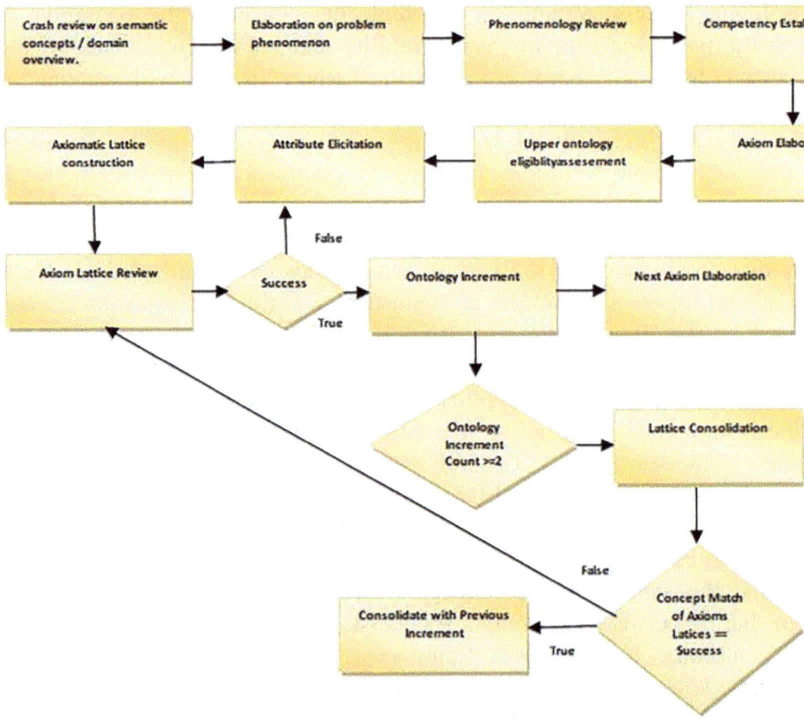

Fig. 2 Executional logic flow of the framework

assigns a specific responsibility to be fulfilled in each step. This emphasizes clarity and the utilization of established brainstorming pipelines such as BMM, Proto Persona, SWOT, Systems Thinking, and Bologna Process contributed to bridging the knowledge gap between domain specialists and ontologists, yielding successful outcomes.

Enforcement of working interoperability and knowledge alliance leading toward collective consensus is the main goal of this framework. Ontology development is a complex task. Hence, the workflow needs to be clearly defined and transparent. This clarity allows heterogeneous stakeholders to work together, as the process is clear to everyone. Therefore, all stakeholders can contribute with a clear understanding. This fosters a knowledge alliance among the group involved in ontology construction. The divide and conquer strategy, explicitly defined established workflows, greatly contributes to overcoming the bottlenecks that hinder working interoperability and collective knowledge alliance. Hence, in the end, all involved stakeholders have proper understanding of the entire pipeline followed. This enforces shared responsibility and comprehension of the created ontology. Accomplishment of this state can be recognized as achieving the goal of collective consensus.

Table 4 Tabular elaboration of the proposed framework's execution

Step	Process/outcome
VoCamp-based crash review	Knowledge—Focused review discussions among the ontologists and the domain specialists. Confirmed technical and domain knowledge sharing among ontologists and the domain specialists
Problem domain elaboration	The Bologna process administered dialectics on the problem to be solved. Once the knowledge sharing is confirmed, the Bologna process assures collaborative problem focus and requirement elicitation
Phenomenology review	Divide and conquer and boundary determination of the problem domain. Locate subtopics as phenomena to be resolved by the ontology to be constructed and their boundaries
Competency establishment	Initiate Bologna process-based dialectics to determine the competency questions to be fulfilled by the ontology. BMM [44], SWOT [45], and Systems thinking [46] notions can be used as tools for methodical brainstorming in finalizing the competency questions
Axiom elaboration	Derive the workflows and associated properties for the recognized competency questions from the previous phase. Utilize the proto persona [47] concept to yield the finalized axiomatic workflows and properties for the determined competency questions
Upper/generic ontology eligibility assessment	Determine the appropriate knowledge fragments associated with the competency questions derived earlier via conducting appropriate dialectics enforced through the Bologna procedures. Asses the existing generic ontological structures via traversing across established ontology pools and determine whether to reuse and existing generic ontological structure or to design from scratch
Attribute elicitation	Utilize agglomerative clustering [48] to inter-relate the derived knowledge fragments. This performs dimensionality reduction, reducing the large amounts of knowledge pieces into more manageable sizes. Henceforth, knowledge seeding [50] is used to enrich the preliminary clusters identified with furthermore coherent proportions. The triadic model suggested by Peirce [49] is used to finally enforce a semantically enriched collective negotiation process to finalize on the terminologies needed to represent the concepts captured
Axiomatic lattice construction	Apply DRSP framework [37] on the findings located from the previous phase. This allows the construction of the initial version of the axiomatic lattice. Subsequently, the primitive lattice defined can be fine-tuned via verifying its structure against John Rowl's "reflective equilibrium" principle [51]. This principle states that high-level meta concepts and low-level domain concepts should be linked via the simplest possible manner, making each knowledge chunk in the lattice act as a self-consistent piece. Business transaction view (BTV) from the UN/CEFACT'S workflow [52] is used to represent a simplified lattice in the form of classes, data, and object properties
Axiomatic lattice review	A primitive version of the lattice adhering to the requirements of the previous step should be created via protégé. Hence, pragmatic review criteria emphasizing domain-independent metadata at the top level and domain-dependent knowledge fragments at a bottom level in the taxonomic structure of the lattice should comply. The next terminological review has to be conducted assessing the validity of the terminology resulting from the attribute elicitation phase
Lattice consolidation	This step performs at least two increments that have been created accurately. A simple case study associated with the experimental domain has to be exercised following the proposed framework's routine. Derived ontology should be verified and validated against a series of queries and responses. Matrices like precision, recall, and F-measure can be utilized to assess the accuracy of the results retrieved for the queries. If the yielded accuracy matrices yield a low value (i.e., less than 60%), axiomatic review or attribute elicitation steps need to be recalled

To determine the operational success of the framework, it has been applied to a university domain to create a university ontology. Deans of faculties, department heads, administrative heads, and student representatives participated as the domain specialists. A team of ontologists, led by a semantic web expert, was liaising with the domain specialists to conduct dialectics as per the Lisbon Strategy and the 4C framework administered by the Bologna Process. Once two ontology increments had been produced, the framework lattice consolidation step was executed. After visual inspection of the constructed ontology increment, a series of Yes/No queries are fired on the ontology to verify the accuracy of domain modeling and it is verified with the involvement of the domain specialists' team. Assessment of the expected outcome results derived from the domain experts, and their comparison with the actual results provided by the ontology via using confusion matrices results (i.e., true positives (TP), false positives (FP), true negatives (TN), false negatives (FN)), was carried out; quantitative test matrices, such as precision, sensitivity and, F-measures were calculated. The results yielded are presented in Fig. 3. Further, qualitative feedback on the framework utilized were derived from the involved stakeholders. The feedback was encoded according to the concepts

N=TP + TN + FP + FN

	Predicted :No	Predicted :Yes	
Actual :No	TN=179	FP=12	(TN+FP)=191
Actual :Yes	FN=27	TP=97	(FN+TP)=124
	(TN+FN)=206	(FP+TP)=109	

Sensitivity = TP / (TP +FN)
=97 / (97 + 27)
= 0.78

Specificity = TN / (TN +FP)
=179 / (179 +12)
=0.93

Precision =TP / (TP +FP)
=97 / (97 +12)
=0.88

Accuracy = (TP + TN) / (TP + TN + FP + FN)
= (97 + 179) / (97 + 179 + 12 + 27)
=0.88

F-Measure = 2 (Precision x Sensitivity) / (Precision + Sensitivity)
= 2(0.88 x 0.78) / (0.88 + 0.78)
=0.82

Fig. 3 Quantitative matrices

Table 5 CCP framework's perspective and question mapping

CCP framework perspectives	"Wh" questionnaire terms	Mapping questions	Encoded facet	Index code
Content	{What}	What are the advantage/disadvantages you experienced in this framework?	Framework goals	FG
		What are the roles you played in this framework?	Framework roles	FR
		What do you think about the contribution you made from your role?	Framework contributions	FC
		What do you think about the operational accuracy of this framework?	Framework accuracy	FA
Context	{Why}	Why do you think this type of framework is needed?	Framework purpose	FP
		Why do you think this framework is accurate/not?	Framework smell	FS
Process	{How}	How to you know your knowledge is represented in the ontology created?	Framework inquiry	FI
		How could this system support applied ontology construction?	Framework benefits	FB

of CCP framework proposed by Joshi [53], which is widely used for opinion segmentation. The CCP framework provides three main perspectives, which can be mapped to "Wh" questions for use in the interviewing process. Controlled interview sessions were conducted with the stakeholders to gather their qualitative opinions about the operation of the proposed framework. These sessions involved the stakeholders who participated in and experienced the entire cycle of the framework. Both quantitative and qualitative metrics confirmed more than an 80% acceptance rate of the defined framework. Afterwards, those questions were used to govern the flow of the interviews conducted. Subsequently, the collected responses were examined and clustered into encoded groups. All responses and opinions were logged/recorded. The CCP framework perspectives, question mappings, and index codes are presented in Table 5. As per the CCP framework, eight different parameters have been yielded to reflect their interaction experience. In the case study of developing the university ontology, the domain specialist's interaction experiences were captured via a controlled interview session.

Subsequently, by following the defined framework's workflows, a practice ontology was developed for a university domain with the involvement of the university's administrative staff and ontologists. Henceforth, the designed ontology increment was queried and the results were verified against the domain knowledge

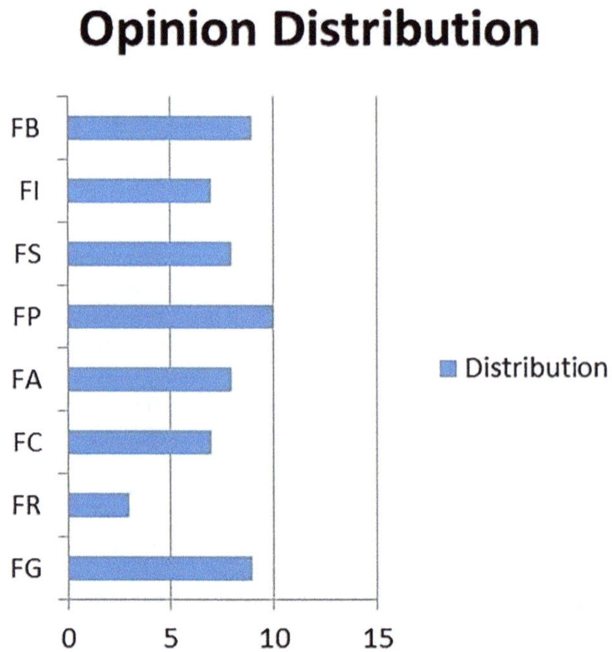

Fig. 4 Stakeholders' response distribution

of the university administration. Depending on the accuracy of the outputs yielded, a confusion matrix has been compiled, as depicted in Fig. 3. Consequently, using the CCP framework, the domain specialists were interviewed about their experience with the framework's operation and the accuracy of the results yielded via it.

Indexed codes were used to label each encoded group, and these index codes were used to plot each cluster's results in the bar chart resulting from the indexed thematic analysis, which is a very effective and methodical approach to convert qualitative data into a quantitative essence, facilitating the analysis [54]. The derived bar chart after the thematic analysis processing is depicted in Fig. 4.

Eventually, the steps specified in the iterative framework are used to methodically relate the outcomes assessed against the framework's success. The iterative framework is a very effective and famous qualitative evaluation framework proposed by Srivastava and Hopwood [55]. It can be used to consolidate all the experimental results toward confirming the accomplishment of the research objective. The information consolidation adhering to the iterative framework's template is depicted in Table 6.

Table 6 Consolidation of experimental results as per iterative framework

Steps in iterative framework	Justification elaborations
01 → What are the data telling me?	Quantitative metrics—As depicted in Fig. 3, multiple quantitative metrics are utilized to validate the effectiveness of the constructed applied ontology's executional outcomes Qualitative assessment—With the involvement of the stakeholders involved in this research, empirical assessment of this ontology's structure, outputs, and framework used to create the ontology is reviewed. As visible in Fig. 4, the stockholder's opinion distribution is plotted in the bar chart As the collective reconciliation, it can be concluded that both the quantitative and qualitative experimental phases have yielded satisfactory results
02 → What do I want to know?	The efficacy of the applied ontology construction mechanism in accomplishing stakeholders' ideology alliance and working interoperability perspectives
03 → Is there a dialectical relationship between steps 1 and 2?	The constructed university ontology has functioned by giving accurate results for the queries fired on it. This implies the domain of concern is modeled appropriately, as it is also evaluated by the domain experts involved in the construction process The total workflow associated with the university ontology construction is fully enforced by the aforementioned proposed framework. This framework has been mainly designed to fulfill the interoperability and alliance goals of the stakeholders, which are strongly enforced by the 4C framework and the Bologna process mediated dialectics Eventually, it can be concluded that the framework has operated successfully, and as a outcome, domain modeling has been properly accomplished. This contributes to the construction of an accurately applied ontology schema, which is capable to accomplishing both qualitative and quantitative validation barriers defined

5 Conclusion

This chapter focuses on a practical problem, related to effective ontology construction. When creating ontologies for applied domains, improper ideological alignment among involved stakeholders triggers numerous cognitive bottlenecks, ultimately ending up with low-quality and erroneous ontologies. Almost all the existing ontology development methodologies do not provide explicit and holistic guidelines to its users. Most of the instructions provided are very abstract and high-level. Therefore, this creates lots of ambiguity and additional problems, as discussed by the Tragedy of the Commons and Rational Choice theory. As the contribution of this research, a novel framework with explicitly defined workflows is proposed to address almost all the problems. Additionally, all salient and significant decisions reached collectively are methodically mapped into numerous work-sheet structures proposed. This improves transparency and traceability of the process in the occurrence of an unexpected issue. Eventually, the completed framework was applied to create an ontology for a university domain, with the involvement of ontologists and multiple domain specialists invited to involve from various faculties and departments of the university. Consequently, the created ontology was practically exercised to validate its efficacy both quantitatively and qualitatively, reflecting around 80% accuracy of the framework proposed in this research. Henceforth, the iterative framework is utilized to empirically assess the outcomes of the research conducted. However, in the future, it is decided to apply this framework to several other domains as well as to integrate tool support to facilitate phase-specific activities.

References

1. M.T. Amith, C. Onye, T.A. Ledoux, G. Xiong, C. Tao, The ontology of fast food facts: Conceptualization of nutritional fast food data for consumers and semantic web applications. BMC Med. Inform. Decis. Mak. **21** (2021)
2. N.S. Abdullah, S.W. Sadiq, *Australasian (ACIS) 2011. A Framework for Industry-Relevant Ontology Development* (2017)
3. K. Vidanage, N.M. Noor, S. Sathsara, Computationally intelligent workflow for improved psychotherapy interventions: An ontology-based approach. Int. J. Inf. Technol. **16**, 4335–4342 (2024)
4. K. Vidanage, R. Mohemad, N.M. Noor, Z.A. Bakar, Fully automated ontology increment's user guide generation. Int. J. Adv. Comput. Sci. Appl. **12**(6), 2021 (2021)
5. K. Vidanage, N.M. Noor, R. Mohemad, Z.A. Bakar, Collaborative ontology construction framework: An attempt to rationalize effective knowledge dissemination. Int. J. Adv. Comput. Sci. Appl. **13**(10), 2022 (2022)
6. A. Alobaid, D. Garijo, M. Poveda-Villalón, I. SantanaPerez, A. Fernández-Izquierdo, O. Corcho, Automating ontology engineering support activities with OnToology. J. Web Semant. **57**, 100472 (2019). https://doi.org/10.1016/j.websem.2018.09.003
7. D. Spoladore, E. Pessot, A. Trombetta, A novel agile ontology engineering methodology for supporting organizations in collaborative ontology development. Comput. Ind. **151**, 103979 (2023)
8. C. Cabanillas, M. Resinas, A. Ruiz-Cortés, A mashup-based framework for business process compliance checking. IEEE Trans. Serv. Comput. **15**, 1564–1577 (2022)
9. A. De Nicola, M. Missikoff, A lightweight methodology for rapid ontology engineering. Commun. ACM **59**(3), 79–86 (2016)
10. J. Ingram, P. Gaskell, Searching for meaning: Co-constructing ontologies with stakeholders for smarter search engines in agriculture. NJAS – Wageningen J. Life Sci. **100300**, 90–91 (2019). https://doi.org/10.1016/j.njas.2019.04.006
11. A. Kocher, L.M. Silva, A. Fay, Modeling and executing production processes with capabilities and skills using ontologies and BPMN, in *2022 IEEE 27th International Conference on Emerging Technologies and Factory Automation (ETFA)*, (IEEE, 2022), pp. 1–8
12. K. Meghraoui, T. Racharak, K.A. Kadi, S. Bensiali, I. Sebari, A quantitative assessment framework for modelling and evaluation using representation learning in smart agriculture ontology. Int. Conf. Agents Artif. Intell. (2024)
13. Y. Huang, S. Dhouib, L.P. Medinacelli, J. Malenfant, Enabling semantic interoperability of asset administration shells through an ontology-based modeling method, in *Proceedings of the 25th International Conference on Model Driven Engineering Languages and Systems: Companion Proceedings*, (2022)

14. M. McDaniel, V.C. Storey, Evaluating domain ontologies. ACM Comput. Surv. **52**(4), 1–44 (2019). https://doi.org/10.1145/3329124
15. R.L. Façanha, M.C. Cavalcanti, M.L. Campos, A systematic approach to review legacy schemas based on ontological analysis. Metadata Semant. Res., 63–75 (2019). https://doi.org/10.1007/978-3-030-14401-2_6
16. K. Vidanage, N.M. Mohamad Noor, R. Mohemad, Z. Abu Bakar, Ontology specific visual canvas generation to facilitate sense-making-an algorithmic approach. Int. J. Electr. Comput. Eng. **12**, 1818–1830 (2022)
17. K. Vidanage, N.M.M. Noor, R. Mohemad, Z.A. Bakar, Semantic web-based knowledge extraction: Upper ontology guided crime knowledge discovery, in *Advanced computing technologies and applications. Algorithms for intelligent systems*, ed. by H. Vasudevan, A. Michalas, N. Shekokar, M. Narvekar, (Springer, Singapore, 2020). https://doi.org/10.1007/978-981-15-32429_30
18. R.F. Calhau, S. Kokkula, D. Cameron, G. Guizzardi, J.P. Almeida, Modeling competence framework elements with an ontology-based approach, in *2023 IEEE25th Conference on Business Informatics (CBI)*, (IEEE, 2023), pp. 1–10
19. S.M. Kim, Antecedents of compliance intention and its impact on waste separation behavior: Based on rational choice theory and deterrence theory. Behav. Sci. **13**(5), 424 (2023)
20. V.L. Johnson, S.M. Esfahani, H. Mohit, Using rational choice theory to explore factors impacting contact tracing application adoption. Inf. Syst. Manag. **41**, 27–43 (2023)
21. V. Martino, G.V. Zani, Green areas: How to avoid the tragedy of the commons. Philosophy City J. **1**(1), 100–111 (2023)
22. K. Vidanage, N.M. Mohamad, R. Mohemad, Upper–level task ontology for generic ontology verbalization, in *2019 2nd International Conference of Intelligent Robotic and Control Engineering (IRCE)*, (IEEE, 2019), pp. 148–155
23. Nagle, Frank, *The Digital Commons: Tragedy or Opportunity? A Reflection on the 50th Anniversary of Hardin's Tragedy of the Commons* (December 13, 2018). Harvard Business School Strategy. Unit, No. 19-060. Available at SSRN: https://ssrn.com/abstract=3301005 or https://doi.org/10.2139/ssrn.3301005
24. J. Tupayachi, H. Xu, O. Omitaomu, M.C. Camur, A. Sharmin, X. Li, Towards next-generation urban decision support systems through AI-powered generation of scientific ontology using large language models—A case in optimizing intermodal freight transportation. ArXiv, abs/2405.19255 (2024)
25. L. Ting, D. Chen, *Mapping IEM to Enterprise Modelling Ontology* (I-ESA Workshops, 2020)
26. L. Li, P. Gu, W. Wang, Ontology modeling of army equipment maintenance support business process under the new system base on IDEF5/Owl, in *2021 IEEE International Conference on Power Electronics, Computer Applications (ICPECA)*, (IEEE, 2021), pp. 331–335
27. Y. Sure, S. Staab, R. Studer, On-ToKnowledge methodology (OTKM), in *Handbook on Ontologies*, (2004), pp. 117–132. https://doi.org/10.1007/978-3-540-24750-0_6
28. D. Vrandecic, S. Pinto, C. Tempich, Y. Sure, The diligent knowledge process. J. Knowl. Manag. **9**(5), 85–96 (2005)
29. S. Auer, H. Herre, RapidOWL—An agile knowledge engineering methodology, in *Ershov Memorial Conference*, Lecture Notes in Computer Science 4378, ed. by I. Virbitskaite, A. Voronkov, (Springer, 2006), pp. 424–430
30. A. Gómez-Pérez, M.C. Suárez-Figueroa, Scenarios for building ontology networks within the NeOn methodology, in *Proceedings of the Fifth International Conference on Knowledge Capture—K-CAP '09*, (2009). https://doi.org/10.1145/1597735.1597773
31. A. Abdelghany, N. Darwish, H. Hefni, An agile methodology for ontology development. Int. J. Intell. Eng. Syst. **12**(2), 170–181 (2019). https://doi.org/10.22266/ijies2019.0430.17
32. E. Toppano, V. Roberto, A semiotic framework for ontology development, in *Proceedings of I-KNOW '09 and I-SEMANTICS '09*, (2009), pp. 738–743
33. K. Kirtiklis, R. Laužikas, I. Kelpšienė, C.J. Dallas, *An Ontology of Semiotic Activity and Epistemic Figuration of Heritage, Memory and Identity Practices on Social Network Sites* (SAGE Open, 2023), p. 13

34. S. Luściński, V. Ivanov, A simulation study of Industry 4.0 factories based on the ontology on flexibility with using FlexSimr software. Manag. Prod. Eng. Rev. **11**(3), 74–83 (2023)
35. I. Sanya, E. Shehab, An ontology framework for developing platform-independent knowledge-based engineering systems in the aerospace industry. Int. J. Prod. Res. **52**(20), 6192–6215 (2014). https://doi.org/10.1080/00207543.2014.919422
36. D. Rousseau, J. Billingham, A systemic framework for exploring worldviews and its generalization as aMultiPurpose inquiry framework. Systems **6**, 27 (2018)
37. D. Cabrera, L. Cabrera, *Systems Thinking Made Simple: New hope for Solving Wicked Problems* (Odyssean Press, Ithaca, NY, USA, 2017)
38. K. Vidanage, N.M. Noor, R. Mohemad, Z.A. Bakar, Arbitrary verification of ontology increments using natural language. Int. J. Adv. Comput. Sci. Appl. **12**(7) (2021)
39. P. Lucian, From the Lisbon strategy to Europe 2020. Stud. Bus. Econ. **10**(2), 53–61 (2015). https://doi.org/10.1515/sbe2015-0020
40. The Bologna Process and the "Process" of Reform and Revolution, *Revolutionizing Global Higher Education Policy* (Routledge, New York, 2016), pp. 159–186. https://doi.org/10.4324/9781315725888-7
41. P.A. Rauschnabel, R. Felix, J. Heller, C. Hinsch, The 4C framework: Towards a holistic understanding of consumer engagement with augmented reality. Comput. Human Behav. **154**, 108105 (2023)
42. A. Fritzsche, Propositional logic, paradox, and Indian dialectics—Towards a deeper ontological approach in paradox theory. Syst. Res. Behav. Sci. **42**, 629–639 (2024)
43. T. Narock, P. Fox, *The Semantic Web in Earth and Space Science. Current Status and Future Directions* (IOS Press, Amsterdam, Netherlands, 2015)
44. M. Pańkowska, Business motivation model for information system architecture development support. J. Soft. Syst. Dev. **2021**, 668927 (2021)
45. M. Haile, Modelling of SWOT analysis using fuzzy integrals, in *ISSC 2016 International Conference on Soft Science*, (2016). https://doi.org/10.15405/epsbs.2016.08.12
46. K.T. Yang, X. Yao, Turning high dimensional optimization into computationally expensive optimization. IEEE Trans. Evolut. Comput. **22**(1), 143–156 (2018)
47. G. Getto, C. Moore, Mapping personas: Designing UX relationships for an online coastal atlas. Compute. Composit. **43**, 15–34 (2017). https://doi.org/10.1016/j.compcom.2016.11.008
48. L. Yang, C. Li, Identification of vulnerable lines in smart grid systems based on improved agglomerative hierarchical clustering. IEEE Access **11**, 13554–13563 (2023)
49. H.R. Cardoso Jr., Peirce and Deleuze in the "Protoplasm of philosophy": Triadic relations and habit as Pragmatic concepts. Rev. Contemp. Philos. **18**, 23 (2018). https://doi.org/10.22381/rcp1820192
50. P. Huang, A. Tafti, S. Mithas, Platform sponsor investments and user contributions in knowledge communities: The role of knowledge seeding. MIS Quarter. **42**, 213–240 (2018)
51. D. Hübner, Three remarks on "reflective equilibrium". Philos. Inq. **41**(1), 11–40 (2017). https://doi.org/10.5840/philinquiry20174112
52. M. Zapletal, R. Schuster, P. Liegl, C. Huemer, B. Hofreiter, The UN/CEFACT modeling methodology UMM 2.0: Choreographing business document exchanges, in *Handbook on Business Process Management 1*, (2014), pp. 625–647. https://doi.org/10.1007/978-3-64245100-3_27
53. M. Joshi, Prioritizing performance evaluation factors of event-based information systems using interpretive structural modeling. Suranaree J. Sci. Technol. **30**(3), 1–10 (2023)
54. F.A. Gildberg, R.L. Wilson, Scientific models for qualitative research: A textual thematic analysis coding system—Part 1. Nurse Res. **31**(3), 36–42 (2023)
55. J.G. Snodgrass, K.R. Clements, W.C. Nixon, C. Ortega, S. Lauth, M. Anderson, An iterative approach to qualitative data analysis: Using theme, cultural models, and content analyses to discover and confirm a grounded theory of how gaming inculcates resilience. Field Methods **32**, 399–415 (2020)

An Exploration of Various Aspects in Opinion Mining and Sentiment Analysis Across Social Media Platforms

S. Vinayaga Vadivu, P. Nagaraj, and M. Raja

1 Introduction

The social media (SM) has emerged as a significant medium for the exchange of information, encompassing a diverse array of subjects. Social multimedia encompasses the assortment of multimedia materials that are shared on SM platforms, to foster individual involvement and collective curation [1]. In the contemporary era of extensive data collection and analysis, a substantial volume of multimedia data is generated within various social networks on a minute-by-minute basis. The exploration of the information included within social multimedia holds significant value due to its crucial role in various prospective applications, including the analysis and prediction of user behavior [2, 3].

One illustrative application of sentiment analysis (SA), also known as opinion mining, involves the extraction of individuals' opinions and sentiments from content generated by users. This analytical technique has the potential to facilitate diverse applications, including the prediction of stock market trends, forecasting box-office performance, projecting political voting outcomes, and monitoring public opinion. The study of sentiment for various forms of multimedia material in social networks has become increasingly relevant in contemporary times. This is due to its significant impact on perception, planning, reasoning, creativity, and decision-making processes.

The concept of SA initially pertains to the objective of identifying, examining, and extracting attitudes, sentiments, and views that are conveyed through written text [4, 5]. The utilization of smart mobile devices and SM platforms, such as Face-

S. V. Vadivu · P. Nagaraj (✉) · M. Raja
Department of Computer Science and Engineering, School of Computing, Kalasalingam Academy of Research and Education, Virudhunagar, Tamil Nadu, India
e-mail: svinayagavadivu@gmail.com; nagu.is.raj@gmail.com; mraja@klu.ac.in

A. Haldorai et al. (eds.), *7th EAI International Conference on Big Data Innovation for Sustainable Cognitive Computing*, EAI/Springer Innovations in Communication and Computing, https://doi.org/10.1007/978-3-032-02979-9_19

book, Twitter, and Weibo, has led to the widespread dissemination of multimedia content, including photographs and videos. This multimedia content has emerged as a significant means of expressing and communicating individuals' thoughts and ideas within social networks, as evidenced by several studies [6]. The study of sentiment in social multimedia has evolved beyond its original scope in natural language processing (NLP), now encompassing interdisciplinary connections with pattern recognition, computer vision, and other facets of artificial intelligence (AI).

Companies must prioritize their clients and allocate substantial financial resources toward research to develop an effective marketing strategy. This encompasses several aspects such as analyzing competition and feedback, as well as ensuring product suitability in expanding markets. The significance of data in formulating policies to differentiate a corporation is evident. Nevertheless, the task of classifying, analyzing, and potentially capitalizing on the vast amount of unstructured data that is readily available seems to be an insurmountable challenge [7].

To flourish, enterprises must possess a comprehensive understanding of the desires, emotions, and actions exhibited by consumers [8]. SA is employed to address the interconnected problems. It makes a considerable contribution to competitive research. Acquiring clients to utilize the services offered by an organization is a prime factor in the progress of any firm. Promoting the firm represents a highly captivating strategy for acquiring consumers. In contemporary times, a revolutionary marketing strategy is being implemented. This marketing strategy employs human emotions to attract and engage clients.

The decision of a buyer to acquire a specific product is influenced by the SA of their feelings. The utilization of SA and machine learning (ML) methodologies has the potential to enhance a brand's overall presentation and deliver satisfactory client experiences [9–12]. The subsequent sections provide an overview of the historical context around the three primary elements examined in this analysis.

This chapter is arranged as follows: The subsequent section provides a background of the study's contextual framework. The third section provides a wide-ranging account of the research methods employed for semantic analysis. The fourth section concludes this chapter.

2 Background

To maintain competitiveness in the current dynamic and complex business environment, businesses must engage in regular and systematic analysis and evaluation of their competitors' strategies, products, and services [13, 14]. Businesses must engage in the monitoring and analysis of their competitors' SM platforms, in addition to their own, due to the widespread adoption of SM as a means of communication between businesses and their existing and prospective clientele. In Ref. [14], it was shown that SM platforms not only provide information about competitors but also present an opportunity to analyze and evaluate client behaviors toward rival organizations. The acquisition and retention of customers are crucial objectives for organizations, and obtaining prompt feedback from consumers is an effective strategy for achieving these goals.

In this regard, SM competitive research has emerged as a significant skill, enabling businesses to gather valuable insights from customers and generate logical reports that contribute to client acquisition and preservation [15–17]. Social data analysis in the context of SM has been extensively investigated in previous studies [18, 19]. The authors of Ref. [20] proposed the utilization of text mining and hierarchical clustering techniques to enhance the management of food supply chains using SM analysis.

A theoretical model is established in Ref. [21], which is aimed at assisting businesses in leveraging data from SM platforms to enhance their customer service capabilities. In Ref. [22], the authors employed support vector machines (SVM) to differentiate the sentiments of Twitter data into +VE, −VE, and NEU to augment decision-making processes. A supervised text classification approach aimed at addressing challenges related to early risk assessment is introduced in Ref. [23]. The proliferation of online SM platforms has provided customers with a unique and noteworthy opportunity to openly articulate their opinions regarding various products and services. This innovation brings about novel insights into customer relationships and brand management, fundamentally transforming the dynamics of consumer-business interactions. SA is employed to study, gather, and assess the influence of consumer perspectives.

SA [24] is a methodology that can be employed to identify ambiguity in words, perspectives, and other elements. SA is a methodology used to determine the emotional stance of both a representative and a user toward a specific subject. The user's choice of tone or emotive style might serve as a reflection of their opinions or sentiments. In recent times, a multitude of algorithms have been established to analyze, forecast, and assess emotions derived from textual data, specifically for product reviews by consumers. The method of recognizing polarity can greatly benefit from the utilization of SA. Furthermore, aside from the aforementioned concerns, the system also encounters issues with fake data/spam, domain heterogeneity, negation, the challenges associated with NLP, the use of bipolar terminology, and the utilization of an extensive vocabulary.

To facilitate SA, preprocessing tasks attempt to systematize the text and pave the way for the effective extraction of significant information, which may subsequently be utilized by text-mining heuristics and ML algorithms [25]. The act of cleansing and making the text ready to facilitate categorization is commonly referred to as data preparation [26]. The objective of preprocessing is to categorize a character set derived from a continuous flow of text, transitioning between different states according to the occurrence of specific letters. The effective management of arbitrary text sequences can be achieved through a meticulous examination of the assortment of characters, including emojis, punctuation marks, white spaces, and emoticons [25].

Preprocessing is an essential step that must be undertaken before inputting unstructured data into an AI system, as computers lack the innate ability possessed by humans to grasp such material. According to the study conducted in Ref. [27],

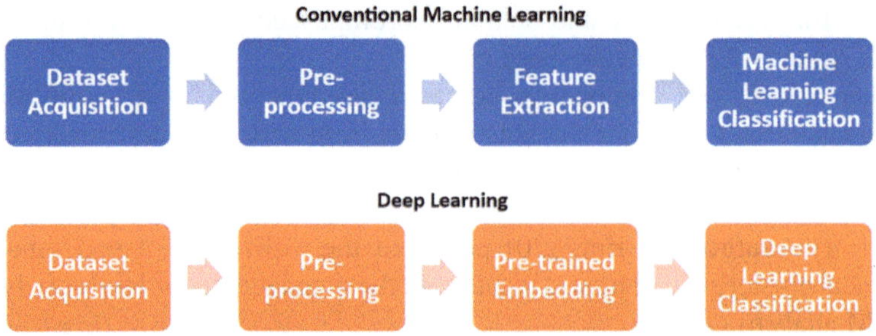

Fig. 1 Key steps involved in ML and DL approaches

it was observed that the implementation of text preparation techniques enhances the accuracy of SA algorithms.

To optimize the effectiveness of the mining process, it is crucial to acknowledge and tackle the aforementioned challenges as documented in Ref. [28]. The study conducted in Ref. [29] revealed that the utilization of semantic sentence analysis has the potential to improve precision, consistency, and methodology. The volume of consumer reviews on digital media has experienced a significant surge. Marketing scholars have employed a range of ways to analyze textual reviews. The study conducted in Ref. [30] showcased the enhanced performance of an NLP technique.

This improvement was achieved by integrating the senses of word and reference into feature vectors and utilizing ensemble classification techniques. The findings of the study revealed that this enhanced NLP approach resulted in improved accuracy in predicting emotions. NLP can analyze, emphasize, and extract semantic information from both textual and vocal data through the utilization of ML techniques and algorithms [31]. At its most basic level, this process entails employing an AI framework to analyze textual content and detect particular words or phrases. Nevertheless, in sophisticated implementations, it empowers computers to accurately discern the emotional state of the speaker, occasionally surpassing human capabilities. Creatively employing advanced AI capabilities could potentially prove to be an effective approach to doing SA.

With this background knowledge, the following section discusses the overview of SA concerning two basic approaches such as ML and deep learning (DL). The conventional steps involved in ML and DL approaches are shown in Fig. 1.

3 SA

This section outlines a thorough review of the existing SA algorithms. SA is a computational procedure that aims to discern and classify the emotional states or sentiments conveyed within written textual content. To accomplish this objective, it is necessary to subject the unprocessed textual data to a series of procedures, which

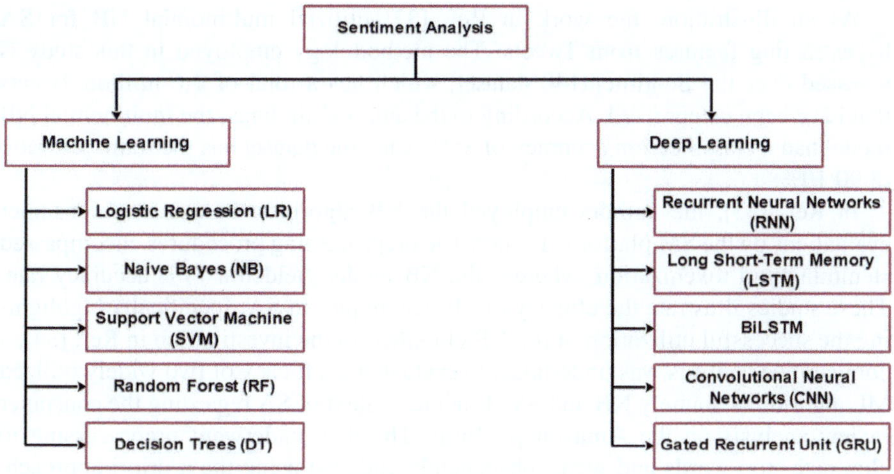

Fig. 2 Categorized approaches to SA

encompass data preprocessing, extraction of text features, and textual classification. The first step performs the most important role in the SA workflow by facilitating the standardization of textual data and eliminating extraneous or disruptive components. This process may involve the utilization of several approaches, such as stemming and lemmatization, in addition to the elimination of special characters and stop words.

The text that has been cleaned is subsequently converted into features, which are then treated by the classification algorithm for making decisions. The classification algorithms employed in SA can be categorized under two main groups: ML and deep learning (DL). ML classification models like SVM, logistic regression (LR), and Naïve Bäyes (NB) apply mathematical models to make predictions on attitudes. Some of the DL classifiers include recurrent neural networks (RNNs) and long short-term memory (LSTM) models, which exploit artificial neural networks (ANN) to do SA [65]. Figure 2 depicts the general classification strategy of SA.

3.1 *ML-Based Algorithms*

The initial step in SA using ML techniques involves the standardization of text input through preprocessing, which entails the elimination of extraneous information. Subsequently, N-grams and term frequency-inverse document frequency (TF-IDF), which are the feature extraction techniques aimed at transforming the textual data into numeral features [66]. These features are then utilized as input for ML classifiers. Several classifiers often employed in SA are SVM, NB, LR, random forest (RF), and decision trees (DTs).

As an illustration, the work in Ref. [32] utilized multinomial NB for SA by extracting features from Tweets. The methodology employed in this study is assessed over the Sentiment140 dataset, which has a total of 1.6 million Tweets that have been categorized. According to the authors' findings, the multinomial NB model had a classification accuracy of 85% when the dataset has the train: test ratio of 90:10%.

In Ref. [33], the authors employed the NB algorithm to examine consumer evaluations on the SM platform Twitter. The preprocessing procedures encompassed stemming and tokenization, whereas the NB model yielded a 77% accuracy rate. These studies illustrate the efficacy of ML techniques in SA, specifically highlighting the successful utilization of the NB classifier. In the investigation in Ref. [34], a comparative analysis was undertaken to evaluate the efficacy of two widely utilized ML algorithms, namely NB and SVM, in the context of SA regarding the consumer review analysis on the Amazon platform. The data underwent preprocessing to eliminate stop-words and were subsequently encoded using the a priori approach. The experimental outcomes indicate that the NB performed better in comparison to the SVM, with an accuracy rate of 90.42% as opposed to 83.42%.

The work presented in Ref. [35] employed three ML techniques, namely NB, SVM, and maximum entropy, to conduct SA experiments. The study involved the utilization of two datasets, namely IMDb and Sentiment140, for conducting tests. Before the trials, several standard preprocessing techniques were applied. The findings indicated that the utilization of the maximum entropy method, in conjunction with the features of unigram and bigram, yielded the most favorable accuracy rates. Specifically, the approach attained an accuracy rate of 88% and 90% on the IMDb and Sentiment140 datasets, respectively.

In Ref. [36], the authors conducted an evaluation of three ML algorithms, including DT, AdaBoost, and SVM, in the domain of SA. The evaluation was done on the Polarity Sentiment 140 and Michigan University datasets. The text underwent preprocessing and TF-IDF is employed for representation. The findings indicate that the SVM exhibited an accuracy of 82%, while the DT model attained an accuracy of 84% and the AdaBoost algorithm yielded an accuracy of 67%.

In a study, the authors of Ref. [37] employed models with regression trees, linear discriminant analysis (LDA), classification, RF, SVM, k-nearest neighbors (k-NN), and C5.0. This work is carried out on 1150 product review Tweets. The text is converted into a term-document matrix and is utilized as input for various models. The findings indicated that the utilization of classification and regression trees, a specific variant of the DT method, yielded the most noteworthy level of accuracy at 88.99%. Table 1 presents a summary of ML algorithms.

An extensive assessment of various ML models for SA was presented in a comparable work in Ref. [38]. The models utilized in the study encompassed AdaBoost, DT, RF, SVM, k-NN, LR, and Gaussian NB. This work is carried out upon a dataset with 14,640 Indian Airlines Tweets, which were categorized. The text data were processed with preprocessing methods such as stopword removal and lemmatization. The analysis shows that 84.5% is achieved as accuracy rates by

Table 1 Summary of ML techniques

Existing works	Classifier	Dataset	Accuracy rate (%)
[32]	MNB	Sentiment-140	85
[33]	NB	Tweets	77
[34]	NB, SVM	Amazon reviews	83.42
[35]	NB, SVM	Sentiment-140	90
[36]	DT, ADABOOST, SVM	Sentiment-140, University of Michigan dataset	846782
[37]	Adaboost	Indian airline reviews	84.5
[38]	Regression tree	Tweets	88.9
[39]	SVM	Indian airline reviews	82.4
	MNB		76.5
[40]	SVM	Tweets	99.5
[41]	NB	Tweets	75.5
[42]	SVM	Indian railway reviews	91.5
[43]	SVM	Agricultural products	92

AdaBoost, which emphasizes that the boosting models outperform other approaches in accurately classifying sentiments.

In Ref. [39], the authors conducted a comparative examination of two ML models, namely multinomial NB and SVM with linear kernels, in the context of SA. The dataset comprised around 10,000 Tweets categorized as favorable, –VE, and NEU, obtained from the SM platform Twitter. The text underwent preprocessing methods including stemming, removal of URLs, and elimination of stop words. The findings indicate that the accuracy rate of SVM (82.48%) is greater than for the multinomial NB model (76.56%).

The work in Ref. [40] introduced the application of an SVM approach for SA of Indonesian Tweets, where preprocessing to exclude stop words in association with stemming and lemmatization techniques is performed. Subsequently, it was denoted using TF-IDF features. The findings indicated that the SVM approach proved a noteworthy accuracy rate of 99.5%.

The work in Ref. [41] undertook a comparative examination of three distinct ML techniques, namely SVM, NB, and k-NN, to evaluate their effectiveness in SA. The dataset comprised Twitter posts about the presidential candidates in Indonesia during the year 2019. The preprocessing procedures encompassed in this study involved text mining, parsing, and tokenization. The findings indicate that the NB algorithm had the greatest accuracy level at 75.58%, followed by the k-NN algorithm at 73.34%, and the SVM algorithm at 63.99%.

The study in Ref. [42] involved a comparison of the performance of four ML techniques, namely C4.5, NB, SVM, and RF. Twitter is used as the dataset and focused on content connected to Indian Railways. The study indicates that the SVM algorithm proves the greatest accuracy rate with 91.5%. The RF algorithm earned a little lower accuracy of 90.5%, while the NB algorithm achieved 89% accuracy rates.

Presented in Ref. [43], a product recommendation method is proposed for rice crop disease products. Sentiment analysis is conducted on the data, encompassing the computation of the rating and the examination of the polarities of the sentiment expressed in the text. Next, the K-means clustering algorithm is used to analyze products and disease-specific product photos. Subsequently, the clusters are utilized to divide the areas of product clusters into four quadrants, and the dataset equipped with labels is then readied for a recommendation. A labeled dataset is evaluated and analyzed using an SVM kernel classifier.

The system has a level of accuracy that is considered to be moderate, which indicates that the datasets that have been labeled are accurate and can be used as the recommendation system. Our method of product recommendations makes recommendations for the top five goods concerning the user's query by utilizing preexisting datasets of products that have been labeled and disease-specific products.

In summary, the primary emphasis of the authors in the examined works revolved around the aspects of preprocessing, feature extraction, and the evaluation of various ML algorithms' performance in the domain of SA. The findings indicated that diverse algorithms exhibited varying degrees of accuracy, spanning from 67% to 99.5%. Several of the most effective algorithms identified in the study were maximum entropy, NB, and SVM.

3.2 DL Approach

The utilization of DL has gained significant traction as a prominent approach for SA, mostly attributed to its capacity to acquire intricate representations of textual input. In the aforementioned approaches, textual input undergoes preprocessing and subsequently undergoes encoding with pretrained embeddings such as word2vec and GloVe. The embeddings are subsequently utilized as inputs for DL architectures, including convolutional neural networks (CNNs), RNNs, LSTM, and gated recurrent unit (GRUs), to facilitate the process of representation learning and categorization.

In the work conducted in Ref. [44], a multilayer perceptron (MLP) model was employed to perform SA Tweets based in Turkey with 3000 tweets containing both +VE and − VE sentiments. The initial phase involved several preprocessing steps, namely Turkish deacidification, tokenization, removal of stop-words and punctuation, as well as stemming. The text was transformed using the pretrained word2vec model. These embeddings were subsequently utilized as inputs for the MLP architecture that involves six and three dense and dropout layers, respectively. The findings indicated that the 81.86% is achieved by the MLP model as classification accuracy.

In Ref. [45], MLP is employed to perform SA on Tweets linked to COVID-19. The dataset comprised one lakh Tweets and underwent preprocessing to exclude HTML tags and nonalphabetic characters, as well as text tokenization and

Table 2 Summary of ML techniques

Existing works	Classifier	Dataset	Accuracy rate (%)
[44]	MLP	Tweets	81.8
[45]	MLP	COVID reviews	93.7
[46]	CNN	IMDb	99.3
[47]	LSTM	US airline tweets Sentiment-140	8266
[48]	LSTM	Tweets	86.3
[49]	BiLSTM	Tweets	94
[50]	CNN and LSTM	Stanford treebank	53.4
[51]	CNN and LSTM	Tweets	59
[52]	CNN and LSTM	Food reviews	75
[53]	CNN and BiLSTM	Sentiment-140	81.2
[54]	CNN and BiLSTM	News articles	90.66
[55]	Hybrid CNN and BiLSTM	IMDb	90.2
[56]	Convolutional BiLSTM	Tweets	77.6
[57]	EGJO-LSTM	Amazon reviews	92
[58]	CNN-LSTM	Tweets	91
[59]	DLMNN	Tweets	91.65
[60]	Hybrid DL	Tweets	93
[61]	BERT	Tweets	92

stemming. The text input was subsequently encoded using two distinct feature extraction techniques, namely the TF-IDF vectorizer and count. These feature extractors were then individually employed in the classification process, utilizing an MLP model based on ReLU activation and five hidden layers. The findings indicated that the MLP-based count vectorizer representation yielded 93.73% accuracy. Table 2 presents a summary of ML techniques.

In Ref. [46], the authors conducted an SA investigation wherein they compared various ML and DL approaches. These approaches encompass NB, SVM, LR, k-NN, CNN, and an ensemble model. The dataset utilized in this study, the IMDb dataset, was labeled as either favorable or −VE. The preprocessing step is intended to eliminate symbols, extraneous characters, stop-words, and redundant words. The count vectorizer is employed for the encoding process and the CNN model has shown 99.33% accuracy.

In Ref. [47], the authors conducted a comparative examination of ML and DL techniques for SA. Sentiment140 and Twitter US Airline Sentiment are the datasets employed for evaluating work performance. The dataset underwent different preprocessing techniques. The compared methods included multinomial NB, LR, SVM, LSTM, and an ensemble approach comprising multinomial NB, LR, and SVM with majority voting. The Sentiment140 dataset yielded an accuracy rate of 82% for the LSTM model, while the SVM scored a maximum accuracy of 68.9%.

In Ref. [48], an investigation centered on SA of 5000 Tweets written in Bangla language was undertaken by employing LSTM techniques. The data underwent preprocessing by the authors, which involved the removal of spaces and punctuation marks Following the hyperparameter optimization, the researchers observed that an LSTM model including five layers with 128 units each, 0.0001 as learning rate and 25 as batch size, attained the maximum accuracy of 86.3%.

In a separate investigation, the authors of Ref. [49] conducted an examination of sentiment in the Saudi dialect by employing LSTM and bidirectional LSTM (BiLSTM) models. The dataset utilized by the authors consisted of 60,000 Tweets written in the Saudi dialect, encompassing both +VE and −VE attitudes. Before analysis, the data underwent preprocessing procedures. A trained word2vec model was employed to encode the data through word embedding. Subsequently, both the LSTM and BiLSTM models were trained and evaluated using these subsets. The findings prove that the BiLSTM model had more accurate performance, with a 94% accuracy rate.

In Ref. [50], the authors introduced a novel approach for SA, utilizing a blend of RNN, CNN, and LSTM models. RNN is utilized as a convolutional filter within the model. The work procedures employed GloVe word embedding for encoding the Stanford Sentiment Treebank dataset. To mitigate the issue of overfitting, the researchers employed the Adam optimizer and implemented early halting techniques; 53.4% is proven as the accuracy of the model.

The hybrid approach described in Ref. [51] presented a proposed approach that involves the integration of CNNs with LSTM models to enhance SA. The dataset included in the research consisted of 32,000 Tweets and lowercase conversion process is done exclusively. Pretrained word embeddings such as word2vec and GloVe are hybridized for text representation. The findings of the study revealed that the LSTM and hybrid CNN model, using GloVe word embeddings, demonstrated a classification accuracy of 59%.

In a similar vein, another hybrid model is discussed in Ref. [52] that amalgamated the utilization of CNN and LSTM models that have been explored for SA. The dataset included in the study consisted of 100 restaurant reviews that were gathered from the Shohoz and Foodpanda applications. The preparation procedures encompassed the elimination of insignificant words and symbols, alongside the utilization of word2Vec transformation. The model that was proposed comprised the model consisting of several layers, namely an embedding layer utilizing the word2vec pretrained model. The model proved a noteworthy accuracy rate of 75.01 on the dataset obtained by the researchers themselves.

In Ref. [53], the authors introduced a model that integrated bidirectional components with CNN upon the Sentiment140 dataset with 1.6 million +VE and −VE preprocessed Tweets. The hybrid model that has been suggested architecture of the model comprises several components. The findings of the study show that this work attained 81.2% precision.

In Ref. [54], the authors conducted a study wherein they proposed a hybrid model that integrated the capabilities of CNNs and bidirectional LSTM (BiLSTM) networks to perform SA. The studies were carried out on a dataset with a total of 2003 articles and news stories, encompassing attitudes that were classified as favorable, −VE, and NEU. The text data underwent preprocessing, to obtain word embeddings with a pretrained doc2vec model. This model consists of multiple layers such as a convolutional, a maxpooling, a BiLSTM, a dropout, and classification layers and is a component that is utilized in various academic disciplines to categorize and assign labels to data or entities based on specific. The experimental results demonstrate that this model attained a noteworthy value of 90.66% as accuracy rate.

The authors proposed improvements to the attention mechanism-based CNN with the BiLSTM model in Ref. [55]. An analysis was performed utilizing the IMDb dataset, which consists of 50,000 reviews classified as either +VE or unfavorable. Before analysis, the text data was encoded using a pretrained word2vec embedding model. To further improve the hybrid model being considered, optimization techniques such as the L2 regularization, Adam optimizer, and dropout were implemented. The results demonstrate that this model has proven a 90.26% accuracy rate and the effectiveness of the attention mechanism in SA is showcased.

In Ref. [56], the authors proposed a model that integrated CNNs with bidirectional LSTM (BiLSTMs) for SA. The construction of the model involved the utilization of various essential elements. These elements encompassed a convolutional layer, which had three filters of size 128 each and employed the ReLU activation function. Furthermore, the possibility of overfitting is reduced by including a maxpooling layer of size 3 and a dropout layer with a 0.2 rate. Before processing, case folding, numerals, special characters, symbols, and superfluous comments removal are carried out. The text data was cleaned, and numerical values were removed. Application of the convolutional BiLSTM model to the dataset resulted in a remarkable accuracy of 77.6%. This outcome serves as evidence of the model's efficacy in performing SA on multilingual data.

A new ML method that has been optimized and given the name enhanced golden jackal optimizer-based long short-term memory (EGJO-LSTM) was presented in Ref. [57] to do SA on e-commerce product reviews. Using a quick cloud dataset obtained from Amazon.com, a performance study was carried out, and the proposed classifier was compared to the most advanced ML models currently available. A comparison of the performance was performed using precision, accuracy, recall, and F1 score measurements. Based on the findings, it is clear that the EGJO-LSTM performs well in contrast to the comparative models concerning sentiment classification. When it comes to precision and accuracy, the proposed strategy is superior to both the conventional and hybrid approaches by a margin of 25% and 32%, respectively.

A CNN-LSTM-based hybrid architecture is utilized in the research presented in Ref. [58] to inquire about people's ideas surrounding the monkeypox disease. On a dataset of Tweets that was freely accessible to the public and relevant to monkeypox, several tests were carried out. Several preprocessing, vectorization, and encoding techniques are applied to the Tweets. Based on the results of our trials, we discovered that the hybrid model offered a higher level of accuracy, which was roughly 91%.

When it comes to Twitter data, the authors of Ref. [59] present a proficient technique for handling SA. Included in the preprocessing of the Twitter database are stemming, tokenization, removal of numbers, stop-words, and other similar operations. Deep learning modified neural networks, also known as DLMNNs, are utilized in the process of sentiment classification procedures. The outcomes of the experiments were analyzed by comparing the output parameters, which included accuracy, F-Score, precision, recall, and execution time, with other traditional techniques, such as artificial neural networks (ANN), SVM, K-Means, and deep convolutional neural networks (DCNN), in conjunction with the suggested model.

The article discussed in Ref. [60] is designed to help app-based meal delivery businesses, undertake competitive research on social broadcasting and convert social broadcasting data into data production for decision-makers [62]. Swiggy, Zomato, and Uber Eats were evaluated in this investigation. R-Studio is employed to acquire customers' Tweets regarding each of these brands, and a sentiment examination based on deep learning is implemented on the retrieved tweets. Based on the statistics, it appears that Zomato received the largest percentage of +VE comments, 29%, followed by Swiggy, 26%, and Uber Eats, 25%. Additionally, Zomato had a lower number of −VE evaluations than Uber Eats and Swiggy, with only 11% of users reporting a −VE experience.

The context and position of words within sentences are used to generate text representations that are determined by the bidirectional encoder representation from the transformer (BERT) model [61]. This work expands upon the previous hybrid deep learning study that employed the BERT representation for SA in Indonesia. We have shown in our simulation that the BERT representation is better than all hybrid architectures.

Recently, the utilization of DL has gained significant traction as a prevalent approach for carrying out sentiment research. Several models, including RNNs, CNNs, GRUs, and LSTM have been employed for representation learning and categorization [63, 64]. GloVe and word2vec have been employed to encode preprocessed text data. Multiple researchers have employed MLP for SA across various datasets and languages. These include English, Korean, Turkish, Bangla, and COVID-19-related Tweets. The MLP model's accuracies are observed between 75% and 93%. Several other methods, including NB, SVM, LR, k-NN, and an ensemble model, were examined in various research. However, the most favorable outcomes were obtained using a CNN model with an accuracy of 99.33%. Furthermore, LSTM and Bi-LSTM models were employed to conduct SA on several datasets, encompassing both the Saudi dialect and English language. In a particular investigation, the Bi-LSTM model had the best level of accuracy, reaching 94%.

4 Discussion

There are still issues that restrict the accuracy and utility of SA research, despite the significant progress that has been made in this field. The following are some of the limitations of this method: its inability to handle texts that are inadequately organized and sardonic texts, the absence of fine-grained SA, annotated data

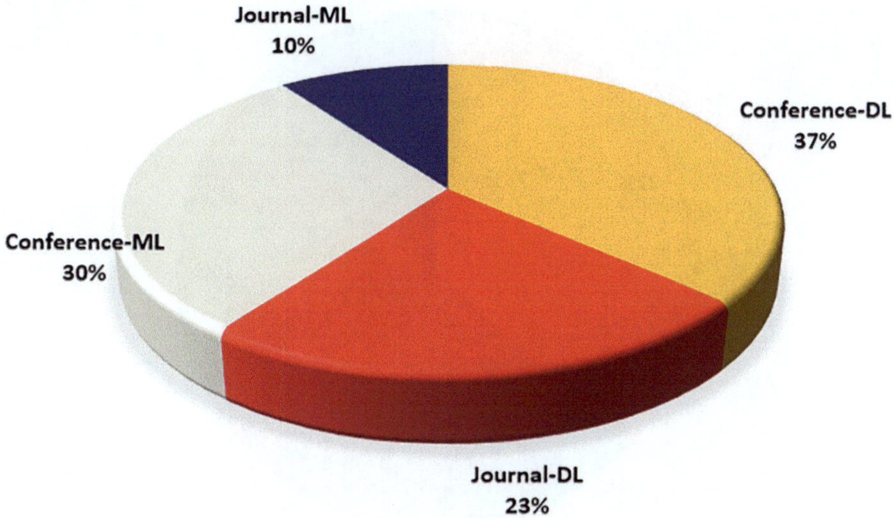

Fig. 3 Distribution of SA-based articles

dependency, word embedding limitations, and the ability to bias the train data. The following graphs analyze the areas in which SA is applied concerning the works discussed here.

Figure 3 illustrates the overall distribution of articles between journals and conferences. Figure 4 shows the domain-wise distribution of the discussed articles.

Language-related SA approaches are greater when compared to the domain-specific articles. Some popular domains, where SA is applicable are healthcare, e-commerce, user review-based, SM, and service-related areas. Future research may concentrate on enhancing the precision and breadth of SA, to counteract these challenges. This may involve the creation of fine-grained SA models that can accurately represent the emotional intensity of a text, the quantification of sentiment distribution across a variety of topics, the management of ambiguous and sarcastic texts, the broadening of the scope of cross-lingual SA, and the enhancement of the applicability of SA to SM platforms.

5 Conclusions

SA holds significant importance in the field of NLP, owing to its extensive array of practical applications. The early studies on SA utilized conventional ML methods, in which the text underwent preprocessing to eliminate stop-words, standardize the language, and denote it through frequency-based features like TF-IDF or bag of words. The preprocessed text was subsequently inputted into ML models, including NB, SVM, and others, to do categorization. Nevertheless, due to the progress

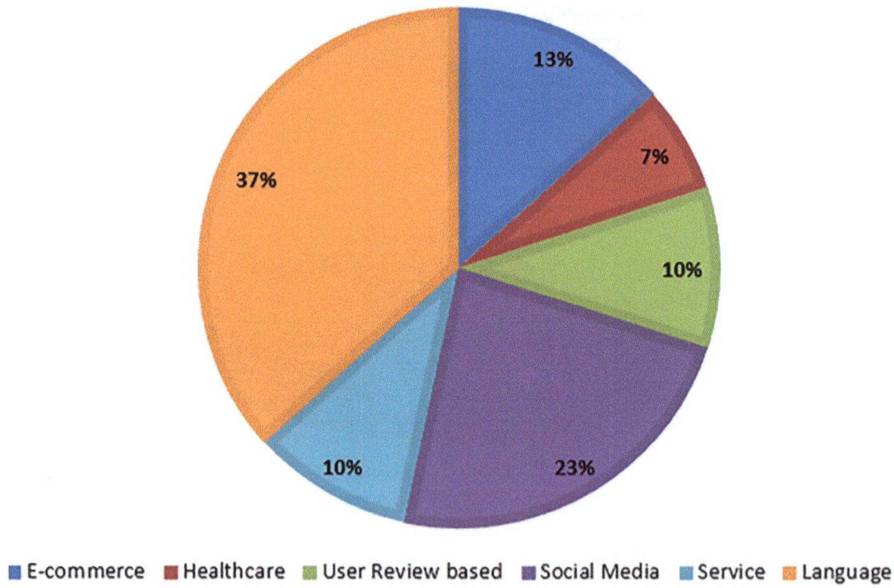

Fig. 4 Domain-wise SA article distribution

made in NLP, scholars have redirected their attention toward the utilization of DL methodologies.

In the field of DL, the initial step involves the encoding of text into pretrained word embeddings, such as GloVe, fastText, or word2vec. The embeddings are capable of capturing the underlying patterns present in the text. These patterns are subsequently utilized as input for DL models, including but not limited to CNN, LSTM, and GRU. Certain SA methodologies also employ an ensemble technique, wherein the predictions generated by numerous ML or DL models are amalgamated to enhance overall performance. Several widely used datasets for SA are the SemEval-2017 Task 4, Internet Movie Database (IMDb), Sentiment140, Twitter US Airline Sentiment datasets, and so on.

Despite the advancements achieved in the field of SA, current methodologies remain susceptible to the challenges posed by inadequately structured and sarcastic textual content. This underscores the imperative for the development of more dependable language models. Furthermore, the majority of existing SA methodologies categorize sentiments into broad classifications, namely +VE, −VE, and NEU. However, future investigations must prioritize the progression of fine-grained SA techniques that encompass a wider range of emotional intensities. These refined classifications should include strongly +VE, +VE, NEU, −VE, and strongly −VE sentiments. The quantification of sentiment, which involves calculating the distribution of polarity for a certain topic, is an additional field of study that holds the potential to inform strategic decision-making.

References

1. M. Naaman, Social multimedia: Highlighting opportunities for search and mining of multimedia data in social media applications. Multimed. Tools Appl. **56**(1), 9–34 (2012)
2. L. Liu, L. Cheng, Y. Liu, et al., Recognizing complex activities by a probabilistic interval-based model, in *Proceedings of the 30th AAAI Conference on Artificial Intelligence, AAAI 2016*, (2016), pp. 1266–1272
3. Y. Liu, L. Nie, L. Liu, et al., From action to activity: Sensor-based activity recognition. Neurocomputing **181**, 108–115 (2016)
4. K. Ravi, V. Ravi, A survey on opinion mining and sentiment analysis: Tasks, approaches, and applications. Knowl.-Based Syst. **89**(C), 14–46 (2015)
5. K. Dashtipour, S. Poria, A. Hussain, et al., Multilingual sentiment analysis: State of the art and independent comparison of techniques. Cogn. Comput. **8**(4), 757–771 (2016)
6. Z. Li, Y. Fan, Survey on visual sentiment analysis. Appl. Res. Comput. **32**(12), 3521–3526 (2015)
7. Z. Cai, D. Cao, D. Lin, et al., A spatial-temporal visual mid-level ontology for GIF sentiment analysis, in *Proceedings of IEEE Congress on Evolutionary Computation*, (2016), pp. 4860–4865
8. E. Cambria, Affective computing and sentiment analysis. IEEE Intell. Syst. **31**(2), 102–107 (2016)
9. V. Campos, A. Salvador, X. Giró-i-Nieto, et al., Diving deep into sentiment: Understanding fine-tuned CNNs for visual sentiment prediction, in *Proceedings of the 1st International Workshop on Affect & Sentiment in Multimedia*, (2015), pp. 57–62
10. S. Ramaswamy, N. DeClerck, Customer perception analysis using deep learning and NLP. Procedia Comput. Sci. **140**, 170–178 (2018)
11. A. Ajmal, H. Aldabbas, R. Amin, S. Ibrar, B. Alouffi, M. Gheisari, Stress-relieving video game and its effects: A POMS case study. Comput. Intell. Neurosci. **2022**, 4239536 (2022)
12. L. Liu, D. Dzyabura, N. Mizik, Visual listening in extracting brand image portrayed on social media. Mark. Sci. **39**, 669–686 (2020)
13. W. He, S. Zha, L. Li, Social media competitive analysis and text mining: A case study in the pizza industry. Int. J. Inf. Manag. **33**, 464–472 (2013)
14. L. Dey, S.M. Haque, A. Khurdiya, G. Shroff, Acquiring competitive intelligence from social media, in *Proceedings of the 2011 Joint Workshop on Multilingual OCR and Analytics for Noisy Unstructured Text Data, Beijing, China, 17 September 2011*, pp. 1–9
15. W. He, G. Yan, Mining blogs and forums to understand the use of social media in customer co-creation. Comput. J. **58**, 1909–1920 (2015)
16. W. He, X. Tian, Y. Chen, D. Chong, Actionable social media competitive analytics for understanding customer experiences. J. Comput. Inf. Syst. **56**, 145–155 (2016)
17. S. Stieglitz, L. Dang-Xuan, Emotions and information diffusion in social media—Sentiment of microblogs and sharing behavior. J. Manag. Inf. Syst. **29**, 217–248 (2013)
18. J. Yang, P. Xiu, L. Sun, L. Ying, B. Muthu, Social media data analytics for business decision-making system to competitive analysis. Inf. Process. Manag. **59**, 102751 (2022)
19. P. Sotiriadou, J. Brouwers, T.-A. Le, Choosing a qualitative data analysis tool: A comparison of NVivo and Leximancer. Ann. Leis. Res. **17**, 218–234 (2014)
20. A. Singh, N. Shukla, N. Mishra, Social media data analytics to improve supply chain management in food industries. Transp. Res. Part E Logist. Transp. Rev. **114**, 398–415 (2018)
21. U. Ramanathan, N. Subramanian, G. Parrott, Role of social media in retail network operations and marketing to enhance customer satisfaction. Int. J. Oper. Prod. Manag. **37**, 105–123 (2017)
22. L.K. Ramasamy, S. Kadry, Y. Nam, M.N. Meqdad, Performance analysis of sentiments in Twitter dataset using SVM models. Int. J. Electr. Comput. Eng. **11**, 2275–2284 (2021)
23. S.G. Burdisso, M. Errecalde, M. Montes-y-Gómez, A text classification framework for simple and effective early depression detection over social media streams. Exper. Syst. **133**, 182–197 (2019)

24. R. Collobert, J. Weston, L. Bottou, M. Karlen, K. Kavukcuoglu, P. Kuksa, Natural language processing (almost) from scratch. J. Mach. Learn. Res. **12**, 2493–2537 (2011)
25. M.A. Palomino, F. Aider, Evaluating the effectiveness of text pre-processing in sentiment analysis. Appl. Sci. **12**, 8765 (2022)
26. E. Haddi, X. Liu, Y. Shi, The role of text pre-processing in sentiment analysis. Procedia Comput. Sci. **17**, 26–32 (2013)
27. V. Nhlabano, P. Lutu, Impact of text pre-processing on the performance of sentiment analysis models for social media data, in *Proceedings of the 2018 International Conference on Advances in Big Data, Computing and Data Communication Systems (icABCD), Durban, South Africa, 6–7 August 2018*, pp. 1–6
28. M.J. Akhtar, Z. Ahmad, R. Amin, S.H. Almotiri, M.A. Al Ghamdi, H. Aldabbas, An efficient mechanism for product data extraction from e-commerce websites. Comput. Mater. Contin. **65**, 2639–2663 (2020)
29. W.G. Mangold, D.J. Faulds, Social media: The new hybrid element of the promotion mix. Bus. Horiz. **52**, 357–365 (2009)
30. M. Kanakaraj, R.M.R. Guddeti, NLP based sentiment analysis on Twitter data using ensemble classifiers, in *Proceedings of the 2015 3Rd International Conference on Signal Processing, Communication and Networking (ICSCN), Chennai, India, 26–28 March 2015*, pp. 1–5
31. M.J. Castillo, H. Taherdoost, The impact of AI technologies on E-business. Encyclopedia **3**, 107–121 (2023)
32. Y.G. Jung, K.T. Kim, B. Lee, H.Y. Youn, Enhanced Naive Bayes classifier for real-time sentiment analysis with SparkR, in *Proceedings of the 2016 IEEE International Conference on Information and Communication Technology Convergence (ICTC), Jeju Island, Republic of Korea, 19–21 October 2016*, pp. 141–146
33. N.S. Athindran, S. Manikandaraj, R. Kamaleshwar, Comparative analysis of customer sentiments on competing brands using hybrid model approach, in *Proceedings of the 2018 IEEE 3rd International Conference on Inventive Computation Technologies (ICICT), Coimbatore, India, 15–16 November 2018*, pp. 348–353
34. S. Vanaja, M. Belwal, Aspect-level sentiment analysis on e-commerce data, in *Proceedings of the 2018 IEEE International Conference on Inventive Research in Computing Applications (ICIRCA), Coimbatore, India, 11–12 July 2018*, pp. 1275–1279
35. N. Iqbal, A.M. Chowdhury, T. Ahsan, Enhancing the performance of sentiment analysis by using different feature combinations, in *Proceedings of the 2018 IEEE International Conference on Computer, Communication, Chemical, Material and Electronic Engineering (IC4ME2), Rajshahi, Bangladesh, 8–9 February 2018*, pp. 1–4
36. M. Rathi, A. Malik, D. Varshney, R. Sharma, S. Mendiratta, Sentiment analysis of tweets using machine learning approach, in *Proceedings of the 2018 IEEE Eleventh International Conference on Contemporary Computing (IC3), Noida, India, 2–4 August 2018*, pp. 1–3
37. A. Tariyal, S. Goyal, N. Tantububay, Sentiment analysis of tweets using various machine learning techniques, in *Proceedings of the 2018 IEEE International Conference on Advanced Computation and Telecommunication (ICACAT), Bhopal, India, 28–29 December 2018*, pp. 1–5
38. T. Hemakala, S. Santhoshkumar, Advanced classification method of twitter data using sentiment analysis for airline service. Int. J. Comput. Sci. Eng. **6**, 331–335 (2018)
39. A.M. Rahat, A. Kahir, A.K.M. Masum, Comparison of Naive Bayes and SVM Algorithm based on sentiment analysis using review dataset, in *Proceedings of the 2019 IEEE 8th International Conference System Modeling and Advancement in Research Trends (SMART), Moradabad, India, 22–23 November 2019*, pp. 266–270
40. U. Makhmudah, S. Bukhori, J.A. Putra, B.A.B. Yudha, Sentiment analysis of Indonesian homosexual tweets using support vector machine method, in *Proceedings of the 2019 IEEE International Conference on Computer Science, Information Technology, and Electrical Engineering (ICOMITEE), Jember, Indonesia, 16–17 October 2019*, pp. 183–186

41. M. Wongkar, A. Angdresey, Sentiment analysis using Naive Bayes Algorithm of the data crawler: Twitter, in *Proceedings of the 2019 IEEE Fourth International Conference on Informatics and Computing (ICIC), Semarang, Indonesia, 16–17 October 2019*, pp. 1–5
42. D.K. Madhuri, A machine learning based framework for sentiment classification: Indian railways case study. Int. J. Innov. Technol. Explor. Eng. **8**, 441–445 (2019)
43. A. Sharma, D. Vora, K. Shaw, S. Patil, Sentiment analysis-based recommendation system for agricultural products. Int. J. Inf. Technol. **16**(2), 761–778 (2024)
44. G.M. Demirci, S.R. Keskin, G. Doğan, Sentiment analysis in Turkish with deep learning, in *Proceedings of the 2019 IEEE International Conference on Big Data, Honolulu, HI, USA, 29–31 May 2019*, pp. 2215–2221
45. G.M. Raza, Z.S. Butt, S. Latif, A. Wahid, Sentiment Analysis on COVID Tweets: An Experimental Analysis on the Impact of Count Vectorizer and TF-IDF on Sentiment Predictions using Deep Learning Models, in *Proceedings of the 2021 IEEE International Conference on Digital Futures and Transformative Technologies (ICoDT2), Islamabad, Pakistan, 20–21 May 2021*, pp. 1–6
46. T. Dholpuria, Y. Rana, C. Agrawal, A sentiment analysis approach through deep learning for a movie review, in *Proceedings of the 2018 IEEE 8th International Conference on Communication Systems and Network Technologies (CSNT), Bhopal, India, 24–26 November 2018*, pp. 173–181
47. P. Harjule, A. Gurjar, H. Seth, P. Thakur, Text classification on Twitter data, in *Proceedings of the 2020 IEEE 3rd International Conference on Emerging Technologies in Computer Engineering: Machine Learning and Internet of Things (ICETCE), Jaipur, India, 7–8 February 2020*, pp. 160–164
48. A.H. Uddin, D. Bapery, A.S.M. Arif, Depression analysis from social media data in Bangla language using long short term memory (LSTM) recurrent neural network technique, in *Proceedings of the 2019 IEEE International Conference on Computer, Communication, Chemical, Materials and Electronic Engineering (IC4ME2), Rajshahi, Bangladesh, 11–12 July 2019*, pp. 1–4
49. R.M. Alahmary, H.Z. Al-Dossari, A.Z. Emam, Sentiment analysis of Saudi dialect using deep learning techniques, in *Proceedings of the 2019 IEEE International Conference on Electronics, Information, and Communication (ICEIC), Auckland, New Zealand, 22–25 January 2019*, pp. 1–6
50. Y. Yang, Convolutional neural networks with recurrent neural filters. arXiv (2018) arXiv:1808.09315
51. D. Goularas, S. Kamis, Evaluation of deep learning techniques in sentiment analysis from Twitter data, in *Proceedings of the 2019 IEEE International Conference on Deep Learning and Machine Learning in Emerging Applications (Deep-ML), Istanbul, Turkey, 26–28 August 2019*, pp. 12–17
52. N. Hossain, M.R. Bhuiyan, Z.N. Tumpa, S.A. Hossain, Sentiment analysis of restaurant reviews using combined CNN-LSTM, in *Proceedings of the 2020 IEEE 11th International Conference on Computing, Communication and Networking Technologies (ICCCNT), Kharagpur, India, 1–3 July 2020*, pp. 1–5
53. V. Tyagi, A. Kumar, S. Das, Sentiment analysis on Twitter data using deep learning approach, in *Proceedings of the 2020 IEEE 2nd International Conference on Advances in Computing, Communication Control and Networking (ICACCCN), Greater Noida, India, 18–19 December 2020*, pp. 187–190
54. M. Rhanoui, M. Mikram, S. Yousfi, S. Barzali, A CNN-BiLSTM model for document-level sentiment analysis. Mach. Learn. Knowl. Extr. **1**, 832–847 (2019)
55. B. Jang, M. Kim, G. Harerimana, S.U. Kang, J.W. Kim, Bi-LSTM model to increase accuracy in text classification: Combining Word2vec CNN and attention mechanism. Appl. Sci. **10**, 5841 (2020)

56. R. Chundi, V.R. Hulipalled, J. Simha, SAEKCS: Sentiment analysis for English–Kannada code switchtext using deep learning techniques, in *Proceedings of the 2020 IEEE International Conference on Smart Technologies in Computing, Electrical and Electronics (ICSTCEE), Bengaluru, India, 10–11 July 2020*, pp. 327–331
57. P. Rasappan, M. Premkumar, G. Sinha, K. Chandrasekaran, Transforming sentiment analysis for e-commerce product reviews: Hybrid deep learning model with an innovative term weighting and feature selection. Inform. Process. Manag. **61**(3), 103654 (2024)
58. K.K. Mohbey, G. Meena, S. Kumar, K. Lokesh, A CNN-LSTM-based hybrid deep learning approach for sentiment analysis on Monkeypox tweets. New Gener. Comput. **42**(1), 89–107 (2024)
59. D. Paulraj, P. Ezhumalai, M. Prakash, A Deep Learning Modified Neural Network (DLMNN) based proficient sentiment analysis technique on Twitter data. J. Exp. Theor. Artif. Intell. **36**(3), 415–434 (2024)
60. R. Vatambeti, S.V. Mantena, K.V.D. Kiran, M. Manohar, C. Manjunath, Twitter sentiment analysis on online food services based on elephant herd optimization with hybrid deep learning technique. Cluster Comput. **27**(1), 655–671 (2024)
61. H. Murfi, T. Gowandi, G. Ardaneswari, S. Nurrohmah, BERT-based combination of convolutional and recurrent neural network for Indonesian sentiment analysis. Appl. Soft Comput. **151**, 111112 (2024)
62. S.V. Vadivu, P. Nagaraj, B.A.S. Murugan, Ensemble machine learning technique-based plagiarism detection over opinions in social media. Automatika **65**(3), 983–991 (2024)
63. S.V. Vadivu, P. Nagaraj, B.S. Murugan, Opinion mining on social media text using optimized deep belief networks. ACM Trans. Asian Low-Resour. Lang. Inform. Process. (2024)
64. P. Nagaraj, V. Muneeswaran, M.C. Prabhu, C.B. Kumar, B.M. Reddy, K.H. Krishna, Instagram-driven social media trend forecasting with learning algorithm using live dataset, in *2023 International Conference on Data Science, Agents & Artificial Intelligence (ICDSAAI)*, (IEEE, 2023), pp. 1–5
65. P. Nagaraj, H.N. Reddy, S.L. Srinivas, M. Jaipal, S.V. Nagendra, Machine learning-based mobile application for store-to-door using sentimental analysis, in *2023 International Conference on Sustainable Computing and Data Communication Systems (ICSCDS)*, (IEEE, 2023), pp. 138–142
66. P. Nagaraj, P.A. Kumar, V.C. Babu, P. Yedukondalu, M.Y. Sai, K.M. Sudar, Movie reviews using sentiment analysis, in *2023 International Conference on Computer Communication and Informatics (ICCCI)*, (IEEE, 2023), pp. 1–6

Leveraging the Power of MediaPipe for Effective Sign Language Detection and Real-Time Audio Conversion

Abhay Krishna, Nandana Manoj, Manni Chellappan, and P. R. Joe Dhanith

1 Introduction

Sign language is an important way for the deaf to communicate their ideas and feelings. However, it is hard for those who know sign language to communicate with those who do not. This shows that better solutions are needed to improve communications between the two groups.

Breakthroughs in machine learning and computer vision are getting people excited about creating systems that can turn sign language into spoken language. Traditionally, most methods view gestures of sign language as individual events, which does not help in the accuracy or understanding of the underlying context behind the communication. In that respect, we have come up with a new approach where the problem stated is solved by motion detection techniques in combination with deep learning through long short-term memory (LSTM) networks.

This model of LSTM is optimal in this task as it could deal with sequences of data. It helps with the complex and sequential nature of sign language gestures. By incorporating motion recognition into the framework of LSTMs, our approach greatly enhances accuracy and contextual understanding in sign language interpretation.

Traditional sign language detection is done by using object detection and image processing techniques to process an image and find out the gesture using features of an image. This is an ineffective method as most words in sign language are not a still gesture but a gesture using the motion of hands and faces in some cases.

A. Krishna (✉) · N. Manoj · M. Chellappan · P. R. J. Dhanith
School of Computer Science and Engineering, Vellore Institute of Technology Chennai Campus, Chennai, India
e-mail: joedhanith.pr@vit.ac.in

A. Haldorai et al. (eds.), *7th EAI International Conference on Big Data Innovation for Sustainable Cognitive Computing*, EAI/Springer Innovations in Communication and Computing, https://doi.org/10.1007/978-3-032-02979-9_20

This chapter solves this issue by adding motion detection to our system and hence improves its accuracy. Many translations in sign language fail because of the fluidity of the language. For example, in most cases, the meaning is closely related to the speed and flow of movement. In this approach, we have captured these dynamic features by motion recognition, resulting in better translation and hence a reliable system that easily grasps the context necessary for conveying the implied meaning behind any gesture [1].

This method assists in breaking the communication barrier by interpreting sign language into text and audio formats so that any person not knowing sign language will easily get what the signer is trying to communicate. Sign language translation thus embodies image processing, pattern recognition, and computer vision.

People with hearing impairments often face challenges with speech due to limited exposure to sound during critical stages of language development. This can create a communication barrier between them and the general public, making daily interactions more difficult. To address this, translating sign language gestures into text provides a crucial bridge, enabling effective communication. However, to make the system even more inclusive, the conversion of text into audio has been integrated. This ensures that individuals with both hearing and visual impairments, as well as those who are blind, can also benefit from the system. Combining sign-to-text and text-to-audio technologies makes for a more accessible and inclusive environment, enabling seamless communication across diverse communities in various settings.

Hand movements, facial expressions, and body postures are the words in sign language. The strong capability of LSTM in handling time-series data makes it very fitting for the task of sign language translation since it can model memory effects due to its special network architecture. An LSTM-based system is therefore capable of remembering longer speech contexts with subtle differences between gestures.

The result of this chapter is way more than the translation of sign language into spoken words, and it builds a bridge between the deaf and the hearing and forms a type of communication. Our approach in providing real-time translation into text and audio enables natural communication between the deaf and the hearing.

2 Review of Literature

There are different methods for recognizing sign language, each with its pros and cons. Multimodal data, which is essentially skeletal and RGB video data, has emerged as a significant enabler for the recognition of American Sign Language (ASL). Skeletal data captures the motion trajectory of different body parts, essential for gesture identification. Such datasets are typically scarce and are obtained in controlled environments and are therefore relatively hard to generalize toward more real-world-based scenarios. Traditional recurrent neural networks would fail to learn long-term dependencies. Long short-term memory (LSTM) networks are able to pass information over long periods by tracking temporal sequences. The movement recognition ability of LSTM network is elevated in a combination with skeletal data. CNNs complement this ability by learning spatiotemporal features effectively. However, the single, isolated systems of RNNs or CNNs usually cannot find the optimal performance state [2].

3D-CNNs extend these capabilities by learning spatiotemporal characteristics from RGB frames, handling problems like temporal misalignments and irrelevant areas in video sequences. KSU-SSL and ArSL datasets demonstrate their effectiveness in addressing hand orientation and position variations. Despite high recognition rates in signer-dependent modes, challenges remain in signer-independent modes. These limitations may be alleviated by integrating LSTMs with 3D-CNNs, improving temporal understanding of gestures [3].

With the translation of sign language to English, NLP is integrated with computer vision algorithms. Techniques such as HSV conversion are applied to feature detection, pseudo-2-dimensional hidden Markov models (P2DHMM) for feature extraction, and part-of-speech (POS) tagging for meaning in the target sentence. These approaches increase the accuracy of transcribable video quality but add complexity in computational requirements. There is further necessity to explore various sign languages [4].

In the case of ISL, segmentation, tracking, and feature extraction techniques involving skin-color-based segmentation, Kalman filter tracking, and scale-invariant feature transform (SIFT) are used. While it is robust in controlled environments, with fluctuations of light intensity or similar-colored garments, performance drops. A proposed enhancement is in segmentation accuracy, sophisticated filtering, and inclusion of facial expression for robustness of recognition and fewer false negatives [5].

Built with palm detection and multi-hand landmark extraction, MediaPipe gives a real-time document to vernacular sign language recognition using images preprocessed and localizing to 21 3D coordinates to enable smooth gesture recognition. Its application of algorithms after acquiring actual data gives impressive accuracy rates as it detects intricate gestures in datasets [6].

Convolutional neural networks (CNNs), VGG-16, and VGG-19 are popular options in sign language detection. VGG-16 is accurate but overfits, whereas VGG-19 underfits and is not very feasible for real-time usage. The computational cost of these models requires architecture optimizations for use in lower-end hardware with a balance of real-time performance and precision [7].

Finally, the American Sign Language (ASL) recognition systems check on 3D hand gesture coordinates for 24 alphabets and exclude movement-based letters like "J" and "Z." Models using algorithms like k-nearest neighbors, random forest, and neural networks have achieved very high accuracy. However, integration with dynamic gestures and word/sentence expansion might enhance its robustness and two-way communication between sign and spoken languages [8].

The COVID-19 pandemic brought out the challenges relating to facial recognition because of the almost universal use of masks. Conventional systems failed while dealing with face differences, and this aspect called for innovative techniques. MediaPipe Facemesh, combined with deep learning, has been used for improved accuracy in masked persons' recognition. Advanced neural networks and hybrid meta-heuristic algorithms have been implemented in the system, so it achieves higher precision even when the face is obstructed. Critical factors include illumination conditions, image quality, and mask variability. Other such efforts include advanced device precision, 3D sensing, and attempts to develop strong recognition models suitable for real-world applications [9].

Deep learning has also helped in the dependable tracking of human facial and body landmarks for linguistic analysis purposes. This chapter aims to compare two models: MediaPipe Holistic (MPH) and OpenFace (OF). Analyzing eyebrow movement in Kazakh-Russian sign language (KRSL) is the goal. MPH shows promise for tracking nonmanual markers, such as eyebrow raises, with fewer distortions than OF in polar and content questions. However, MPH's 3D modeling introduces challenges during head movements, particularly in upward and downward head pitches. Future improvements may involve integrating correction models to address these distortions and enhance robustness for broader applications [10].

An important application of face detection technology is deepfake detection, especially in terms of security and content authenticity in the digital world. This review analyzes MTCNN, Dlib, and MediaPipe compared with FFHQ and Celeb-A datasets. The test for MediaPipe proved to be better when it comes to accuracy and performance—99.3% better than its competitors. Nonetheless, noisy dataset scenarios and multi-face detection present challenges, notably in the OpenForensics dataset. Addressing these issues will require the need for more sophisticated correction algorithms to improve robustness and performance in challenging environments [11].

3 Proposed Architecture

For this chapter we will be using Google's MediaPipe for detecting the pose and faces for sign language detection. MediaPipe offers pretrained models for real-time detection of human features such as hand landmarks, facial expressions, and pose estimation. Its lightweight, cross-platform framework enables efficient processing on edge devices, ensuring minimal latency and computational overhead. Additionally, MediaPipe's robustness to varying lighting and background conditions enhances its applicability in diverse environments.

Given below are the functionalities taken from MediaPipe's library:

3.1 Holistic Landmarks by MediaPipe

To analyze full-body gestures and poses, we used Google MediaPipe library, best known for having good models in computer vision and machine learning to create a complete system. We then imported the holistic and drawing modules of MediaPipe, essential to combine face-hand-pose landmarkers with each module complementing one another in a unified model. The `mp.holistic` framework was able to capture and analyze a total of *543 landmarks* in real time which includes:

1. Pose—33 landmarks
2. Face—468 landmarks
3. Hand—up to per hand (21 × 2) landmarks per hand

Combining face, hand, and body poses data into a single holistic function allows us to process complicated gestures/activities all-conjointly, more accurately, and with better performance:

1. **Face Landmark Detection**:
   - Face landmarks and expressions in photos and videos can be identified using the MediaPipe Face Landmarker task. This task allows you to create virtual avatars, apply facial filters and effects, and recognize human facial expressions. Machine learning (ML) models are used for this task; these models can operate on a single image or on an ongoing stream of images. In order to infer detailed facial surfaces in real time, the task generates three-dimensional face landmarks, blendshape scores (coefficients representing facial expression), and transformation matrices to carry out the necessary transformations for effects rendering.
   - To precisely localize facial landmarks on images and videos, we employed the MediaPipe Face Landmark model, enabling it to capture detailed three-dimensional points for all faces in real time.
   - We had 30 videos for each word and used MediaPipe to precisely recognize all facial points in the data, which were then saved individually.
   - The landmarks it detected as shown in Fig. 1 were stored in an organized fashion, making them a good dataset for further tasks like expression recognition and motion detection [12].
2. **Pose Landmark Detection**:
   - One can identify landmarks of human bodies in an image or video by using the MediaPipe Pose Landmarker task. This exercise can be used to classify movements, assess posture, and pinpoint important body parts. Machine learning (ML) models that operate on individual images or videos are used in this task. Body pose landmarks in image coordinates and 3-dimensional world coordinates are output by the task.
   - It recognizes body landmarks as shown in Fig. 2 from images or videos accurately, including key points for human body locations, posture prediction, and category classification.
   - This task was used to extract body pose landmarks with original 2D image coordinates and corresponding world-centered 3D coordinates for accurate examination.

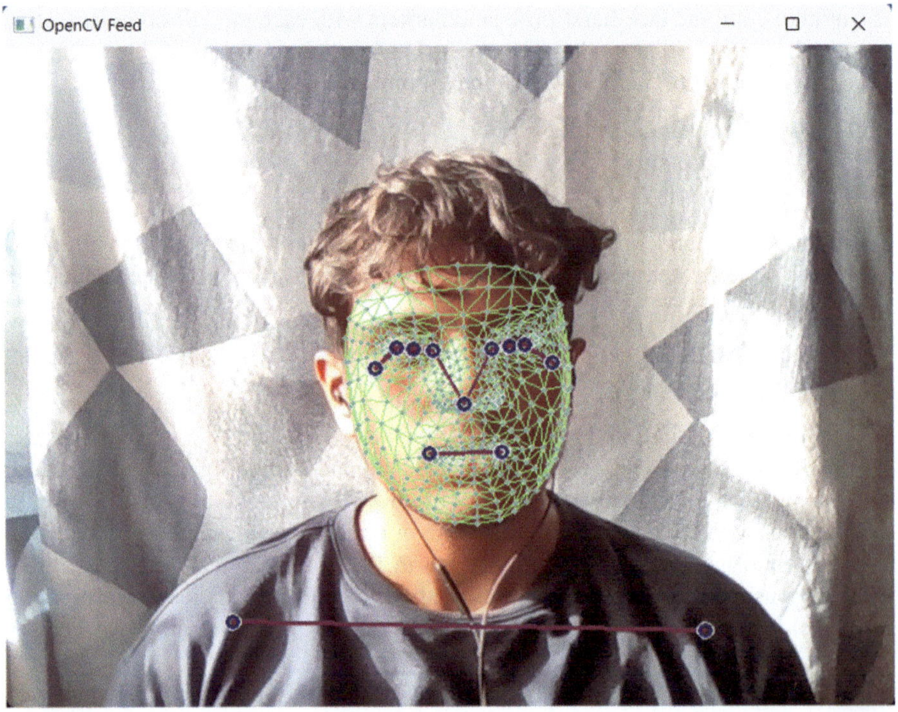

Fig. 1 Face landmarks

- In our paper, we successfully detected and stored all critical points such as the elbow, shoulder, chest, and fingers, ensuring a comprehensive dataset for further tasks like postural assessment and movement tracking [13].

3. **Hand Landmark Detection**:

- One can identify the hand's landmarks in an image by using the MediaPipe Hand Landmarker task. This task can be used to locate hands' important points and apply visual effects to them. The task uses a machine learning (ML) model to process image data, either as static data or as a continuous stream. It outputs hand landmarks in world coordinates, hand landmarks in image coordinates, and the handedness (left/right hand) of several hands that are detected.
- MediaPipe Hand Landmark detection deals with recognizing hand landmarks in images—the user identification of key points on a hand.
- Basic task is to detect both right and left hands by identifying the points with high precision and accuracy.
- For our paper, we extracted and saved every significant hand point as shown in Fig. 3 over time—individual landmarks for each finger were also tracked in detail [14].

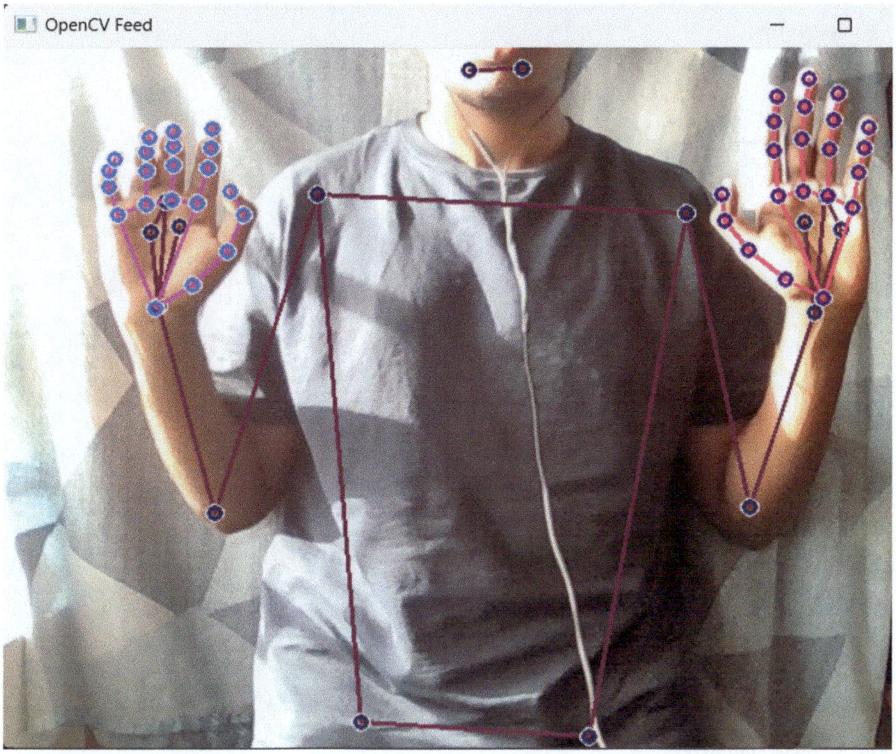

Fig. 2 Pose landmarks

3.2 Architecture of MobileNetV2 Model

A convolutional neural network architecture, MobileNetV2, aims to function well on mobile devices. Its foundation is an inverted residual structure, in which the bottleneck layers are connected by residuals. Lightweight depth-wise convolutions are used by the intermediate expansion layer to filter features as a source of nonlinearity. The first fully convolution layer with 32 filters makes up the entirety of MobileNetV2's architecture. It is followed by 19 residual bottleneck layers as shown in Fig. 4 [15].

MobileNetV2 greatly reduces the computational cost of such networks while maintaining high accuracy through weight factorization among others to build lighter but powerful deep neural network stack of layers using depth-wise separable convolutions. This design is ideal for resource-limited environments, such as mobile devices. The architecture tries to keep the computational complexity as low possible by applying depth-wise and point-wise convolution instead of standard convolutions [16].

Fig. 3 Hand landmark

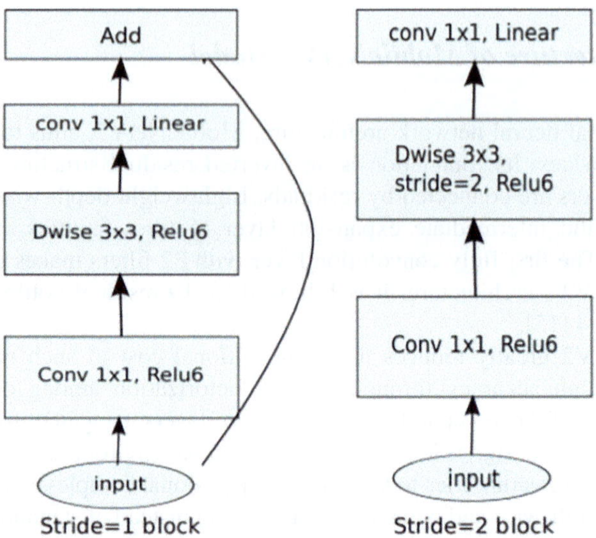

Fig. 4 MobileNetV2 architecture

Total Multiplications (Standard Convolution)

$$\text{Total multiplications} = M \times D_k^2 \times D_p^2 \times D_p \times N, \tag{1}$$

where:

- M is the number of input channels.
- D_k is the kernel size.
- D_p is the output size.
- N is the number of filters.

Multiplications per Channel

$$\text{Mults per channel} = D_k^2. \tag{2}$$

Total Depth-Wise Multiplications

$$\text{DC Mults} = M \times D_k^2. \tag{3}$$

Total Point-Wise Multiplications

$$\text{PC Mults} = N \times D_p^2 \times M, \tag{4}$$

where:

- N is the number of filters.
- D_p is the output size.
- M is the number of input channels [17].

3.3 *Architecture of BlazePose GHUM Model*

This system is a statistical 3D human shape modeling and articulated process that meshes with the next iteration of fully trainable, modular deep learning frameworks. Along with detailed 3D full-body scans of individuals in a variety of poses (sitting, crouching, or lying), it includes close-up shots capturing their heads and facial expressions as well as hand gestures, leveraging artist-designed, gender-neutral rigged quad-meshes as a dataset and starting from the same parametrization of model blocks that incorporates nonlinear shape spaces via variational autoencoders (VAEs), pose-space deformation correctives, skeleton joint center predictors, and blend skinning functions within one unified learning pipeline. The models are trained on an extensive dataset of over 60,000 human poses that represent the complete range of configurations expected in daily life (87), ensuring component consistency and capturing correlations. These models support facial expression analysis, as well as detailed hand and body shape and pose estimation. The framework offers fully trainable human models in different resolutions—GHUM

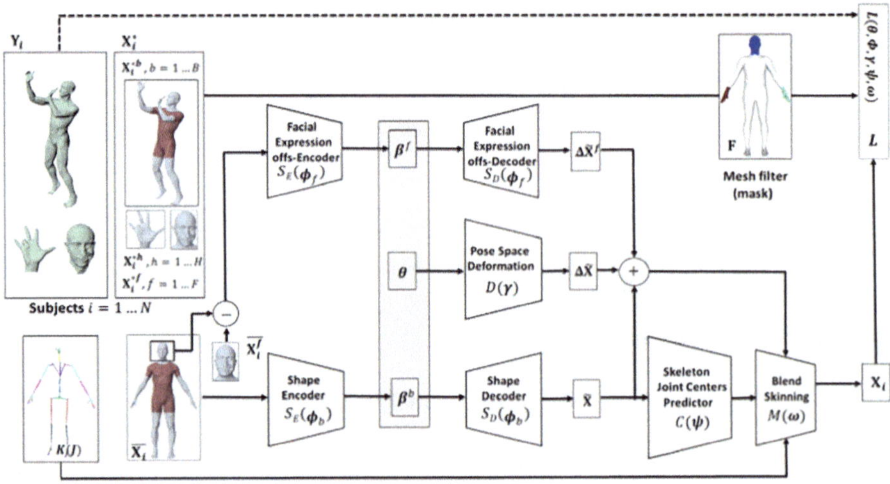

Fig. 5 Architecture of GHUM model

with 10,168 vertices and GHUMLite with 3,194 vertices—and compares their performance, analyzing various components and demonstrating reconstruction from image data. The architecture is shown in Fig. 5 [18].

The primary difficulty working with the 3D aspect of our pose estimator has been to find accurate, in-the-wild data. Despite 2D data can typically be labeled by human annotations, it is difficult for labeling the 3D data. This usually involves a controlled lab environment or specialized hardware fitted with depth sensors, making the process more complicated from maintaining diversity in human subjects and environmental settings within dataset. One alternative is to create a completely synthetic dataset; however, this also has the issue of domain adaptation when using real-world images [19].

We use a statistical 3D human body model, GHUM, which is learnt from millions of images data and motion traces in our approach. We used a fitting process to create 3D human body pose ground truth for our dataset based on aligning the GHUM model with our existing 2D pose annotations. This alignment leads to the dataset being significantly expanded with 3D key point coordinates in a real-world metric space. To this end, the fitting process minimized them with respect to GHUM shape and pose variables in order for reconstructed model to best fit image data. The process covered aligning 2D key points, silhouette semantic segmentation, and maintaining the shape as well as pose regularization terms.

3.4 LSTM

The use of long short-term memory (LSTM) networks has a major advantage of effectively recognizing continuous motion. LSTMs are different from traditional

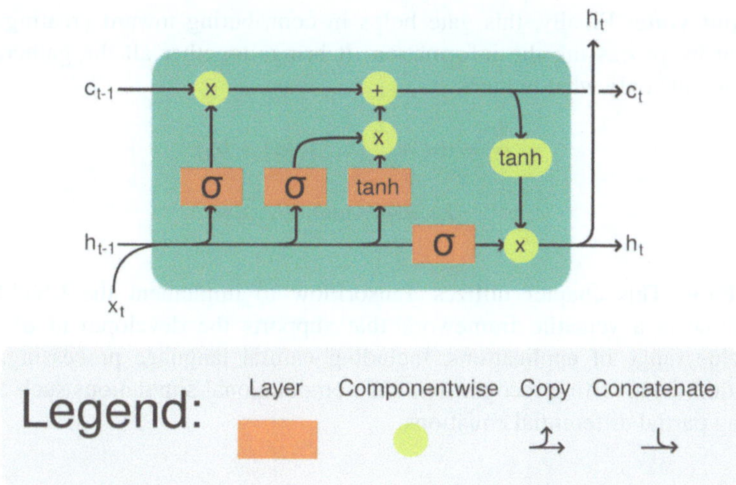

Fig. 6 Architecture of LSTM model

models that mainly concentrate on individual data points. They can manipulate and learn from whole sequences of data thus proving useful in tasks requiring sequential information. Classification derived by the LSTM model makes use of training set in order to take into account temporal dynamics inherent in the data.

To address certain limitations in traditional neural networks, scientists developed long short-term memory (LSTM) networks. LSTM is designed to retain long-term information and overcome the vanishing gradient problem, which affects many standard models. The architecture is shown in Fig. 6 [20].

LSTM networks capture the temporal dependencies between sequential hand and pose landmarks, improving the model's ability to distinguish between gestures that rely on dynamic movements. This enhances recognition accuracy and ensures efficient processing for continuous gestures in real-time applications:

1. **Forget Gate:** Either keeps or discards information. It uses a sigmoid activation function that outputs values between zero and one. Values close to 0 signify that the information is forgotten, while those around 1 suggest that it remains.

$$f_t = \sigma\left(W_f \cdot [h_{t-1}, x_t] + b_f\right). \tag{5}$$

2. **Input Gate:** The new data's relevance is controlled by the input gate, which has two activation functions: sigmoid and tanh.

$$i_t = \sigma\left(W_i \cdot [h_{t-1}, x_t] + b_i\right) \tag{6}$$

$$\tilde{C}_t = \tanh\left(W_C \cdot [h_{t-1}, x_t] + b_C\right). \tag{7}$$

3. **Output Gate:** Finally, this gate helps in contributing toward creating a final output by processing the information. It brings together all the gathered data and outputs only what matters.

$$o_t = \sigma\left(W_o \cdot [h_{t-1}, x_t] + b_o\right) \tag{8}$$

$$h_t = o_t \cdot \tanh(C_t). \tag{9}$$

TensorFlow This chapter utilizes TensorFlow to implement the LSTM layer. TensorFlow is a versatile framework that supports the development of models for a wide range of applications, including natural language processing, image recognition, handwriting recognition, and computational simulations such as those involving partial differential equations.

3.5 SkLearn

Sklearn (short for Scikit-learn) is the best library python has to offer when it comes down on machine learning and analysis of data. This includes middle-level methods for building and testing models such as classification, regression, clustering, or dimensionality reduction. Scikit-learn provides an easy-to-use API and has extensive documentation, which eases employing various algorithms as well as model performance evaluation. It also provides some of the essential metrics like model accuracy, precision, recall, etc. out-of-box so that you can get a good glimpse into your trained model performance before applying any form of training data augmentation or regularization steps and figure-out what obviously wrong with base models architecture choices then only go to add-on regularization methods to pull-off hi-fi results from troublesome architectures [21].

To calculate the performance metrics of our model, we have used various scikit-learn libraries and functions. These tools help us evaluate and draw conclusions on how well our machine learning models are performing using various evaluation metrics.

3.6 Text-to-Speech Conversion Using pyttsx3

The proposed system incorporates a text-to-speech (TTS) module for real-time conversion of recognized sign language gestures into audible speech. This functionality is achieved using Python's pyttsx3 library, an offline TTS engine that provides a flexible and efficient mechanism for speech synthesis. Unlike cloud-based TTS services, pyttsx3 operates locally, ensuring privacy, speed, and independence from Internet connectivity, which is critical for accessibility in diverse environments.

It supports multiple TTS engines, including those provided by SAPI5 (Microsoft Windows) and NSSpeechSynthesizer (macOS), and offers robust control over speech properties such as voice selection, rate, and volume. In the proposed system, the library is integrated to convert the recognized text, corresponding to sign language gestures, into spoken audio. This feature extends the system's accessibility to the visually impaired community, bridging communication gaps and enhancing inclusivity.

The integration of pyttsx3 ensures that the system remains lightweight and platform-independent, making it deployable on edge devices. This design choice aligns with the overarching goal of creating a portable, user-friendly tool that caters to individuals with hearing, speech, and visual impairments. Through this approach, the system not only translates gestures into text but also provides a seamless auditory representation, fostering effective communication in real-world scenarios.

3.7 OpenCV for Real-Time Visualization

The OpenCV (Open-Source Computer Vision Library) is vital toward the continuous implementation of computer vision applications in real time. This is because it possesses powerful interest functions that capture videos, process images and visualize them, thus making it more unique with respect to some other applications that allow human beings to interactively or actively deal with them. The key things we did with OpenCV during our paper include visualization in real time and dataset creation [22].

Read the Camera OpenCV captures video frames from the camera in real time. This continuous stream of live images is essential for processing and analyzing dynamic visual data as it happens.

Dataset Creation Script This script automates the process of dataset creation. It captures 30 videos per class—and "hello," "sorry," and "thank you"—and processes these videos to detect hands and extract relevant features. The data is then stored as numpy files, each containing sequences of hand points and actions. This process ensures that the dataset is well-structured and ready for training and testing machine learning models.

3.8 Architecture of Our Paper

For our research we have come with an architecture as shown in Fig. 7.

Input and Preprocessing

- **Image Acquiring:** This is the initial step where images are captured, using the webcam.

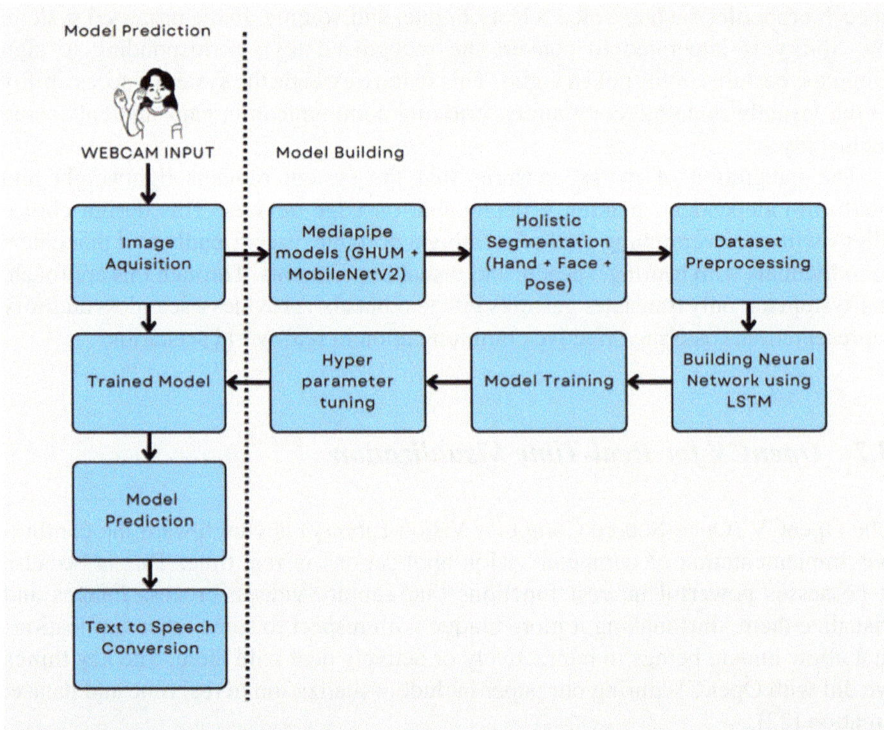

Fig. 7 Block diagram for this paper

- **Holistic Detection:** This stage involves identifying and marking the parts of the hand, face, and body using MediaPipe functionality.
- **Image Segmentation:** Once the hand is detected, the image is segmented to isolate the hand region from the background, and coordinates of the region are noted down.

Model Training and Feature Extraction

- **Model Training:** A machine learning model is trained using LSTM network using the dataset of hand coordinates. This model learns to recognize patterns and features associated with different hand gestures.

Recognition and Output

- **Recognition:** The trained model is used to predict the gesture or sign represented by a new hand image.

- **Text-to-Speech Conversion:** The recognized gesture or sign is converted into text, which is then synthesized into speech using a text-to-speech engine.

Database

This component is used to store the training data, the trained model, and possible intermediate results.

4 Experimental Configuration

4.1 Dataset

The experimental configuration includes a dataset containing image coordinates for various sign language words, represented as NumPy arrays and used as the training dataset. The careful selection of words in the dataset was crucial to ensure a diverse range of signs for effective training and testing of the system. One of the main features of this study is the ability of the system to identify and recognize the words included in our customized dataset (Fig. 8).

Fig. 8 Four basic sign languages

4.2 Data Collection and Representation

The dataset was created considering natural conditions without extra brightness, orientation, background adjustments, gloves, etc. In this chapter, we utilized a dataset comprising three distinct actions: "thank you," "sorry," and "hello." Each action category was represented by 30 video samples, with each video consisting of 30 frames. The frames were stored in a NumPy array format, where each frame captured the spatial coordinates of the actions being performed. This approach allowed for precise numerical representation and analysis of the movement patterns associated with each action across the dataset. The examples of dataset collection have been shown in Figs. 9, 10, and 11.

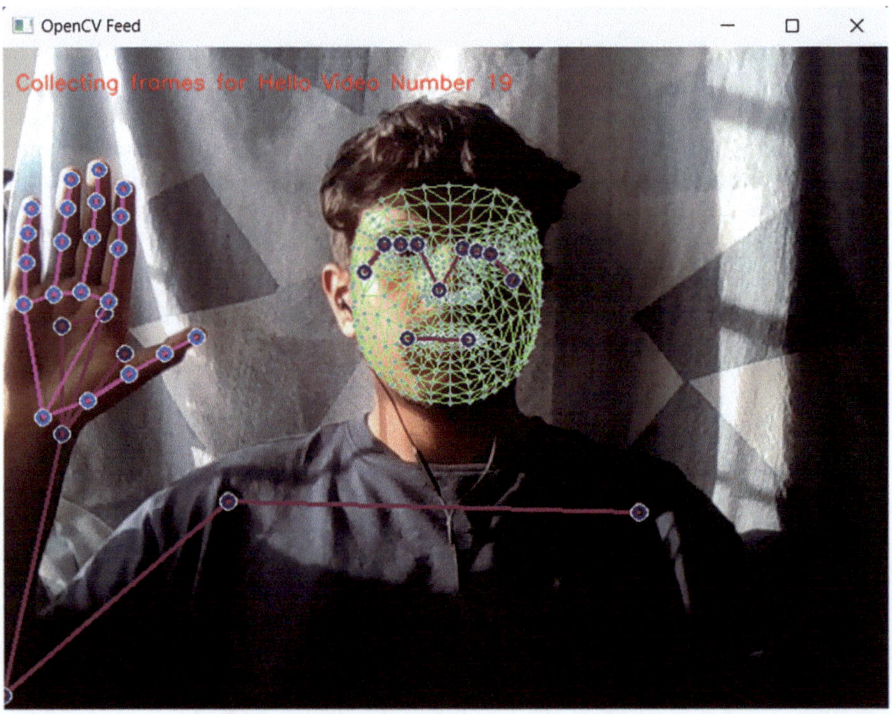

Fig. 9 Dataset collection for action: hello

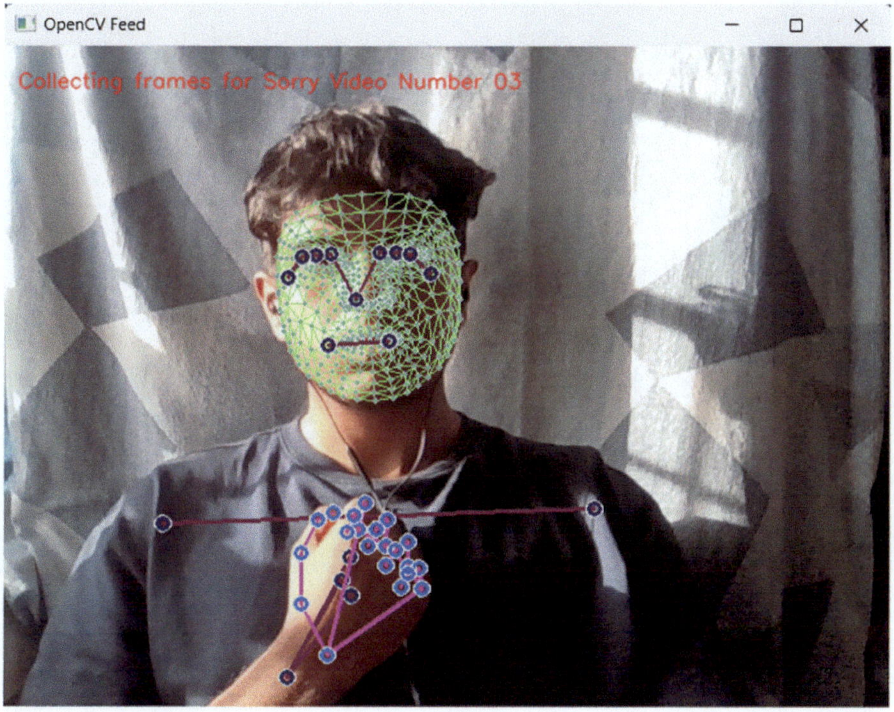

Fig. 10 Dataset collection for action: sorry

4.3 Hardware and Software

We used a computer with an 16GB RAM and 6GB graphics card to train our model. Visual Studio Code (VSCode) was used as the integrated development environment (IDE) for coding and testing the system (Fig. 12).

4.4 Model Visualization and Monitoring

We made use of TensorBoard as a tool for visualizing and monitoring various metrics such as loss and accuracy. This proved to be an invaluable resource in tracking the performance of our model through the generation of diverse plots and histograms. The insights gained from these visualizations have been instrumental in refining and enhancing our sign language detection system. Furthermore, TensorBoard facilitated the visualization of the model graph, allowing for a comprehensive understanding of its architecture. It also provided the capability to display training data, including images and audio, which was essential for gaining a deeper understanding of the dataset and model performance (Fig. 13).

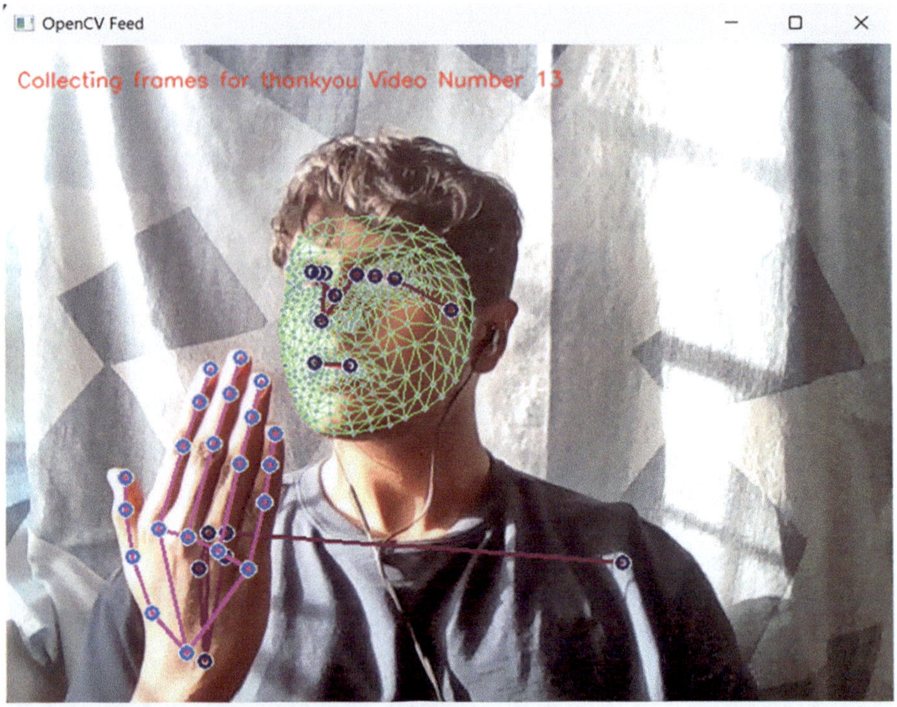

Fig. 11 Dataset collection for action: thank you

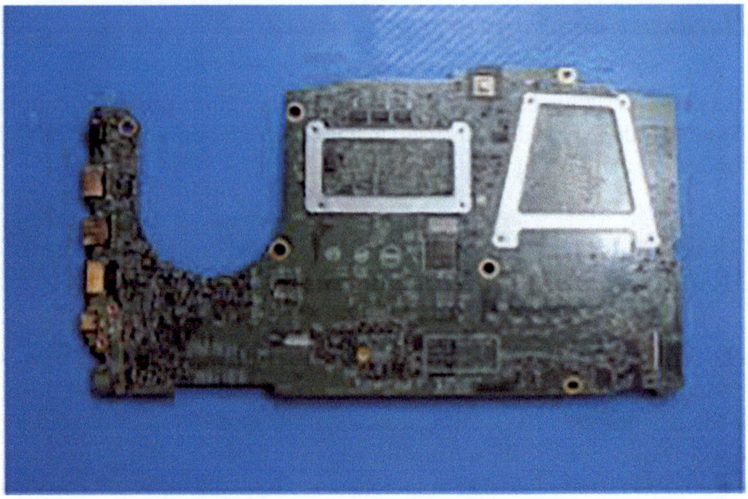

Fig. 12 Hardware setup: GPU used in the system for model training

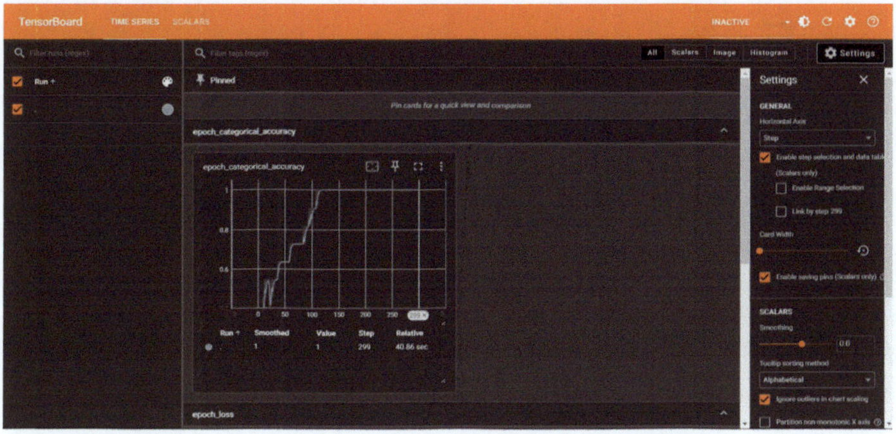

Fig. 13 Visualization of performance using Tensorboard

5 Results and Analysis

5.1 *Model Performance*

The results obtained at epoch 300 provide a comprehensive insight into the model's performance and robustness. The key metrics are summarized in Table 1.

5.2 *Accuracy Analysis*

The training accuracy achieved is 1.0000, indicating that the model has perfectly learned the patterns in the training data, with every instance in the training set classified correctly. Similarly, the validation accuracy of 1.0000 is a remarkable outcome, suggesting that the model is equally effective when applied to unseen data in the validation set. These results indicate that the model is not only highly accurate but also capable of generalizing well beyond the training data (Figs. 14, 15, 16).

Table 1 Key metrics at epoch 300

Metric	Value
Training accuracy	1.0000
Validation accuracy	1.0000
Training loss	0.0020
Validation loss	0.0085

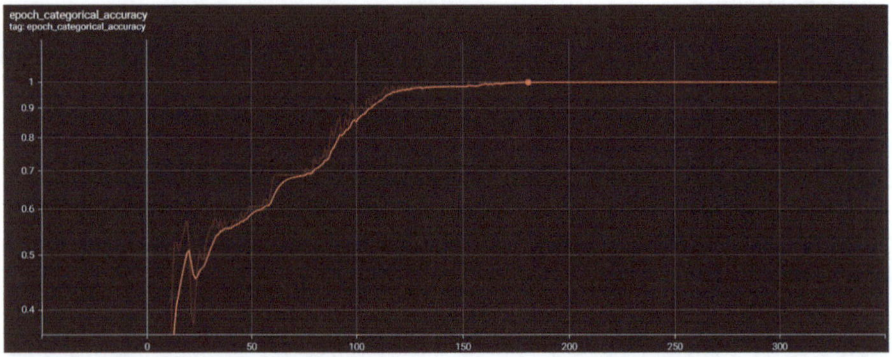

Fig. 14 Training accuracy

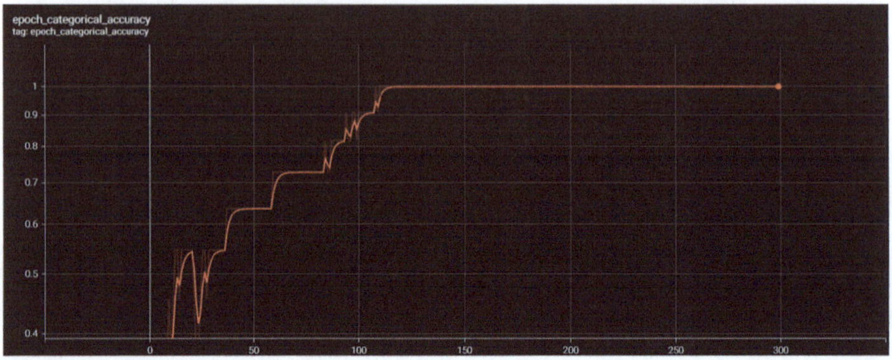

Fig. 15 Validation accuracy

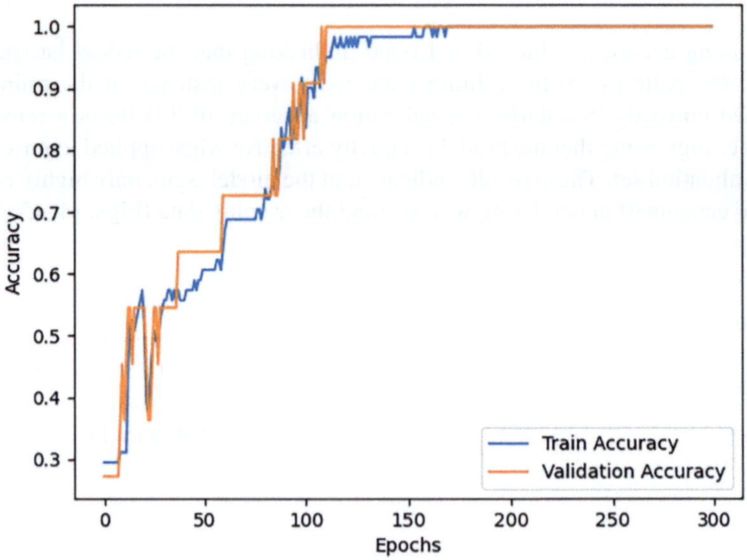

Fig. 16 Training and validation accuracy over epochs

5.3 Loss Analysis

In terms of loss, the training loss stands at 0.0020, indicating that the model's predictions on the training data are very accurate with minimal errors. The validation loss is also low at 0.0085, showing that the model maintains its predictive power when tested on validation data. The small gap between training and validation loss further reinforces that the model is learning effectively without overfitting, which is crucial for its generalization to new data (Figs. 17, 18, 19).

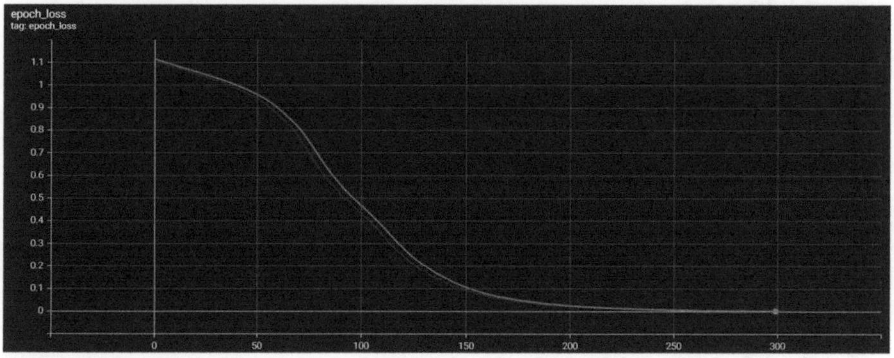

Fig. 17 Training loss

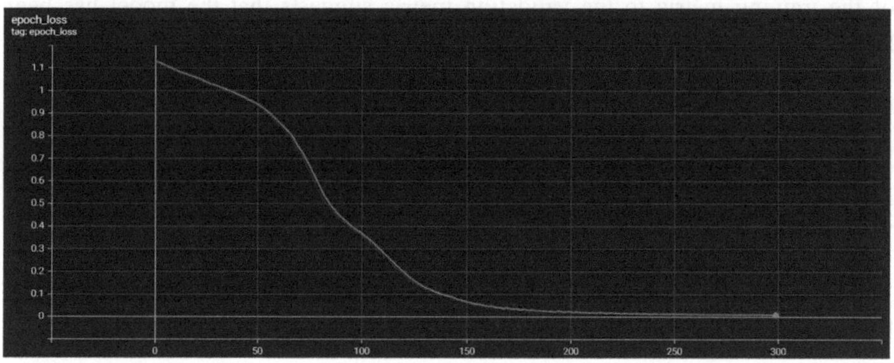

Fig. 18 Validation loss

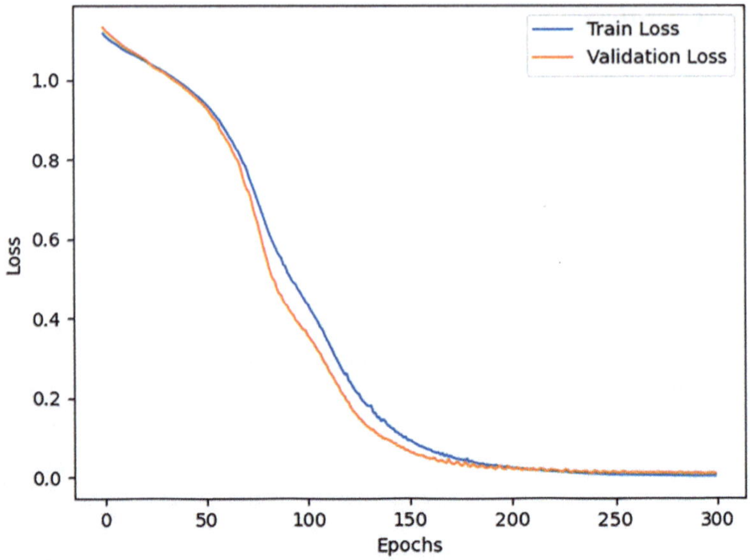

Fig. 19 Training and validation loss over epochs

5.4 Overfitting and Regularization

Compared to most models, which are often very accurate during training but less so during validation, this model demonstrates a balanced performance. The closeness of the training metric to the validation metric suggests that the model has been properly regularized and optimized to avoid common pitfalls such as overfitting. This balance is achieved using several effective techniques, including dropout, batch normalization, and careful tuning of hyperparameters.

5.5 Model Summary

The architecture of the model is summarized in Table 2, providing an overview of the layers, output shapes, and number of parameters:

- **Total params:** 1,332,875 (5.08 MB)
- **Trainable params:** 444,291 (1.69 MB)
- **Non-trainable params:** 0 (0.00 B)
- **Optimizer params:** 888,584 (3.39 MB)

Table 2 Model summary

Layer (type)	Output shape	Number of parameters
LSTM	(None, 64)	442,112
Dense	(None, 32)	2080
Dense_1	(None, 3)	99

5.6 Multifunctional Model Capabilities

In addition, our model has the ability to automatically translate the detected actions into their equivalent audio outputs. Using the TTS (text to speech) library from python, we were able to convert the text output into an audio output.

6 Conclusion

This chapter showed the usability of Google MediaPipe library for full-body gestures and poses analysis; in total, 543 landmarks were captured in real time. The holistic framework processes the body, face, and hand pose data simultaneously into one model supporting improved accuracies in processing complex gestures and activities. Besides, the implementation of the MobileNetV2 architecture hugely reduces computation cost with almost similar high accuracy, making the implementation particularly fit for the resource constraint. The confluence of these technologies has phenomenally expedited computer vision and machine learning by dispensing a solid dataset for the recognition of expressions, the creation of virtual avatars, postural assessment, movement tracking, etc. This kind of approach will not only make the system more efficient and precise but will also contribute to the knowledge applicable with regard to technology in a valuable and innovative way. This model can be more extended among the various other datasets using others, having actions in large amounts to make this model widely applicable and vast.

References

1. N. Alleema, S. Chandrasekaran, Recognition of American sign language using modified deep residual CNN with modified Canny edge segmentation (2022)
2. P.S. Santhalingam, P. Pathak, J. Košecká, H. Rangwala, Sign language recognition analysis using multimodal data, in *2019 IEEE International Conference on Data Science and Advanced Analytics (DSAA)* (IEEE, 2019, October), pp. 203–210
3. M. Al-Hammadi, G. Muhammad, W. Abdul, M. Alsulaiman, M.A. Bencherif, M.A. Mekhtiche, Hand gesture recognition for sign language using 3DCNN. IEEE Access **8**, 79491–79509 (2020)
4. S.S. Wazalwar, U. Shrawankar, Interpretation of sign language into English using NLP techniques. J. Inf. Optim. Sci. **38**(6), 895–910 (2017)

5. S. Divya, S. Kiruthika, A.N. Anton, S. Padmavathi, Segmentation, tracking and feature extraction for Indian sign language recognition. Int. J. Comput. Sci. Appl. **4**(2), 57–72 (2014)
6. A. Halder, A. Tayade, Real-time vernacular sign language recognition using MediaPipe and machine learning, 7421 (2021). Journal homepage: www.ijrpr.com. ISSN:2582
7. W.M.Y.W. Bejuri, Sign language detection using convolutional neural network for teaching and learning application. Indonesian J. Electric. Eng. Comput. Sci. (IJEECS) **28**(1), 358–364 (2022)
8. Y. Bajaj, P. Malhotra, American sign language identification using hand trackpoint analysis, in *International Conference on Innovative Computing and Communications: Proceedings of ICICC 2021*, vol. 1 (Springer Singapore, 2022), pp. 159–171
9. M.H. Moghaddam, H.R. Ghaffary, Facial recognition of people with masks using MediaPipe Facemesh and Deep Learning Algorithms. https://doi.org/10.2139/ssrn.5022821
10. A. Kuznetsova, V. Kimmelman, Testing MediaPipe holistic for linguistic analysis of nonmanual markers in sign languages. arXiv preprint arXiv:2403.10367 (2024)
11. M.T. Abdullah, N.H.M. Ali, Facial deepfake performance evaluation based on three detection tools: MTCNN, Dlib, and MediaPipe, in *AIP Conference Proceedings*, vol. 3097. No. 1 (AIP Publishing, 2024)
12. Google AI, *Face Landmarking*. Google. Available at: https://ai.google.dev/edge/mediapipe/solutions/vision/face_landmarker
13. Google AI, *Pose Landmarking*. Google. Retrieved from https://ai.google.dev/edge/mediapipe/solutions/vision/pose_landmarker
14. Google AI, *Hand Landmarking*. Google. Retrieved from https://ai.google.dev/edge/mediapipe/solutions/vision/hand_landmarker
15. MobileNetV2, Inverted Residuals and Linear bottlenecks. Retrieved from: https://paperswithcode.com/method/mobilenetv2
16. K. Dong, C. Zhou, Y. Ruan, Y.Li, MobileNetV2 model for image classification, in *2020 2nd International Conference on Information Technology and Computer Application (ITCA)* (IEEE, 2020, December), pp. 476-480
17. GeeksforGeeks, *Image Recognition with MobileNet*. Retrieved from https://www.geeksforgeeks.org/image-recognition-with-mobilenet/
18. *GHUM & GHUML: Generative 3D Human Shape and Articulated Pose Models*. Taken from : https://research.google/pubs/ghum-ghuml-generative-3d-human-shape-and-articulated-pose-models/
19. TensorFlow Blog, *3D Pose Detection with MediaPipe BlazePose and GHUM in TensorFlow.js*. Retrieved from https://blog.tensorflow.org/2021/08/3d-pose-detection-with-mediapipe-blazepose-ghum-tfjs.html
20. Wikipedia, Long short-term memory. Retrieved from https://en.wikipedia.org/wiki/Long_short-term_memory#:~:text=Long
21. CodenX, *Machine Learning using Scikit-learn (sklearn): Evaluating Classification Model using Metrics*. Medium. Retrieved from https://medium.com/codenx/machine-learning-using-scikit-learn-sklearn-evaluating-classification-model-using-metrics-018857ea4b69
22. D.M. Prasanna, Ch. Ganapathy Reddy, Development of real time face recognition system using OpenCV. Development **4**(12), 791 (2017)

Index

A. Haldorai et al. (eds.), *7th EAI International Conference on Big Data Innovation for Sustainable Cognitive Computing*, EAI/Springer Innovations in Communication and Computing, https://doi.org/10.1007/978-3-032-02979-9

MIX
Papier aus verantwortungsvollen Quellen
Paper from responsible sources
FSC® C105338

If you have any concerns about our products,
you can contact us on
ProductSafety@springernature.com

In case Publisher is established outside the EU,
the EU authorized representative is:
Springer Nature Customer Service Center GmbH
Europaplatz 3, 69115 Heidelberg, Germany

Printed by Libri Plureos GmbH
in Hamburg, Germany